Semiconductor Interfaces and Microstructures

Semiconductor Interfaces and Microstructures

Editor

Zhe Chuan Feng

Dept. of Physics, Emory University, USA

World Scientific
Singapore • New Jersey • London • Hong Kong

Published by

World Scientific Publishing Co. Pte. Ltd.
P O Box 128, Farrer Road, Singapore 9128
USA office: Suite 1B, 1060 Main Street, River Edge, NJ 07661
UK office: 73 Lynton Mead, Totteridge, London N20 8DH

Library of Congress Cataloging-in-Publication data is available.

SEMICONDUCTOR INTERFACES AND MICROSTRUCTURES

Copyright © 1992 by World Scientific Publishing Co. Pte. Ltd.

All rights reserved. This book, or parts thereof, may not be reproduced in any form or by any means, electronic or mechanical, including photocopying, recording or any information storage and retrieval system now known or to be invented, without written permission from the Publisher.

ISBN 981-02-0864-2
 981-02-0988-6 (pbk)

Printed in Singapore by General Printing Services Pte. Ltd.

Preface

In the middle of 1990, the American Physical Society (APS) - Materials Physics Group invited me and Drs. L. C. Feldman (AT&T Bell Lab.), F. K. Le Goues (IBM Thomas J. Watson Research Center), and H. H. Wieder (University of California, San Diego) to organize a focused session on *Semiconductor Interfaces and Microstructures* for the 1991 APS March meeting. Due to the attractive topics, the good work from four organizers and the support of APS, we had 75 speakers for this topical session during 18-22 March 1991 in Cincinnati. About the same time, the World Scientific Publishing (WSP) and the Institute of Physics Publishing (IOPP) asked me to edit the review books in this field. Because of my special interest and great achievements on the semiconductor interfaces and microstructures in recent years, I considered it a good plan. I contacted many scientists and professors working in the frontiers of this field. Many of them were willing to review their fruitful work. Due to so many important achievements and developments that have appeared in recent years, it is worthy to publish two books in this active field.

The present book consists of 12 review chapters. Each one is a comprehensive review on the recent achievements and creative contributions in the corresponding field. This book emphasize more on the interface problems for various semiconductor materials and systems, explored from different concepts or perspectives, and studied by various techniques and methods.

The book is grouped into four sections: *General, III-V, II-VI, and IV-IV*. It includes new developments in growth methods; electric, optical, magnetic and structural characterization and properties; relative theories; devices and applications. Each chapter focuses on a special subject and is prepared by one or more experts in the field. Some introductory materials are also included to make the book suitable for graduates and new scientists in the field. We paid more attention to organize those materials that have not appeared in existing books.

The *general section* consists of 3 chapters. In Chapter 1, *carrier transport in artificially structured two-dimensional semiconductor systems*, W. Walukiewicz from Lawrence Berkeley Laboratory reviews the basic concept of electron and hole transport in two-dimensional and quasi-three-dimensional semiconductor systems. Modulation doping has provided structures with extremely high mobilities. Contributions of different scattering processes to the total electron and hole mobilities in various

types of modulation doped heterostructures are calculated and compared with experiments.

A. Sibille, F. F. Palmier and C. Minot in chapter 2 review the *miniband conduction in semiconductor superlattices* (SL), based on the available experimental data and existing theoretical models of transport. An impressive progress has been made recently in this field with in particular the conclusive observation of tailorable perpendicular miniband conduction by adjusting SL parameters, the experimental verification of the pertinence of Wannier-Stark quantization related to Bloch oscillations in biased SLs, and the experimental validation of the 20 years old ideas of Esaki and Tsu on nonlinear miniband transport.

Chapter 3 by J. J. Song, J. F. Zhou and J. M. Jacob focuses on various optical properties of superlattices, emphasizing those which depend on the barrier widths, L_b's. The well-to-well couping of electronic wave functions in thinner L_b SLs leads to subband dispersion. They discuss many interesting optical properties associated with subband dispersion and mixing, and mostly observed in the excitonic spectra of III-V SLs.

There are 5 chapters in the *III-V section*, which is the most widely studied compound system. Chapter 4 by P. O. Holtz, B. Monemar and J. L. Merz deals with *radiative processes in GaAs/AlGaAs heterostructures*. The first observation of luminescence from a semiconductor heterointerface was observed only a few years ago. The authors review a large number of experiments related to the 2D electrons in the interface notch, theoretical aspects, the so-called H-bands, dynamical studies, magneto-optical and photoluminescence excitation (PLE) spectroscopy, and photoluminescence (PL) with an applied electric field.

Chapter 5 by Guo-Hua Li focuses on the *type-I-type-II transitions in GaAs superlattices*. Theoretical and experimental studies on the electronic structures and optical properties of GaAs/AlAs SLs are reviewed. The $(GaAs)_n(AlAs)_n$ SL is type-I SL in which the electrons and holes are confined in the same layers, when GaAs and AlAs layers are thick, while it is type-II SL with the electrons and holes confined in adjacent layers, as the layers are thin enough. The transition point between type-I and type-II is in the range of $8 \leq n \leq 14$. Theoretical calculations by various models and experimental results by PL, PLE, PL under pressure, absorption and other spectroscopies are shown. Γ-X mixing and X-valley splitting are discussed.

Chapter 6 written by D. Gammon, B. V. Shanabrook and D. S. Katzer is concerned with *photoluminescence studies of interface roughness in*

GaAs/AlAs quantum well structures. They review the use of PL to characterize the structural disorder, which is existed as monolayer-high islands, of the interfaces (IFs) in GaAs/AlAs quantum well structures. Excitons measured by PL is a powerful probe of the IF roughness. Differently-sized islands on the top and bottom interfaces lead to different types of PL spectra.

D. C. Reynolds and K. R. Evans contributed the chapter 7, *optical and magneto-optical properties of narrow $In_xGa_{1-x}As$-GaAs quantum wells*. High quality pseudomorphic $In_xGa_{1-x}As$-GaAs (lattice-mismatched) quantum wells (QWs) and superlattices have been grown without misfit dislocations on GaAs substrate by molecular beam epitaxy (MBE). The details of MBE growth, band alignment and interface characteristics are presented. The authors also review the PL studies on excitonic and band-to-band transitions from narrow QWs; direct coupling of GaAs barrier excitons with InGaAs well excitons, magneto-optical effects and coupled QWs.

J. B. Webb and R. Rousina review the *growth and studies of antimony based III-V compounds by magnetron sputter epitaxy using metalorganic and solid elemental sources* in chapter 8. Two forms of magnetron sputtering, magnetron sputter epitaxy (MSE) and metalorganic magnetron sputtering (MOMS) have been used to the epitaxial deposition of InSb, GaSb, $In_{1-x}Ga_xSb$, $In_{1-x}Al_xSb$, and $In_{1-x}Al_xSb$/InSb strained layer SLs. Their structural and opto-electronic properties are discussed.

The *II-VI section* includes 2 chapters. J. M. Wrobel and J. J. Dubowski review the structural and optical *properties of CdMnTe films and CdMnTe-CdTe superlattices grown by pulsed laser evaporation and epitaxy* (PLEE) in chapter 9. A brief discussion is also given concerning the fundamentals of PLEE: a technique that has been especially developed for the deposition of thin films, QWs and SLs of semiconductor materials.

Chapter 10 by R. E. Nahory, M. J. S. P. Brasil and M. C. Tamargo is concerned with $Zn_{1-y}Cd_ySe_{1-x}Te_x$ *quaternary II-VI wide bandgap alloys and heterostructures* grown by MBE. They model the bandgap vs composition (x,y) for this system and the results are in good agreement with values measured by photoconductivity and PL techniques. The band-offset engineering with respective to ZnSe for the system ZnSe-ZnCdSeTe-ZnCdSe is discussed. Lattice-matched growth of the quaternary on InAs is potentially expected to produce visible-light lasers in the red.

We have two chapters also in the last *IV-IV section*. R. P. G. Karunasiri, K. L. Wang and J. S. Park review the *intersubband transitions in SiGe/Si quantum structures*. The physics of intersubband transitions is discussed using a parabolic QW as an example. An intersubband infrared detector for the 8-14 μm atmospheric window by using $Si_{1-x}Ge_x$/Si multiple quantum well is demonstrated. The peak responsivity and detectivity of the device can be improved by optimizing parameters.

M. M. Anikin et al. from Ioffe Institute contribute the chapter 12, *high-temperature discrete devices in 6H-SiC: sublimation epitaxial growth, device technology and electrical performance*. They review the 14 years research and development work in Ioffe Institute on 6H-SiC devices, including schottky and rectifier diodes, UV photodetectors and junction gate field-effect transistors, that are suitable for the performance in high temperature and cruel environments. Because these authors didn't have a laser printer and a good word processor to prepare their paper, I recompiled the whole article for them. Many figures are hand-drawn by them. Chapter 5 is another example without using a laser printer. These two chapters are very good in science. Contributors' hard work are complimented by us. The lack of ordinary tools and equipments for science in high level institutes shows the necessity of reformation for both countries in some sense.

No one book can cover the tremendous achievements and developments. I recommend to the interested readers another sister book, *<<Semiconductor Interfaces, Microstructures and Devices: Properties and Applications>>*, edited also by me and published by Institute of Physics Publishing, Inc. These two books do not overlap in material, however, they complement each other for a full appreciation of the field to date.

Finally, I'd like to acknowledge all the authors of this book and editors Tony Moore, Ph. Tham and S. H. Gan, who made this book successful. I'd like also compliment all the scientists and engineers who are working in the frontiers of this field. Because of their achievements and contributions, this book is possible to appear. We are looking forward to further developments in the future. We hope that this book may benefit scientists, engineers, professors and students in further exploration and understanding of semiconductor interfaces and microstructures.

Zhe Chuan Feng
Emory University
Department of Physics
Atlanta, GA 30322

CONTENTS

Preface — Z. C. Feng (Emory University) v

General Section

(1) Carrier transport in artificially structured two-dimensional semiconductor systems 1
 — W. Walukiewicz (Lawrence Berkeley Laboratory)
 1. Introduction
 2. Electronic structure
  2.1 Electronic structure of n-type MDH
  2.2 Electronic structure of p-type MDH
 3. Charge scattering mechanisms in 2D MDH
  3.1 Scattering by charged centers
  3.2 Acoustic phonon scattering
  3.3 Optical phonon scattering
  3.4 Alloy disorder scattering
  3.5 Interface roughness scattering
  3.6 Intersubband scattering
  3.7 Scattering mechanisms in p-type MDH
 4. Scattering mechanisms in wide parabolic wells
 5. Calculated mobility and comparison with experiment
  5.1 N-type $Al_xGa_{1-x}As$/GaAs MDH
  5.2 N-type $In_{0.53}Ga_{0.47}As$ based heterostructures
  5.3 P-type AlGaAs/GaAs MDH
  5.4 Wide parabolic AlGaAs/GaAs wells
 6. Conclusions

(2) Miniband conduction in semiconductor superlattices 31
 — A. Sibille, J. F. Palmier, C. Minot
  (Centre National d'Etudes Télécommunications, FRANCE)
 1. Introduction
 2. Miniband structure
 3. Experimental observations of low field miniband conduction
 4. Non-linear miniband transport
 5. Theory and models of semi-classical miniband conduction
 6. Limits of semiclassical miniband transport
 7. Conclusions

(3) Barrier width dependence of optical properties
in semiconductor superlattices 63
- J. J. Song, J. F. Zhou and J. M. Jacob
(Oklahoma State University)
1. Introduction
2. Exciton peak shift
   2.1 Binding energy
   2.2 The shift S_n
   2.3 Subband width Wn
   2.4 Seperation between 11H and 11L excitons
3. Effects of subband dispersion and saddle points on excitonic spectra
4. Unconfined states
5. Line narrowing
6. Subband mixing in SL's
7. Tunneling phenomena on optical properties of superlattices
8. Stark effect
9. Confinement and coupling of LO phonons in superlattices
10. Concluding remarks

III-V Section

(4) Radiative processes in GaAs/AlGaAs
heterostructures 93
- P. O. Holtz, B. Monemar (Linkoping University, SWEDEN)
and J. Merz (University of California at Santa Barbara)
1. Introduction
2. Theoretical aspects
3. Samples
4. Photoluminescence measurements on GaAs/AlGAs heterostructures
5. Dynamical studies
6. Magneto-optical spectroscopy
7. Photoluminescence excitation spectroscopy
8. Photoluminescence with an applied electric field
9. Conclusions

(5) Type-I-Type-II transition in GaAs/AlAs
superlattices 120
- G. H. Li (Semiconductor Institute, CHINA)
1. Introduction
2. Structure of GaAs/AlAs superlattice
3. Theoretical calculation
   3.1 Effective mass approximation

    3.2   Tight-binding method
    3.3   Wannier orbital model
    3.4   Empirical pseudopotential method
    3.5   First-principle calculation
4. Experimental measurements
    4.1   Photoluminescence and photoluminescence excitation
    4.2   Absorption and other spectroscopies
    4.3   Photoluminescence under hydrostatic pressure
5. Γ-X mixing and X-valley splitting
    5.1   Order of X_z and X_{xy} splitting
    5.2 Γ-X mixing
6. Conclusion

(6) Photoluminescence studies of interface roughness in GaAs/AlAs quantum well structures 149
    - *D. Gammon, B. V. Shanabrook and D. S. Katzer*
      (Naval Research Laboratory)
1. Introduction
2. Growth details
3. Excitons as probes of the IF
4. Catagories of IF's according to island sizes
5. Absolute exciton energies and microroughness
6. Conclusion

(7) Optical and magneto-optical properties of narrow $In_xGa_{1-x}As$-GaAs quantum wells 168
    - *D. C. Reynolds (Wright State University)*
      and K. R. Evans (WL/ELRA)
1. Introduction
2. Pseudomorphic $In_xGa_{1-x}As$-GaAs heterostructures
    A.   Molecular beam epitaxy growth
    B.   Band alignment
    C.   Interface Characteristics
3. Photoluminescence properties of narrow quantum wells
    A.   Excitonic transitions
    B.   Bound-to-bound transitions
4. Direct coupling of GaAs barrier excitons with InGaAs well excitons
5. Magneto-optical effects
6. Coupled quantum wells

(8) **Growth and studies of antimony based III-V compounds by magnetron sputter epitaxy using metalorganic and solid elemental sources** 199
- *J. B. Webb and R. Rousina*
  (National Research Council Canada)
1. Introduction
2. The magnetron sputter cathode and epitaxy system
3. Epitaxy of III-V Compounds by magnetron sputtering
   3.1 Metalorganic magnetron sputtering
   3.2 Carbon incorporation in MOMS
   3.3 Magnetron sputter epitaxy
4. Properties of InSb, $In_{1-x}Ga_xSb$ and $In_{1-x}Al_xSb$, InSb/InSb, InSb/GaAs
   4.1 $In_{1-x}Ga_xSb$ on GaAs
   4.2 $In_{1-x}Al_xSb$ strained layer superlattices
5. Conclusions

II-VI Section

(9) **Properties of $Cd_{1-x}Mn_xTe$ films and $Cd_{1-x}Mn_xTe$-CdTe superlattices grown by pulsed laser evaporation and epitaxy** 216
- *J. M. Wrobel (University of Missouri - Kansas City) and J. J. Dubowski (National Research Council Canada)*
1. Introduction
2. Pulsed laser evaporation and epitaxy
3. Epitaxial layers and microstructures of $Cd_{1-x}Mn_xTe$
   3.1 Substrate preparation
   3.2 Composition
   3.3 Surface morphology and crystal structure
   3.4 Film orientation
4. Optical properties
   4.1 Photoluminescence
   4.2 Raman spectra
   4.3 Photoreflectance
   4.4 Photovoltaic effect
5. Conclusions

(10) **$Zn_{1-y}Cd_ySe_{1-x}Te_x$ quaternary II-VI wide bandgap alloys and heterostructures** 238
- *R.E. Nahory, M.J.S.P. Brasil and M.C. Tamargo*

(Bell Communications Research)
1. Introduction
2. Ternary $Zn_{1-y}Cd_ySe$ and $ZnSe_{1-x}Te_x$
3. Bandgap of the quatenary $Zn_{1-y}Cd_ySe_{1-x}Te_x$
4. Growth and characterization of $Zn_{1-y}Cd_ySe_{1-x}Te_x$ epilayers
5. Discussion of band-offsets
6. $Zn_{1-y}Cd_ySe_{1-x}Te_x$ on InAs or InP substrates
7. Summary

IV-IV Section

(11) Intersubband transitions in SiGe/Si quantum structures 252
- R. P. G. Karunasiri, K. L. Wang and J. S. Park
  (University of California at Los Angles)
1. Introduction
2. Physics of intersubbband transitions
   2.1 Transition in a quantum well
   2.2 Transition between superlattice minibands
   2.3 Intervalence band transition
3. Intersubband transition in SiGe/Si quantum wells
4. Intersubband transitions in δ-doped quantum wells
   4.1 Doping dependence of intersubband absorption
   4.2 Many-body effects
5. Detector applications
   5.1 Detector characterization
   5.2 Detectivity
6. Summary

(12) High-temperature discrete devices in 6H-SiC: sublimation epitaxial growth, device technology and electrical performance 280
- M. M. Anikin, P. A. Ivanov, A. A. Lebedev, S. N. Pytko,
  A. M. Strel'chuk, and A. L. Syrkin *(Ioffe Institute, RUSSIA)*
1. Epitaxial growth of pn structures by open sublimation method
2. Fabrication of devices based on epitaxial structures: profiling and application of ohmic contacts
3. 6H-SiC schottky diodes
   3.1 The electrical characteristics of SiC surface-barrier structures
   3.2 Forward current mechanism in 6H-SiC surface-barrier structures

4. Rectifier diodes
   4.1 Capacitance-voltage characteristics and deep centers
   4.2 Forward current-voltage characteristic of the pn structures
      4.2a. Low currents
      4.2b. High currents
   4.3 Reverse current-voltage characteristics of the pn structures
      4.3a. The breakdown voltage
      4.3b. Current in the pn structures prior to breakdown
   4.4 Characteristics of radiationless recombination processes
   4.5 Electroluminescence spectra
5. UV photodetectors in 6H-SiC
6. Junction gate field-effect transistors

CARRIER TRANSPORT IN ARTIFICIALLY STRUCTURED TWO-DIMENSIONAL SEMICONDUCTOR SYSTEMS

W. Walukiewicz
Center for Advanced Materials, Materials Sciences Division
Lawrence Berkeley Laboratory, University of California
1 Cyclotron Road, Berkeley, California 94720 USA

ABSTRACT

Transport of electrons and holes in two-dimensional and quasi-three-dimensional semiconductor systems is reviewed. Contributions of different scattering processes to the total electron and hole mobilities in various types of modulation doped heterostructures are calculated. It is shown that in a wide temperature range phonon scattering is the principal scattering mechanism limiting electron and hole mobilities in high quality AlGaAs/GaAs modulation doped heterostructures. The importance of nonspecular scattering from rough walls in wide parabolic wells is emphasized. Also, several unresolved or poorly understood aspects of charge transport in two-dimensional semiconductor systems are discussed.

1. INTRODUCTION

The last decade has witnessed an unprecedented growth of research on low dimensional semiconductor systems. Sophisticated epitaxial techniques such as Molecular Beam Epitaxy, Metalorganic Chemical Vapor Deposition and numerous variations of these two growth methods allow atomic scale control of the the growth process and are widely used to grow these complex semiconductor structures. Since all the epitaxial techniques provide excellent control of composition and doping only along the growth direction, they are used most successfully to fabricate two dimensional structures. Additional confinement in the growth plane is much more difficult to achieve and control. Therefore, although substantial progress has been made in last few years, studies of truly one- and zero-dimensional semiconductor systems are still quite rare.

Low dimensional semiconductor systems have become fertile ground for basic research. The discovery of the Quantum Hall Effect [1] and the Fractional Quantum Hall Effect [2] were major developments in fundamental solid state physics in the last several years. Flexibility with which various structures could be designed and practically realized has led to new concepts of charge transport

in mesoscopic systems. Also, a wealth of new linear and nonlinear optical effects has been observed in semiconductor quantum wells [3].

One of the most basic characteristics of a semiconductor system is its response to the external fields. This response is determined by the properties of the semiconductor material as well as by specifics of the interaction of the charge carriers with collective excitations and imperfections of the crystal lattice. Studies of the electronic transport in three-dimensional (3D) semiconductors have provided invaluable information on a variety of semiconductor band structure parameters [4]. In addition, in many instances an analysis of the free carrier mobility has been used to determine the strength of the charge carrier scattering potentials [5,6].

It was evident from the beginning that the introduction of artificially structured 2D systems would open an opportunity to study basic transport phenomena in a new and in many respects unusual material systems [7]. Using the concept of modulation or selective doping it has been possible, for the first time, to separate charge carriers from the parent impurities [7,8]. This concept led to a very substantial reduction of impurity scattering and allowed the study of charge transport in almost perfectly pure semiconductors where phonon scattering plays a dominant role down to very low temperatures.

There are several review papers on properties of 2D systems. Ando *et al.* [9] have reviewed the progress in the field of 2D systems until the early 1980's. Understandably, at that time most of the work was limited to studies of 2D inversion layers in Si and InAs. However, the main theoretical concepts presented in the review can be easily adopted to any 2D system. Recently, various aspects of modulation doped semiconductor systems were reviewed in a series of articles covering basic properties of modulation doped structures [10] as well as their applications for electronic [11] and optoelectronic devices [12,13].

The main objective of this review is to present recent theoretical and experimental results on electronic transport in 2D and quasi-3D modulation doped heterostructures. We will focus our attention on 2D electron or hole gases confined at heterointerfaces of different semiconductors. However, many of the concepts and results will be directly applicable to charge transport in any two dimensional system. The basic theoretical concepts will be illustrated with experimental results on the most extensively studied material systems. Due to the limited scope of the review, we will refrain from discussing lattice mismatched systems where strain plays a significant role in determining electronic structure and transport properties [14].

The material in this chapter is structured as follows: The background information on the electronic structure of electron and hole gases confined in thin 2D films is presented in section 2. In section 3 and 4 the scattering processes limiting carrier mobilities are introduced and the differences in carrier scattering in 2D and quasi-3D systems are discussed. Calculated electron and hole mobilities in various modulation doped materials systems are compared with representative experimental data in section 5. Finally the discussion of currently

actively studied, and possible future research areas are discussed in the closing section 6.

2. ELECTRONIC STRUCTURE

Quantum confinement of electrons and holes in 2D planes can be achieved by several methods. Charge carriers can be confined in an accumulation or inversion layer at a semiconductor surface [15]. The carriers can also be confined by an attractive electrostatic potential created by δ or planar doping [16]. However, the method most widely used to produce 2D systems utilizes band offsets at semiconductor heterointerfaces [17]. The band offsets can be used to confine electrons or holes in a thin square quantum well or at a interface between two different semiconductors. Modern epitaxial techniques provide an atomic scale control over the thickness of the films and the smoothness of the interfaces. The value of the available band offset depends on the semiconductor materials and for III-V compounds may be as large as 1.3 eV [18].

In order to achieve 2D confinement the system has to satisfy certain conditions. First, the thickness of the well has to be smaller than the electron or hole de Broglie wavelength. For a 3D electron or hole gas of concentration n the de Broglie wavelength at the Fermi energy is $\lambda_F = 2\pi/k_F$, where $k_F = (3\pi^2 n)^{1/3}$. Hence the condition to observe 2D confinement of such gas is that the thickness w of the well satisfies the condition

$$w < 2\left(\frac{\pi}{3n}\right)^{1/3} \qquad (1)$$

It should be emphasized that this is a universal condition which does not depend on any semiconductor material parameters.

Another characteristic length for an electron or hole gas is the mean free path, i.e., the distance an electron or a hole can travel between momentum randomizing scattering events. For a 2D gas the mean free path l_p has to satisfy the condition,

$$l_p = v_F \cdot \tau = \hbar k_F \mu/e = \hbar (3\pi^2 n)^{1/3} \mu/e > w \qquad (2)$$

where v_F is the carrier velocity at the Fermi level, τ is the relaxation time and μ the mobility. The above condition has a simple physical interpretation: it states that in a 2D system electrons or holes interact with the confinement walls more frequently than with random scattering centers.

In general, for low effective mass high mobility carriers, condition (1) is more restrictive than condition (2). For example, to observe a 2D gas in GaAs with 10^{17} cm^{-3} charge carriers $\lambda_F \cong 440$ Å and a plane confinement of a width smaller than 400 Å is required. On the other hand, for a carrier concentration of 10^{17} cm^{-3} the low temperature electron mobility easily exceeds 10^4 cm^2/Vs. Therefore, in this case the condition (2) is satisfied for w<1000 Å. It can be

shown that in most semiconductor systems a confinement of the order of 100 Å is necessary to clearly observe quantum effects. Such a degree of confinement is easily accessible and can be practically realized in many semiconductor systems.

Recently a new type of structure which connects 2D with 3D limits has been proposed and practically realized. In so-called parabolic quantum wells a wide well can be obtained by a combination of proper design of the alloy composition in the well with remote doping [19,20]. The width of such wells can be of the order of 1000 Å, i.e., it is larger than the de Broglie wavelength. Therefore, although the carrier motion is not quantized the mean free path in such structure can be larger than the width of the well. Consequently, one has to account for an interaction of free carriers with the confinement walls.

Among the variety of possible 2D semiconductor systems the single quantum well modulation doped heterostructures (SQW-MDH) are the semiconductor structures most frequently used to study transport of 2D electrons and/or hole gases. In such structures, schematically shown in Fig. 1, the shallow donors or acceptors are located in the barrier forming semiconductor S2. The carriers are transferred from the parent dopants into the well forming semiconductor S1. The resulting electric field confines the mobile charges in a narrow quasi-triangular well at the heterointerface. By changing the distance d separating the dopants from the quantum well one can control the concentration of the electrons or holes in the well. Also, by increasing the separation one is able to very substantially enhance the carrier mobility by reducing the effectiveness of the ionized impurity scattering. Practical implementation of this idea has resulted in n-type AlGaAs/GaAs MDHs with low temperature electron mobilities exceeding 10^7 cm^2/V·s [21,22] and p-type structures with hole mobilities approaching 4×10^5 cm^2/V·s [23]. The principal objective of all theoretical calculations of the 2D transport is to understand the scattering mechanisms limiting electron and hole mobilities in such structures. The starting point

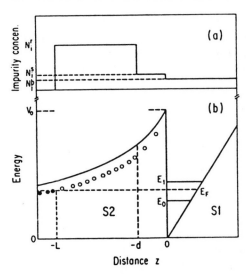

Fig. 1: Schematic representation of (a) the doping profile and (b) the energy configuration for a n-type, single quantum well modulation doped heterostructure. (After ref. [26])

of the theory is to provide an accurate description of the electronic structure of electrons or holes confined in a quasi-triangular quantum well at the heterointerface.

2.1 Electronic Structure of n-type MDH

The wavefunction of an electron gas confined in the z-direction can be written in the form,

$$\Psi_{n,\vec{k}}(\vec{r},z) = \frac{1}{L^2} \chi_n(z) \exp(i\vec{k}\cdot\vec{r}) \qquad (3)$$

where $\vec{R} = (\vec{r},z)$, $\vec{r} = (x, y)$ and $\vec{k} = (k_x, k_y)$. The wavefunction $\chi_n(z)$ is given by the solution of coupled Schrödinger and Poisson equations [24,25],

$$-\frac{\hbar^2}{2m^*}\frac{d^2}{dz^2}\chi_n(z) + \left[V_0\theta(-z) - e\phi(z)\right]\chi_n(z) = E_n\chi_n(z) \qquad (4)$$

$$\frac{d^2}{dz^2}\phi(z) = \frac{4\pi e^2}{\varepsilon_0}\left[\sum_{i=0}^{m} N_i\chi_i(z) + N_A(z) - N_D(z)\right] \qquad (5)$$

Here $E_n(\vec{k}) = E_n + \hbar^2\vec{k}^2/2m^*$ is the dispersion relation for the n-th subband. $V_0\theta(-z)$ represents a step-like band offset at z=0. $N_D(z)$ and $N_A(z)$ are the functions representing the distribution of shallow donors and acceptors respectively, N_i is the density of electrons in the i-th subband. In the present approximation the exchange-correlation potential has been neglected. It has been shown that this potential does not play a significant role in AlGaAs/GaAs like MDHs [24]. The distribution of localized charges is schematically represented in Fig. 1 and is given by,

$$N_D(z) - N_A(z) = \begin{cases} N_i^b & z > 0 \\ N_i^s & -d < z < 0 \\ N_i^r & -L < z < 0 \end{cases} \qquad (6)$$

Here we assume a constant, z-independent concentration of dopants in each of the considered regions. In general, solutions of Eqs. (4) and (5) require numerical calculations. However, it has been shown that in the case of heterostructures shown in Fig. 1, an approximate variational solution can be obtained. The wave function of the ground subband can be represented by [25],

$$\chi_0(z) = \begin{cases} Bb^{1/2}(bz+\beta)\exp\left(-bz/2\right) & z>0 \\ B'b'^{1/2}\exp\left(+b'z/2\right) & z<0 \end{cases} \quad (7)$$

where B, b, B', b' and β are variational parameters related to each other through the boundary conditions at the heterointerface, and the normalization condition for $\chi_0(z)$. Therefore the parameters B, B' and β can be expressed in terms of b and b' [27]. The corresponding energy E_0 of the bottom of the ground subband depends in a complex way on the variational parameters and the details of the carrier and impurity distribution in the MDH [27].

The part of the wavefunction for z < 0 represents the finite penetration of the electron gas into the barrier. Incorporation of the penetration is important for calculations of the alloy disorder and surface roughness scatterings in GaAlAs/GaAs MDHs. The variational wavefunction (7) is very frequently used in the calculations of the electronic transport at low electron densities when only the ground subband is occupied [25,27]. At higher electron densities or at elevated temperatures one needs to consider also the transport in higher lying excited subbands [26]. An exact treatment of such a problem is rather difficult and requires extensive numerical calculations. An approximate description of the two band transport will be discussed later.

In numerous attempts to calculate the electron mobility in MDHs, an even simpler approach has been used in which the penetration of the wavefunction into the barrier is ignored. In such a case the ground state subband wavefunction takes the form [9,28],

$$\chi_0(z) = \begin{cases} \left(\frac{b_0^3}{2}\right)^{1/2} z\exp\left(-b_0 z/2\right) & z>0 \\ 0 & z<0 \end{cases} \quad (8)$$

The parameter b_0 has a simple form [28],

$$b_0^3 = \frac{33\pi}{2}\frac{m^* e^2}{\varepsilon_0 \hbar}N_{eff}$$

where $N_{eff} = N_s + (32/11)N_{dep}$, N_s is the electron density in the well, $N_{dep} = N_i^b / W_d$, and W_d is the width of the depletion region in the well-forming semiconductor. For very high purity well-forming semiconductors, the concentration of dopants is very small and therefore very often the term 32/11 N_{dep} can be neglected in comparison with N_s.

In thermal equilibrium the charge transfer across the interface is governed by the condition [25],

$$E_0 + E_F = V_0 - E_b - \frac{4\pi e^2}{2\varepsilon_0} \frac{(N_s + N_{dep})^2}{N_i^r} - \frac{4\pi e^2}{\varepsilon_0}(N_s + N_{dep}) \cdot d + \frac{4\pi e^2}{\varepsilon_0} N_s \frac{B'^2}{b'} \quad (9)$$

where E_b is the donor binding energy in the doped region of the barrier-forming semiconductor. For large enough band offsets the last term, representing the charge penetrating the barrier, is very small and can be safely ignored. Eq. (9) has been obtained using the neutrality condition,

$$N_s + N_{dep} + N_i^r(L - d) \quad (10)$$

where $L - d$ is the depletion width of the intentionally doped region of the barrier.

2.2 Electronic Structure of p-type MDH

Transport of a 2D hole gas has been much less extensively studied both theoretically and experimentally. The complex nature of the valence Γ_8 band maxima in group III-V semiconductors makes a detailed analysis of the electronic structure of 2D holes difficult and numerically involved. The main difference in the description of electron and hole electronic structures is that in the case of holes one has to consider the four-fold degenerate valence band. Therefore, the Schrödinger equation (Eq. (4)) is replaced by,

$$\left[H\left(\vec{k}, \frac{1}{i}\frac{\partial}{\partial z}\right) + V(z)\right] \chi_{nk}(z) = E_{nk}(z) \quad (11)$$

where H is a 4x4 $\vec{k} \cdot \vec{p}$ matrix with k_z replaced by the operator $\partial/i\partial z$, V (z) is the potential including band offsets and electrostatic interactions with charged impurities. Eq. (11) has to be solved for each pair of (k_x, k_y).

Numerical calculations of the band structure for specific p-type MDHs have been performed by several groups [29-31]. The main conclusion of these calculations was that the electronic structure of the four-fold degenerate Γ_8 valence band splits into a number of subbands. For $k_x = k_y = 0$ these subbands are two-fold spin degenerate. Also, as is shown in Fig. 2, for the case of a p-type AlGaAs/GaAs MDH, the energy dispersion relations for the light and heavy holes are described by nonparabolic energy dependent effective masses [31]. Also, it has been demonstrated that the constant energy surfaces are warped, therefore the effective masses depend on a crystallographic direction. A detailed description of the hole transport in such systems would require extensive numerical calculations. However, under certain conditions, one can use an approximation which can significantly simplify the problem. As is seen in

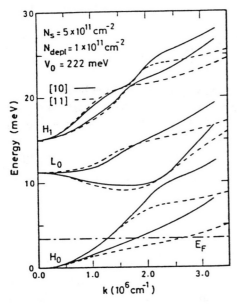

Fig. 2:Calculated hole subband energies of p-type AlGaAs/GaAs MDH as functions of the k wavevector for two different directions, [10] and [11]. Note that the sign of the hole energy is reversed. (After ref. [31])

Fig. 2, the first excited subband is located at about 10 meV below the ground subband. Therefore, for all practically achievable hole concentrations, the Fermi energy is low enough so that only the ground subbands are occupied. Consequently, at low temperatures one has to consider transport within the ground spin-up and spin-down subbands only. Under such approximations one can use an approach previously described for electrons in n-type MDHs. Therefore, we can write the hole wave function in the form [32],

$$\chi_h(z) = \left(\frac{b_h^3}{2}\right)^{1/2} z \exp(-b_h z/2) \quad (12)$$

where

$$b_h^3 = b_{1h}^3 + b_{2h}^3 \quad (13)$$

$$b_{ih}^3 = \frac{33 \pi e^2}{2\varepsilon_0 \hbar^2} P_i \quad (14)$$

P_i (i=1,2) is the concentration of holes in spin-up and spin-down subbands, respectively. The wavefunction (12) is used to calculate hole scattering rates for different scattering potentials.

3. CHARGE SCATTERING MECHANISMS IN 2D MDH

Most of the theoretical descriptions of charge transport are based on the solution of the Boltzman equation [4,6,33,34]. For elastic, momentum randomizing scattering processes, the solution of the Boltzman equation can be expressed in terms of an energy dependent relaxation time. For inelastic scattering processes, i.e., the processes in which the electron or hole loses or gains energy comparable or larger than $k_B T$, a different method to solve the Boltzman equation has to be used. The most frequently used approach is based on either the variational principle [33] or an iterative method used e.g. by Rode [34].

All the major scattering processes limiting the electron and hole mobilities in compound semiconductors are now well established [4,34]. In 3D compound semiconductors the low temperature mobility is determined by ionized impurity scattering and in the case of alloys also by the alloy disorder scattering. In ultra-high purity materials additional contributions from neutral impurities have to be taken into account. With increasing temperature, scattering by phonons comes into play. This scattering mechanism is especially important in lightly doped semiconductors. There are several different ways phonons can scatter charge carriers. The acoustic phonons can scatter electrons or holes via the deformation potential mechanism. Also, since compound semiconductors lack the center of inversion symmetry, the carriers can be scattered by acoustic phonons via the piezoelectric mechanism. Finally, the carriers can be scattered by optical phonons. The major contribution to the scattering of electrons by optical phonons comes from the interaction of electrons with electrostatic Fröhlich potential. In addition to the polar interaction the holes can be scattered by optical phonons via the optical phonon deformation potential.

Among those different scattering mechanisms, only the scattering by ionized impurities and by alloy disorder can be considered as strictly elastic processes. Scattering by phonons always results in an exchange of energy. However, in the case of acoustic phonons the energy of phonons for small phonon wavevectors is very low. Therefore, for all practical purposes the scattering by acoustic phonons can be treated as an elastic process for the temperatures higher than ~4 K. A high wavevector independent optical phonon energy leads to strongly inelastic scattering in the whole, practically important temperature range. It should be noticed, however, that the optical phonon scattering is strongly reduced at low temperatures. Therefore, in most instances this scattering process can be neglected at temperatures below ~40 K.

There are certain features which distinguish electron transport in two and three dimensions. In the case of ionized impurities there are two distinctly different types of scattering in the 2D case: Electrons can be scattered by remote impurities located within the doped region of the S2 semiconductor as well as by residual impurities and/or charged defects in the S1 semiconductor. Also, there is the possibility of charge scattering by structurally rough interfaces.

For elastic scattering processes the electron or hole mobility can be expressed in terms of energy dependent relaxation times. In general, the total relaxation time is given by the expression,

$$\frac{1}{\tau_{tot}(E)} = \sum_i \frac{1}{\tau_i(E)} \qquad (15)$$

where $\tau_i(E)$ are microscopic relaxation times associated with the different scattering processes. The macroscopic relaxation time, and thus also macroscopic mobility, is given by averaging of Eq. (15) over the energy E.

$$\overline{\tau(E)} = \frac{\int \tau(E)\, E\,(\partial f_o/\partial E)\, dE}{\int E\,(\partial f_o/\partial E)\, dE} \qquad (16)$$

where $f_o(E) = [1 + \exp[(E - E_F)/k_B T]]^{-1}$ is the Fermi-Dirac distribution function and E_F is the Fermi energy. In many instances Eq. (16) can be significantly simplified. At low temperatures, when the electron or hole gases are degenerate, $E_F > k_B T$, one has to consider only the electrons or holes on the Fermi surface, and the integration in (16) gives a simple result,

$$\overline{\tau(E)} \cong \tau(E_F) \qquad (17)$$

For example, in the case of n-type AlGaAs/GaAs, MDH with $N_s = 3 \times 10^{11}$ cm^{-2} the condition $k_B T < E_F$ is satisfied for $T < 70$ K. However, for very lightly doped heterostructures or for p-type MDHs with larger effective masses the degeneracy condition is satisfied at much lower temperatures and therefore a numerical evaluation of expression (16) is necessary.

3.1. Scattering by charged centers

Although the objective of modulation doping is to reduce the scattering from Coulomb potentials associated with ionized impurities, nevertheless, the scattering by remote impurities in the barrier and residual background impurities and defects in the well still play a major role in limiting the electron or hole mobilities at low temperatures. The relaxation time for the scattering of charge carriers from a Coulomb potential is given by [9,25],

$$\tau_C^{-1} = \frac{2\pi}{\hbar} \int dz\, N_i(z) \sum_q \left[\frac{2\pi e^2}{q\,\varepsilon(q)}\right]^2 |F_C(q,z)|^2 \times (1 - \cos\theta)\, \delta(E_{\vec{k}}^- - E_{\vec{k}-\vec{q}}^-) \qquad (18)$$

where $q = 2k \sin(\theta/2)$, θ is the scattering angle, $\varepsilon(q)$ is the static dielectric function,

$$\varepsilon(q) = 1 + \frac{2e^2 m^*}{\varepsilon_0 q \hbar^2} F(q) \qquad (19)$$

and ε_0 is the static dielectric constant. The form factors are given by,

$$F_C(q,z) = \int dz'\, |\chi_0(z')|^2 \exp(-q|z-z'|) \qquad (20)$$

and

$$F(q) = \int dz' \int dz'\, |\chi_0(z)|^2 |\chi_0(z')|^2 \exp(-q|z-z'|) \qquad (21)$$

Expression (19) for the static dielectric function is valid only for $T = 0$. At $T \neq 0$ a more complex equation including the polarizability of 2D gas at finite temperatures has to be included. A simple expression for the temperature dependent dielectric function has been obtained and used to calculate the mobility in AlGaAs/GaAs MDHs [26],

$$\varepsilon(q) = \left[1 + \frac{q_s}{q}\right] \quad (22)$$

where

$$q_s = \frac{2\pi e^2}{\varepsilon_0} \frac{N_s}{k_B T} \left\{\left[1 + \exp(-E_F/k_B T)\right] \ln\left[1 + \exp(E_F/k_B T)\right]\right\}^{-1} \quad (23)$$

An inspection of Eqs. (18) and (20) provides an insight into how the remote impurity scattering is reduced by the separation of the impurities from the quantum well. For the impurities located in the $z' < -d$ region of the barrier and for 2D gas located at $z=0$ the scattering rate is reduced by an exponential factor $\exp(-2k_F d)$. Therefore, for typical Fermi wavevectors of 10^6 cm^{-1}, a large enhancement of the remote impurity mobility is expected for $d > 10^{-6}$ cm.

3.2. Acoustic phonon scattering

An acoustic wave propagating in a crystal can couple to electrons and holes by deforming the crystal and affecting the positions of the conduction and valence band edges. In binary, partly ionic compounds an additional coupling between acoustic phonons and the carriers stems from the piezoelectric effect produced by the deformation of the unit cell. Both deformation potential and piezoelectric coupling of charge carriers to acoustic phonons are well known and were extensively studied in 3D compound semiconductors [4].

In 2D MDHs it is customarily assumed that the phonons can freely propagate in all directions. Therefore one considers a confined electron or hole gas which interacts with phonons which are 3D in nature [35-41]. Under such assumptions the relaxation times due to deformation potential and piezoelectric acoustic phonon scattering are given by [40],

$$\tau_{DP}^{-1} = \frac{3 m_e^* a_c^2 b_0 k_B T}{16 \pi \hbar^3 C_l} \int_0^\pi S^2 (1 - \cos\theta) d\theta \quad (24)$$

and

$$\tau_{PE}^{-1} = \tau_L^{-1} + 2\tau_T^{-1} \quad (25)$$

where

$$\tau_{L,T}^{-1} = \frac{\hbar \alpha_{L,T}}{k_B T k^2} \int qS^2 f_{L,T}(q) d\theta \quad (26)$$

The screening factor S has the form,

$$S^2 = \frac{q^2}{(q + q_s H)^2} \qquad (27)$$

where

$$H = \frac{1 + \tfrac{9}{8}r + \tfrac{3}{8}r^2}{(1+r)^3}, \quad r = \frac{q}{b_0} \qquad (28)$$

a_c is the deformation potential constant and c_l is the elastic constant. The functions $f_{L,T}(Q)$ and constants $\alpha_{L,T}$ are given in ref. [40]. The factor S in Eqs. (24) and (26) represents the screening by free electrons. When the screening is neglected for the short range deformation potential coupling, one obtains the following simple result,

$$\tau_{DP}^{-1} = \frac{3 m_e^* a_c^2 k_B T b_0}{16 C_l \hbar^3} \qquad (29)$$

From this expression one finds that the electron mobility is inversely proportional to the temperature T and the parameter b_0. Therefore, for the deformation potential limited mobility one expects a very weak dependence on the electron density,

$$\tau_{DP} \sim b_0^{-1} \sim \left(N_S + \frac{32}{11} N_{dep} \right)^{-1/3} \qquad (30)$$

The piezoelectric coupling limited mobility shows a more complex dependence on the electron density. However, in most instances the deformation potential is the predominant scattering process which determines the total acoustic phonon scattering.

The expressions (24) and (26) were obtained under the assumption that the thermal energy $k_B T$ is larger than the energy of acoustic phonons participating in the scattering process, $k_B T > \hbar \omega_q$. Under such conditions the phonon distribution function can be approximated by,

$$N_q = \frac{1}{\exp\left(\dfrac{\hbar \omega_q}{k_B T}\right) - 1} \simeq \frac{k_B T}{\hbar \omega_q} \qquad (31)$$

At low temperatures this equipartition condition is not satisfied. Also, at such low temperatures the scattering by acoustic phonons cannot be treated as an elastic process. This temperature range corresponds to the so-called Bloch - Grüneisen (B-G) regime. [42,43]. Theoretical studies have shown that in this regime the acoustic phonon mobility is very rapidly increasing with decreasing temperature [43,44]. Recently, a strong increase of the acoustic phonon limited

mobility has been experimentally observed by subtracting the temperature independent part of the electron mobility in high quality AlGaAs/GaAs MDHs [43]. It should be noted, however, that at very low temperatures the scattering by charged centers is the dominant process limiting electron and hole mobilities and that the B-G regime cannot be clearly distinguished from direct measurements of the electron or hole mobilities.

3.3. Optical phonon scattering

A proper treatment of optical phonon scattering is by far the most complex problem of charge transport in 2D systems. The large optical phonon energy (30 meV to 40 meV) makes the scattering process highly inelastic even at room temperature. In addition, since the typical energy separation between electric subbands is of the order of 10 meV, an optical phonon can couple several subbands. Any theoretical treatment of this problem requires an accurate knowledge of the energies and wave-functions for all those subbands [45]. Numerical calculation of the optical phonon scattering rates for AlGaAs/GaAs MDH have shown that although the 2D confinement leads to reduction of the optical phonon mobility at 77 K, the room temperature mobilities in the 2D and 3D cases are practically the same [45]. This finding is consistent with the experimental results which demonstrated that the optical phonon limited mobility does not depend on the size of the 2D confinement or on the density of 2D electrons [46].

These results provide justification for another approach proposed in Ref. (26) where it has been argued that since the optical phonons probe a wide range of subbands, therefore, the density of states participating in transport is an average over all those subbands. It can be shown that the total density of states for several subbands resemble the density of state of 3D gas. Consequently, one can use a 3D approximation to calculate the optical phonon scattering in 2D MDHs.

3.4. Alloy disorder scattering

For MDH involving ternary compounds, an additional scattering from random alloy-disorder potential has to be included [47]. Two distinctly different types of MDHs containing ternary compounds are possible [48,26]. In the first one the well is formed by a binary compound and the barrier is formed by a ternary alloy. In such MDH there is no significant alloy scattering. The only contribution to this scattering process is through the scattering of the charge carriers penetrating into the barrier. In the other type of structure with the well formed by the ternary alloy the charge carriers are very efficiently scattered by the disordered alloy in the well.

Since the alloy disorder scattering results from a potential, localized to within one unit cell, there is a formal similarity between this scattering

mechanism and the deformation potential, acoustic phonon scattering. In the case of the quantum well formed by the ternary compound the relaxation time is given by Eq. (24) with $a_C^2 k_B T/C_l$ replaced with $x(1-x)\Omega \langle V \rangle^2$ where x is the alloy composition, Ω is the unit cell volume and $\langle V \rangle$ is the alloy disorder parameter. In order to evaluate the carrier mobility for 2D gas confined within the binary semiconductor, the scattering rate has to be reduced by the additional factor [26],

$$\frac{64 \pi^2 e^4 \hbar (N_s/2 + N_{dep})^2}{3 \varepsilon_0^2 (2 m^*)^{1/2} V_0^{5/2} b_0} \qquad (32)$$

For a very large band offset, V_0, the penetration of the wavefunction into the barrier is very small and, as is seen from Eq. (32), the contribution of alloy-disorder scattering is very low.

3.5. Interface roughness scattering

Despite the atomic scale precision of the modern epitaxial growth techniques, most of the epitaxially grown interfaces exhibit some degree of structural roughness. A rough interface can scatter the carriers confined in the well. The interface roughness plays a very significant role in scattering of electrons in Si metal-oxide-semiconductor structures [49]. The scattering process is especially significant at higher electron densities. The theoretical expression originally obtained for Si inversion layers [9] has been adopted to 2D gas in MDHs [25],

$$\tau_{IR}^{-1} = \frac{2\pi}{\hbar} \sum_{\vec{q}} \left[\frac{\Delta \Lambda F_{eff}}{\varepsilon(q)}\right]^2 \exp\left(-\frac{1}{4} q^2 \Lambda^2\right) \times (1 - \cos\theta) \delta(E_{\vec{k}} - E_{\vec{k}-\vec{q}}) \qquad (33)$$

Here, Δ represents the mean square deviation of the height and Λ is a measurement of the lateral spatial decay rate of the roughness, respectively. The effective field is given by

$$F_{eff} = \frac{4 \pi e^2}{\varepsilon_0} \left(\frac{1}{2} N_s + N_{dep}\right) \qquad (34)$$

One finds from Eq. (33) that the interface-roughness scattering rate is greatly reduced for Λ greater than the carrier wavelength, $2\pi/k_F$. This can be contrasted with the case of nonspecular scattering of a quasi-3D electron gas in wide-parabolic quantum wells [50], where the largest contribution comes from large size fluctuations. The interface-roughness has been frequently invoked to explain lower than usual mobilities in MDHs [51,52]. However, since there is no independent way to measure the parameters Δ and Λ, it is difficult to ascertain the relative importance of this scattering compared with the other low temperature scattering processes.

3.6. Intersubband scattering

All the above considerations of the elastic scattering process were restricted to the case of carrier transport within a ground electric subband. This approximation is strictly applicable only to the transport at low temperatures and low carrier densities when only the lowest subband is occupied. In heavily doped MDHs one has to account for the occupation of higher lying subbands [45, 53]. A proper treatment of multisubband transport is difficult since one has to consider the momentum randomizing events within all occupied subbands, as well as the scattering between the subbands. Such a problem, in general, requires solving of a set of coupled Boltzman equations which can be done only numerically. Numerical solutions of multisubband transport was presented in Refs. [45, 53, 54].

In Ref. [26] a simplified approach to the transport in the two lowest subbands has been proposed. Instead of solving two coupled Boltzman equations it has been assumed that since the concentration and momentum of the carriers in the upper subband is much smaller than those in the ground subband, therefore, one can neglect the contribution of the upper subband to the total conductivity. The only effect of this subband is that it provides additional channel for the carrier scattering. Since there is a large momentum change for the carriers which are scattered from the ground to the excited subband, therefore, such scattering processes will lead to a substantial reduction of mobility of the carriers. The intersubband scattering rates for various scattering potentials were derived in Ref. [26] and were used to calculate scattering rates in AlGaAs/GaAs and InAlAs/InGaAs MDHs.

3.7. Scattering mechanisms in p-type MDH

As has been discussed in Section 2, the transport of holes in p-type MDHs has attracted much less attention. The available experimental data is restricted to AlGaAs/GaAs MDHs [23,55-57]. Also, the only existing theoretical treatment of the 2D hole transport is limited to a simple approach proposed by the present author and based on the assumption of decoupled spin-up and spin-down subbands [32,38,58]. It has been argued that since the light holes in the spin-down subband are very efficiently scattered by the heavy holes in spin-up subband, the mobility of holes in spin-down states is much lower. This, combined with a much smaller concentration of light holes in spin-down band, provides justification for neglecting the light hole contribution to the total conductivity. Under such circumstances the mobility of a 2D hole gas can be described in terms of carrier transport in a single subband. This allows use of the methods previously developed for a single parabolic electronic subband with one important modification which takes into account the p-type symmetry of the valence band Bloch wavefunctions. The scattering rates for long range

interactions (Coulomb and piezoelectric acoustic phonon scatterings) given by Eqs. (18) and (26) have to be multiplied by a factor of $1/4(1+3\cos^2\theta)$.

In the case of the deformation potential interaction the deformation potential constant a_c in Eq. (24) is replaced by an effective valence band deformation potential, E_{dp}, which has been shown to have the form [38],

$$E_{dp}^2 = E_{intra}^2 + E_{inter}^2 \qquad (35)$$

where

$$E_{intra}^2 = \frac{C_l + C_t}{4C_L}\left[(l+m)^2 + n^2\right] + \frac{C_l}{4C_t}m^2 \qquad (36)$$

and

$$E_{inter}^2 = \frac{m_2^*}{3m_1^*}\frac{C_l}{C_t}m^2 \qquad (37)$$

represent intra and inter subband scattering, respectively, C_t is the transverse elastic constant, m_1^* and m_2^* are heavy (spin-up) and light (spin-down) hole effective masses, and l, m and n are the valence band deformation potential parameters [59].

As is shown in section 3.4., the strength of the alloy-disorder scattering is determined by the parameter, $\langle V \rangle$. For the conduction band electrons $\langle V \rangle = \langle S | V | S \rangle$ where $\langle S |$ is the conduction band Bloch amplitude and V is the difference of the core potentials of two elements in the alloy. For the valence band hole the alloy-disorder parameter $\langle V \rangle$ is replaced by $\langle V_v \rangle$, given by,

$$\langle V_v \rangle = \langle X | V | X \rangle \qquad (38)$$

where X is the Bloch amplitude for the Γ_8 valence band. The parameters $\langle V \rangle$ and $\langle V_v \rangle$ can be related to the conduction band and valence band offsets for the binary compounds forming the alloy.

4. SCATTERING MECHANISMS IN WIDE PARABOLIC WELLS

So far we have considered 2D systems in which charge carriers are confined in one direction, and can freely move in a 2D plane. Such systems satisfy both conditions (1) and (2), necessary for a fully quantized 2D motion of the charge. Also, modulation (or remote) doping allows for a dramatic reduction of ionized impurity scattering. In standard 3D structures the ionized impurities and charge carriers are not spatially separated and ionized impurity scattering is the dominant process limiting the mobility at low temperatures.

Recent progress in the epitaxial growth of AlGaAs films allowed a realization of a new type of structure which has all the basic features of 3D

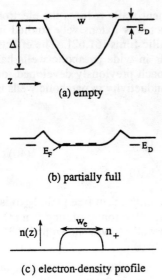

Fig. 3: Schematic illustration of the conduction-band edge in (a) an empty parabolic well, (b) a partially full well. The electron density profile is shown in (c). (After ref [60]).

systems and still has impurities removed from the immediate vicinity of the charge carriers [19,20]. In modulation (or remotely) doped wide parabolic wells the composition of $Al_xGa_{1-x}As$ in the well is parabolically varied with the distance. As shown in Fig. 3, for the case n-type AlGaAs/GaAs structure, this creates a well with the conduction band edge parabolically dependent on the distance [60]. If such a structure is doped in the regions outside the well, the electrons are transferred to the well. The resulting electrostatic potential flattens the bottom of the well. Consequently, one obtains a structure in which electrons are confined in a wide well and are separated from their parent donors. This design allows preparation of wells with a thickness of the order of 1000 Å and with the electron concentration of about 10^{16} cm^{-3} [60].

From condition (1) we find that the de Broglie wavelength of the electron gas in such systems is smaller than the well width. Therefore, the electron motion in the well is not quantized. Such system can be considered a model of modulation doped 3D electron gas, which should exhibit an enhanced low temperature electron mobility. In fact, mobilities as high as 3×10^5 cm^2/V·s were observed in AlGaAs/GaAs wide parabolic wells [50]. Although such mobilities are much higher than those observed in standard 3D semiconductors, they are still much lower than mobilities in 2D MDH or what one would expect in any modulation doped system.

The reason for this lower than expected mobility lies in nonspecular scattering of the electrons from rough confinement walls. As has been discussed in Section 1, in addition to the condition (1), one also has to consider condition (2), which in general is less restrictive than Eq. (2), and is easily satisfied in 2D MDHs. In wide parabolic wells with $\simeq 3\times10^5$ cm^2/V·s and n $\simeq 3\times10^{16}$ cm^{-3}/V·s,

$$l_p \simeq 10^4 \text{ Å} > w_e = 10^3 \text{ Å}$$

This means that although the electron motion is not quantized in the parabolic well the mean free path of the electron is much larger than the well width. The transport of electrons in the well will be affected by the interaction with the confining walls.

The conditions resemble the situation frequently observed in thin metallic films characterized by a small de Broglie wavelength and high electron mobility.

The interaction of the electrons with the confinement walls reduces the electron conductivity in the plane of the well. This effect was extensively studied in metals and is known as the size effect in thin metallic films [61,62]. The effects of electron interaction with the confinement walls in wide parabolic wells has been considered in Ref. [50]. Adopting the approach previously developed for thin metallic films [63], it can be shown that the conductivity in parabolic wells is given by,

$$\sigma = \frac{3\sigma_b}{4} \int_0^\pi d\theta \frac{\sin^3(\theta)}{1 + l_b/l_s(\theta)} \qquad (40)$$

where $\sigma_b = ne\mu_b$ is the bulk conductivity, l_b is the bulk mean free path, $l_s(\theta)$ is the effective electron mean free path associated with electron scattering from the confining wells, and θ is the angle of incidence at a wall. It was shown that [63],

$$l_s(\theta) = -w_e/\{|\cos(\theta)| \ln[p(\theta)]\} \qquad (41)$$

where $p(\theta)$ is the specularity parameter representing the probability that an electron will be specularly reflected from a wall, and w_e is the well width.

For randomly rough walls the specularity parameter takes the form [64],

$$p(\theta) = \exp\left[-\left[\frac{4\pi\alpha}{\lambda}\right]^2 \cos^2\theta\right] \qquad (42)$$

where α is the roughness parameter, and λ is the electron wavelength. It has been argued that the parameter α should be of the order of Thomas-Fermi screening length [50],

$$\alpha = R_s = \frac{2\sqrt{2}\, e^2 m^{*3/2} (k_B T)^{1/2}}{\pi \hbar^3 \varepsilon_0} F_{-1/2}\left(E_F/k_B T\right) \qquad (43)$$

where $F_{-1/2}(\cdot)$ is the Fermi-Dirac integral of the order $-1/2$. Substitution of Eqs. (41) and (42) into Eq. (40), and integration over θ yields an analytic expression for the electron mobility in the parabolic quantum well,

$$\mu = \mu_b\, G(c) \qquad (44)$$

where

$$G(c) = c^{-1/3} \frac{1}{4} \ln\left[\frac{\left(1+c^{1/3}\right)^3}{(1+c)}\right] = \frac{\sqrt{3}}{2}\left[\tan^{-1}\left[\frac{2c^{1/3}-1}{\sqrt{3}}\right] + \frac{\pi}{6}\right] \quad (45)$$

where $c = 4\, l_b\, k_F^2\, \alpha^2 / w_e$. For perfectly smooth walls $\alpha \to 0$ or for a very wide well $w_e \to \infty$, $c \to 0$ and, as expected, $\mu \to \mu_b$.

5. CALCULATED MOBILITY AND COMPARISON WITH EXPERIMENT

Lattice matching at heterointerfaces is one of the principal requirements to achieve high quality, high mobility MDHs. To date the $Al_xGa_{1-x}As/GaAs$ heterointerface is the system most often used to fabricate MDHs. Since AlAs is quite well lattice matched to GaAs, therefore, there is the possibility of preparing MDHs with variable composition of Al in AlGaAs barriers. This adds a degree of freedom in creating MDHs with variable band offsets or to taylor the barrier height for specific applications. Other possible lattice matched systems do not offer such flexibility. For example, $In_xGa_{1-x}As$ is lattice matched to InP only for x = 0.53. Also, $In_yAl_{1-y}As$ is lattice matched to either InP or $In_{0.53}Ga_{0.47}As$ only for y = 0.48. Recently, there has been significant progress in improving the quality of the lattice mismatched pseudomorphic heterosystems. However, the electron or hole mobilities in strained systems are still much lower than those in the lattice matched MDHs [14].

In this section we shall present results of calculations of electron and hole mobilities in various MDHs and in wide parabolic wells. The results will be compared with existing experimental data. We will also discuss several aspects of carrier transport in artificially structured systems which are still controversial and not fully understood. Here we present results of calculations of electron mobility in MDHs in which the electron gas is confined in GaAs or in $In_{0.53}Ga_{0.47}As$ with the barrier formed by either AlGaAs, in the former, or by InP or $In_{0.48}Al_{0.52}As$ in the latter case.

5.1. N-type $Al_xGa_{1-x}As/GaAs$ MDH

The parameters describing the band structure of $Al_xGa_{1-x}As/GaAs$ MDHs, along with the coupling parameters for electron-phonon interactions, are known from studies of electron transport in bulk GaAs. Here we adopt the parameters listed in Ref. [38]. We also assume a value of 0.26 eV for the conduction band offset at the $Al_{0.35}Ga_{0.65}As/GaAs$ interface and a value of $\langle V \rangle = 0.5$ eV for the alloy disorder parameter in AlGaAs. This value of $\langle V \rangle$ is smaller than the previously used $\langle V \rangle = 1$ eV {25,26]. However, it should be

pointed out that the alloy disorder scattering contribution to the total scattering is negligibly small with the possible exception of the MDHs with high 2D electron densities, approaching 10^{12} cm^{-2} [26].

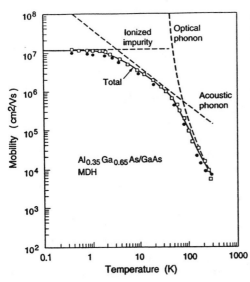

Fig. 4: Theoretical temperature dependence of the electron mobility in AlGaAs/GaAs MDH with a spacer width of 700 Å. Points are experimental data of ref. [21] (□) and ref. [22] (•) for the spacer width 700 Å and 750 Å, respectively.

The calculated temperature dependence of the electron mobility in Al$_{0.3}$Ga$_{0.7}$As/GaAs MDH is shown in Fig. 4. Comparison of these calculations with available experimental data [21,22] on high quality MDHs indicates that in the wide temperature range of 4 K to 300 K the electron mobility is determined by phonon scattering. The optical phonons dominate at T > 70 K and the acoustic phonon scattering plays a major role in the range 4 K to 40 K. At even lower temperatures the mobility levels off due to the scattering from remote impurities in the barrier and residual impurities or charged defects in nominally undoped quantum wells. The alloy disorder scattering of the electrons penetrating the AlGaAs barrier is very small and amounts to about 4% of the total scattering in MDHs with an electron mobility of 10^7 cm^2/V·s.

The mobility in Fig. 4 has been obtained assuming the equipartition condition for the acoustic phonons. As has been shown previously [43] this condition is valid only for T ≳ 4 K. At lower temperatures the acoustic phonon mobility drastically increases with decreasing temperature, $\mu_{ac} \sim T^{-s}$, where the exponent s depends on the type of coupling. Mobility calculations in the B-G regime show that s=5 for piezoelectric scattering [43]. Also, it has been found that for the screened deformation potential, s=7, whereas it is reduced to s=5 when the screening is not included [43]. As is seen in Fig. 4, the B-G regime cannot be clearly observed in the experimental data because of the onset of temperature independent scattering from charged centers.

The electron mobility due to optical phonons was calculated using the 3D approximation [26]. The justification for the approximation has been discussed in Section 4.4. It is also seen in Fig. 4 that this approximation quite well describes the 2D electron mobility for T > 100 K, supporting the conclusion

about insensitivity of optical phonon scattering to electron gas confinement [46]. Since the ionized impurity scattering contributes only at low temperatures where the 2D electron gas is degenerate, the mobility due to this scattering is independent of the temperature for $T < 40$ K. In high-quality MDHs the ionized impurity mobility is temperature dependent only at higher temperatures where the ionized impurity contribution to the total scattering is insignificant.

The calculated total mobility in Fig. 4 has been obtained using the Matthiessen's rule, i.e., instead of averaging the total microscopic relaxation time over energy using equation (15), one performs the averaging for each of the scattering mechanisms separately and then combines all the relaxation times to obtain the total mobility. Using both methods we find that for the MDH shown in Fig. 4, the error of using Matthiessen's rule does not exceed a few percent.

As seen in Fig. 4, simple model calculations account extremely well for the temperature dependence of the electron mobility in high-quality AlGaAs/GaAs MDHs. In the structures reported in Refs. [21] and [22] the low temperature mobility exceeds 10^7 cm^2/V · s and is mostly determined by scattering from remote impurities. In the MDH measured in Ref. [22] with a spacer width of 750 Å additional scattering from the background impurities with concentrations of only 2×10^{13} cm^{-3} is required to account for the maximum mobility of 1.05×10^7 cm^2/V·s.

The early studies of AlGaAs/GaAs MDH have demonstrated that the mobility of 2D electrons depends on the growth sequence of the epitaxial layers forming the heterostructure [65]. The very high mobilities can be achieved only in so-called normal-MDH (N-MDH) in which high purity well-forming GaAs is grown first, followed by the growth of the barrier forming, doped AlGaAs. In an inverted-MDHs (I-MDH) in which the growth sequence is reversed the mobilities are typically about one order of magnitude lower than in a normal MDH.

Several explanations have been put forward to understand the reduced mobility in I-MDHs. It has been proposed that the inverted heterointerfaces are less structurally perfect [65]. These imperfections would scatter the electrons resulting in lower mobility [65]. It has also been suggested that a diffusion of impurities from heavily doped regions in AlGaAs towards the well can be responsible for an increased ionized impurity scattering [66].

Most recently a new explanation for the mobility reduction in I-MDH has been advanced [67]. It has been shown that there is a significant difference in the native defect incorporation in I- and N-MDHs. In I-MDH the well-forming GaAs layer is grown in the presence of electrons transferred from heavily doped barriers forming AlGaAs. The electrons enhance the incorporation of native charged defects [68]. These defects act as scattering centers reducing the electron mobility in the well. The effect of scattering by the charged defects in the well of a typical AlGaAs/GaAs I-MDH is shown in Fig. 5. The electron mobility is found to be at least one order of magnitude lower than that expected in an

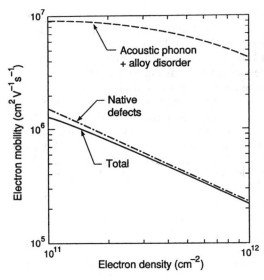

Fig. 5: Electron mobility as a function of 2D electron gas density in inverted AlGaAs/GaAs MDH at 4K. The native defect scattering is the dominant mechanism limiting the mobility.

equivalent N-MDH. The proposed explanation is supported by studies which indicate that a long range Coulomb potential, rather than short range interface roughness, is responsible for the low mobilities observed in I-MDHs [69].

In most high-quality MDHs, the 2D electron density is very low when the structure is measured in the dark. However, one can increase the density by illumination of heavily doped AlGaAs with infrared light of energy ~ 1 eV [70]. The illumination transfers the deep DX-like donors into a shallow donor configuration leading to much more efficient charge transfer to the well. A typical light induced 2D density is ~ 2×10^{11} cm^{-2} to 3×10^{11} cm^{-2} [21,22]. It has been also shown that the concentration of electrons can be controlled with an external electric field applied to the backside of MDH [71]. By increasing the electron density one can reach conditions where the higher lying excited subbands are occupied. Experimental results indicate that the onset of the occupation of the first excited subband is always associated with a decrease of the measured electron mobility. Systematic studies have shown that in AlGaAs/GaAs MDH, the mobility decreases by about 30% when the first excited subband is occupied [71,73]. Simple theoretical calculations in Ref. [26] are consistent with these results. It has been found that the onset of the intersubband scattering reduces the electron mobility due to acoustic phonons and background ionized impurities by about 28%. It has been also shown that the remote ionized impurities contribute very little to the intersubband scattering due to an exponential reduction of the scattering rates at large momentum transfers.

5.2. N-type $In_{0.53}Ga_{0.47}As$ based heterostructures

In InP/InGaAs and InAlAs/InGaAs MDHs the well-forming $In_xGa_{1-x}As$ is lattice matched to InP and InAlAs for x=0.53. There is one important difference between the charge transport in GaAs and in InGaAs. In the latter case the

electrons are very efficiently scattered by the alloy-disorder potential in the well. Early experimental results on the 2D electron gas mobilities have shown that the mobility is independent of temperature for T < 50 K [74]. Also, the highest reported mobilities were limited to about 10^5 cm^2/V·s. These results were confirmed on both InAlAs/InGaAs [75] and on InP/InGaAs [76] MDHs.

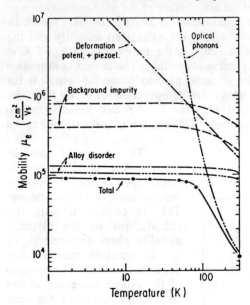

Fig. 6: Electron mobility vs. temperature in $In_{0.48}Al_{0.52}As/In_{0.53}Ga_{0.47}As$ MDH with 80 Å spacer. Experimental points are from ref. [74]. The upper and lower background impurity mobilities correspond to $N_i^b = 5 \times 10^{15}$ cm^{-3} and 10^{16} cm^{-3}, respectively. Alloy disorder potential $\langle V \rangle$ was used as a fitting parameter. (After ref. 26)

Theoretical calculations have demonstrated that the mobility at these low temperatures is determined by alloy disorder scattering in the well [26,48,77]. Fig. 6 shows the comparison of the theoretical calculation with the experimental data of Ref. [74]. The material parameters used in the calculations are given in Ref. [26]. Because of high electron density in the MDH studied [74], the calculations included the effects of the first excited subband using the approximate approach outlined in Section 3.6. An analysis of the low temperature mobility data was used to determine the alloy-disorder parameter $\langle V \rangle$. It has been found that the experimental results can be explained assuming $\langle V \rangle$ to be in the range of 0.55 eV to 0.63 eV. This value very favorably compares with the independent determination of $\langle V \rangle = 0.6$ eV reported in Ref. [78].

5.3. P-type AlGaAs/GaAs MDH

As has been discussed in Section 3.7, there is only a very limited amount of experimental data on transport properties of 2D holes in MDHs. Practically all of the data is restricted to the AlGaAs/GaAs system [23,55,56]. Similarly, as in the case of electrons, modulation doping leads to a large increase of the 2D hole mobilities. A low temperature mobility as high as 3.8×10^5 cm^2/V·s was measured at T = 0.3 K [23]. The only theoretical calculations of the 2D hole

mobilities are based on the model of decoupled electric subbands [32,38,58]. It has been shown in these calculations that only the heavy holes contribute significantly to the total conductivity in a MDH. The contribution of the light holes is much smaller since they are very efficiently scattered by the heavy holes. Therefore, the transport in p-type MDHs can be described in terms of single heavy hole electric subband. Theoretical calculations of the hole mobility in a p-type AlGaAs/GaAs MDH with the hole density of 2×10^{11} cm^{-2} are shown in Fig. 7 [58]. Very good agreement between the calculated mobility and the experimental data of Ref. [57] is obtained. In the temperature range 4 K to 80 K the mobility is determined by phonon scattering. The largest contribution to the scattering is coming from the acoustic phonon scattering which is the dominant scattering mechanisms up to the temperature of ~ 60 K. The discrepancy between theory and experiment for T < 4 K can be attributed to the background ionized impurity scattering. This scattering mechanism has not been included in the calculations.

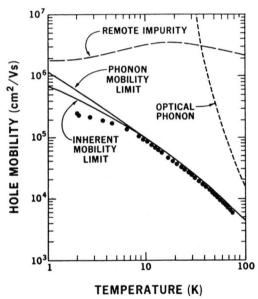

Fig. 7:Temperature dependent hole mobility in p-type Al$_{0.5}$Ga$_{0.5}$As/GaAs MDH. The points represent experimental data of ref. [57] for the hole density $P = 2 \times 10^{11}$ cm^{-2}. (After ref.[58])

The theoretical results in Fig. 7 were obtained for an unscreened acoustic phonon deformation potential scattering. This is consistent with the calculations of the electron mobility where the screening of the deformation potential has not been included. The problem of the proper screening of the short range potentials has been discussed in several papers. It has been shown recently that the electron energy loss rates [27], as well as phonon-drag contribution to the thermoelectric power [39] in 2D electron systems, can be explained assuming the unscreened deformation potential. However, it has also been argued by others that the dependence of the phonon scattering on the electron density can be better understood when the screening is included in the calculations [79,80].

It has been found that to explain experimentaly observed mobilities with a screened deformation potential very high values of deformation potential

constants are required. Thus, in AlGaAs/GaAs MDH, deformation potentials as high as 13.5 eV were used to explain the value of the acoustic phonon limited mobility [80]. Such a high value is at variance with the results on electron mobility in high purity GaAs [81] and is much higher than the value of $a_C = -9.3$ eV directly determined by an independent experiment [82]. Also, theoretical calculations has provided even a lower value of $a_C = -7.3$ eV [83].

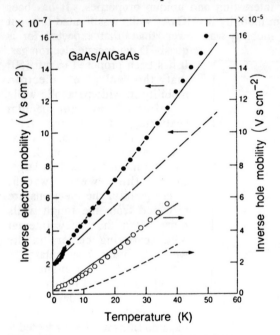

Fig. 8: Temperature dependence of the inverse electron and hole mobilities in AlGaAs/GaAs MDHs. The experimental points are from refs [57]; (o) and ref. [84]; (•). The solid and broken lines represent the calculated mobilities for unscreened and screened deformation potentials, respectively. (After ref. [38])

The analysis of electron and hole mobilities in high quality MDH provides an independent test for the values of the conduction and valence band acoustic phonon deformation potentials [58]. As is shown in Fig. 8, the temperature dependences of inverse electron and hole mobilities are well explained by calculations in which acoustic phonon deformation potential interaction is not screened by free carriers. The best fit is obtained for $a_C = -9.3$ eV and $a_V = -0.7$ eV. These values of the deformation potentials were independently determined from other experiments [82]. It is also seen in Fig. 8 that incorporation of screening very significantly reduces the acoustic phonon scattering rates which are much lower than those determined experimentally. Therefore, the results in Fig. 8 support the notion that a simple Fermi-Thomas screening of the short range deformation potential cannot be used in 2D systems.

5.4. Wide parabolic AlGaAs/GaAs wells

The wide parabolic wells were designed to simulate quasi-3D systems with reduced ionized impurity scattering. It has been proposed that such systems could exhibit Wigner crystallization at very low temperatures. Although the crystallization has not ever been clearly observed, it has been shown that wide-parabolic wells exhibit other interesting and unique properties. It has been demonstrated that the maximum mobility in AlGaAs wells was limited to about 3×10^5 cm^2/Vs. This value of mobility was lower than that expected for a quasi-3D degenerate electron gas. It has been proposed in ref. [50] that the value of electron mobility in wide parabolic wells is affected by nonspecular scattering of electrons from rough confining walls. As has been shown in Section 3, the electron mean free path is larger than the well width. Consequently, the electrons are reflected from the rough walls more often than they interact with scattering centers in the well. In order to calculate the effect of the nonspecular reflection, a model theory which has been previously developed to calculate the size effect in thin metallic films has been adopted to evaluate electron mobility in parabolic wells. Results of the calculations are shown in Fig. 9. The results indicate that the nonspecular scattering reduces the electron mobility by a factor of 2 to 3. It is also seen in Fig. 9 that incorporation of nonspecular scattering accounts for the experimentally observed temperature dependence of the mobility.

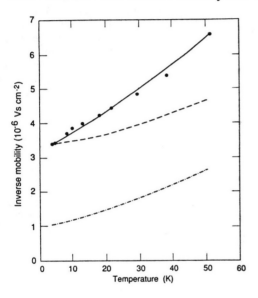

Fig. 9: Temperature-dependent inverse mobility in wide-parabolic AlGaAs/GaAs wells. (___), scattering from rough walls included, ionized impurity concentration $N_i^b = 2.3\times10^{14}$ cm^{-3}; (-----), no scattering from the walls, $N_i^b = 2.3\times10^{14}$ cm^{-3}; (-----) no scattering from the walls, $N_i^b = 10^{15}$ cm^{-3}. (After ref. [50])

6. CONCLUSIONS

In this chapter we have reviewed basic aspects of electron and hole transport in artificially structured 2D systems. The concept of modulation (or selective) doping has provided structures with extremely high electron and hole mobilities. Very significant reductions of impurity scattering in modulation doped systems allowed extensive studies of phonon scattering down to very low temperatures. These studies raise several issues concerning the free carrier screening of acoustic phonon deformation potential. It is now quite evident that the reduced dimensionality is clearly observed only at low temperatures. At higher temperatures of $T > 100$ K charge transport in 2D systems is very similar to that observed in 3D high purity semiconductors. For example, it has been demonstrated that charge confinement does not affect highly inelastic scattering by optical phonons.

Although this review was restricted to lattice matched semiconductor systems, most of the concepts and methods can be easily adopted to lattice mismatched and pseudomorphic systems. At strained semiconductor interfaces an important complication arises from the fact that one has to consider the effect of planar strain on the electronic structure. The effects of the strain are especially complex in the case of the degenerate valence bands. Also, the electronic transport depends on whether the strained layer is unrelaxed, partially, or fully relaxed, since it can be affected by structural defects at the interfaces. All these issues are of great importance for strained GaAs/InGaAs and Si/SiGe heterostructures which are considered the most promising systems for a high performance heterjunction bipolar transistors.

Spectacular improvements in electron and hole mobilities found in AlGaAs/GaAs MDHs are not matched by other semiconductor systems. Interesting and potentialy very promising heterostructures are based on III-V and II-VI narrow band gap compounds Because of very small electron effective mass in these materials one should, in principle, be able to achieve electron mobilities much higher than those in GaAs based heterostructures. The progress in this area will, however, depend on further improvements in the quality of the epitaxial layers of these materials. One of the main problems is still much lower purity of the narrow gap materials compared with GaAs or AlGaAs thin films. In narrow gap II-VI semiconductors, lower than expected mobilities can be also attributed to the difficulty in preparing high quality abrupt heterointerfaces. Reducing of the heterointerface intermixing will be necessary to take full advantage of HgTe or HgSe based modulation doped heterostructures.

Charge transport in 2D systems is now a mature research area with well developed experimental and theoretical methods. Progress in material preparation techniques have made possible to study charge transport in 1D (quantum wires), as well as unusual properties of 0D (quantum dot) systems. An intense effort in several laboratories into development of these new methods of preparation of strongly confined low-dimensional systems has already produced a wealth of new results and certainly will be the research area actively pursued in the future.

ACKNOWLEDGEMENTS

The author would like to gratefully acknowledge helpful discussions with E. E. Haller. This work was supported by the Director, Office of Energy Research, Office of Basic Energy Sciences, Materials Science Division of the U.S. Department of Energy under Contract No. DE-AC03-76SF00098.

REFERENCES

[1] K.v. Klitzing, G. Dorda, and M. Pepper, Phys. Rev. Lett. **45**, 494 (1980).
[2] D. Tsui, H. L. Störmer and A.C. Gossard, Phys. Rev. Lett. **48**, 1559 (1982).
[3] S. Schmitt-Rink, D.S. Chemla, and D.A.B. Miller, Adv. Phys. **38**, 89 (1989).
[4] W. Zawadzki, in *Handbook of Semiconductors*, Vol. **1**, edited by W. Paul (North Holland, Amsterdam, 1982) p. 713.
[5] G.E. Stillman and C.M. Wolfe, Thin Solid Films **37**, 69 (1976).
[6] W. Walukiewicz, J. Lagowski, L. Jastrzebski, M. Lichtensteiger, and H.C. Gatos, J. Appl. Phys. **50**, 899 (1979).
[7] R. Dingle, H.L. Störmer, A.C. Gossard, and W. Wiegmann, Appl. Phys. Lett. **33**, 665 (1978).
[8] T.J. Drummond, H. Morkoc, and A.Y. Cho, J. Appl. Phys. **52**, 1380 (1981).
[9] T. Ando, A.B. Fowler, and F. Stern, Rev. Mod. Phys. **54**, 437 (1982).
[10] S. Hiyamizu, in *Semiconductors and Semimetals*, Vol. **30**, edited by R.K. Wiliardson and A.C. Beer, (Academic Press, Inc., Harcourt Brace Jovanovich, Publishers, 1990), p. 53.
[11] H. Morkoc and H. Unlu, in *Semiconductors and Semimetals*, Vol. **24**, edited by R.K. Wiliardson and A.C. Beer, (Academic Press, Inc., Harcourt Brace Jovanovich, Publishers, 1987), p. 135.
[12] D.S. Chemla, D.A.B. Miller, and P.W. Smith, *ibid*, Vol. **24**, p. 279.
[13] W.T. Tseng, *ibid*, Vol. **24**, p.397.
[14] G.C. Osbourn, P.L. Gourley, I.J. Fritz, R.M. Biefeld, L.R. Dawson, and T.E. Zipperian, *ibid*, Vol. **24**, p. 459.
[15] See e.g., F. Stern, and W.E. Howard, Phys. Rev. **163**, 816 (1967).
[16] A. Zrenner, H. Reisinger, F. Koch, K. Ploog, and J.C. Maan, Phys. Rev. B **33**, 5607 (1986).
[17] L. Esaki and K. Tsu, IBM J. Res. Dev. **14**, 61 (1970).
[18] G. Tuttle, H. Kroemer, and J. English, Appl. Phys. Lett. **65**, 5239 (1989).
[19] M. Sundram, A.C. Gossard, J.H. English, and R.M. Westervelt, Superlattices and Microstructures **4**, 683 (1988).
[20] M. Shayegan, T. Sajoto, J. Jo, M. Santos, and C. Silvestre, Appl. Phys. Lett. **53**, 791 (1988).
[21] L. Pfeiffer, K.W. West, H.L. Störmer, and K.W. Baldwin, Appl. Phys. Lett. **55**, 1888 (1989).
[22] T. Saku, Y. Kirayama, and Y. Horikoshi, Jap. J. Appl. Phys. **30**, 902 (1991).
[23] W.I. Wang, E.E. Mendez, Y. Iye, B. Lee, M.H. Kim, and G.E. Stillman, Appl. Phys. Lett. **60**, 1834 (1986).
[24] T. Ando, J. Phys. Soc. Jap. **51**, 3893 (1982).
[25] T. Ando, *ibid*, **51**, 3900 (1982).
[26] W. Walukiewicz, H.E. Ruda, J. Lagowski, and H.C. Gatos, Phys. Rev. B **30**, 4571 (1984).

[27] Y. Okuyama and N. Tokuda, Phys. Rev. B **40**, 9744 (1989).
[28] F. Stern and W.E. Howard, Phys. Rev. **163**, 816 (1967).
[29] U. Ekenberg and M. Altarelli, Phys. Rev. B **30**, 3569 (1984).
[30] D.A. Broido and L.J. Sham, Phys. Rev. B **31**, 888 (1985).
[31] T. Ando, J. Phys. Soc. Japan **54**, 1528 (1985).
[32] W. Walukiewicz, Phys. Rev. B **31**, 5557 (1985).
[33] See e.g., D.J. Howarth and E.H. Sondheimer, Proc. Phys. Soc. London A **219**, 53 (1953).
[34] D.L. Rode, in *Semiconductors and Semimetals*, Vol. **10**, edited by R.K Wiliardson and A.C. Beer, (Academic Press, Inc., Harcourt Brace Jovanovich, Publishers, 1975), Chap. 1.
[35] P.J. Price, Annals of Physics **133**, 217 (1981).
[36] B.K. Ridley, J. Phys. C: Solid State Phys. **15**, 5899 (1982).
[37] V. Karpus, Sov. Phys. Semicon. **20**, 6 (1986).
[38] W. Walukiewicz, Phys. Rev. B **37**, 8530 (1988).
[39] Y. Okuyama and N. Tokuda, Phys. Rev. B **42**, 7028 (1990).
[40] P.J. Price, Surf. Sci. **143**, 145 (1984).
[41] T. Kawamura and S. Das Sarma, Phys. Rev. B **45**, 3612 (1992).
[42] P.J. Price, Solid State Comm. **51**, 607 (1984).
[43] H.L. Stromer, L.N. Prfieer, K.W. Baldwin, and K. West, Phys. Rev. B **41**, 1278 (1990).
[44] V. Karpus, Semicon. Sci. Technol. **5**, 691 (1990).
[45] B. Vinter, Appl. Phys. Lett. **45**, 581 (1984).
[46] H.J. Polland, W.W. Rühle, K. Ploog, and C.W. Tu, Phys. Rev. B **36**, 7722 (1987).
[47] L. Makowski and M. Glicksman, J. Phys. Chem. Solids **34**, 487 (1973).
[48] G. Bastard, Appl. Phys. Lett. **43**, 591 (1983).
[49] A. Hartstein, T.H. Ning, and A.B. Fowler, Surf. Sci. 58, 178 (1976).
[50] W. Walukiewicz, P.F. Hopkins, M. Sundaram,. and A.C. Gossard, Phys. Rev. B **44**, 10 909 (1991).
[51] H. Sakaki, T. Noda, K. Hirakawa, M. Tanaka, and T. Matsusne, Appl. Phys. Lett. **51**, 1934 (1987).
[52] H. Morkoc, T.J. Drummond, and R. Fischer, Appl. Phys. **53**, 1030 (1982).
[53] S. Mori and T. Ando, Phys. Rev. B **19**, 6433 (1979).
[54] N.T. Thang, G. Fishman, and B. Vinter, Surf. Sci. **142**, 266 (1984).
[55] H.L. Stormer, K. Baldwin, A.C. Gossard, and W. Wiegmann, Appl. Phys. Lett. **44**, 1062 (1984).
[56] W.I. Wang, E.E. Mendez, and F. Stern, Appl. Phys. Lett. **45**, 639 (1984).
[57] E.E. Mendez and W.I. Wang, Appl. Phys. Lett. **46**, 639 (1985).
[58] W. Walukiewicz, J. Appl. Phys. **53**, 3577 (1986).
[59] G.E. Pikus and G.L. Bir, Sov. Phys., Solid State Phys. **1**, 1502 (1958).
[60] E. Gwinn, P.F. Hopkins, A.J. Rimberg, K.M. Westervelt, M. Sundaram, and A.C. Gossard, Phys. Rev. B **41**, l0700 (1990).
[61] K. Fuchs, Proc. Cambridge Philos. Soc. **34**, 100 (1938).
[62] J.R. Schrieffer, Phys. Rev. **97**, 641 (1955).
[63] A.A. Cottey, Thin Solid Films **1**, 297 (1967).
[64] S.B. Soffer, J. Appl. Phys. **38**, 1710 (1967).
[65] H. Morkoc, T.J. Drummond, and R. Fischer, J. Appl. Phys. **53**, 1030 (1982).
[66] S. Sasa, J. Saito, K. Nanbu, T. Ishikawa, and S. Hiyamizu, Jpn. J. Appl. Phys. **23**, L573 (1984).
[67] W. Walukiewicz and E.E. Haller, Appl. Phys. Lett. **58**, 1638 (1991).

[68] W. Walukiewicz, Appl. Phys. Lett. **54**, 2094 (1989).
[69] U. Merirav, M. Heiblum, and F. Stern, Appl. Phys. Lett. **52**, 1268 (1988).
[70] For recent review see e.g. D.V. Lang in *Deep Levels in Semiconductors*, edited by S.T. Pandelides (Gordon and Breach Science Publishers, New York, 1986) p. 489.
[71] H.L. Stormer, A.C. Gossard, and W. Wiegmann, Solid State Comm. **41**, 707 (1982).
[72] D.R. Leadley, R.J. Nicholas, J.J. Harris, and C.T. Foxon, Semicon. Sci. Techn. **5**, 1081 (1990).
[73] R. Fletcher, E. Zaremba, M. DiIorio, C.T. Foxon, and J.J. Harris, Phys. Rev. B **38**, 7866 (1988).
[74] A. Kastalsky, R. Dingle, K.Y. Cheng, and A.Y. Cho, Appl. Phys. Lett. **41**, 279 (1982).
[75] M.A. Tischler and B.D. Parker, Appl. Phys. Lett. **58**, 1614 (1991).
[76] M. Takikawa, J. Komeno, and M. Ozeki, Appl. Phys. Lett. **43**, 280 (1983).
[77] Y. Takeda and A. Sakaki, Jap. J. Appl. Phys. **24**, 1307 (1985).
[78] J.R. Hayes, A.R. Adams, and P.D. Greene in *GaInAsP Alloy Semiconductors*, edited by T.P. Pearsall (Wiley, New York, 1982).
[79] E.E. Mendez, P.J. Price, and M. Heiblum, Appl. Phys. Lett. **45**, 294 (1984).
[80] P.J. Price, Phys. Rev. B **32**, 2643 (1985).
[81] W. Walukiewicz, H.E. Ruda, J. Lagowski, and H.C. Gatos, Phys. Rev. B **32**, 2465 (1985).
[82] D.D. Nolte, W. Walukiewicz, and E.E. Haller, Phys. Rev. Lett. **59**, 501 (1984).
[83] C.G. Van de Walle and R.M. Martin, Phys. Rev. Lett. **62**, 2028 (1989).
[84] J. H. English, A.C. Gossard, H. L. Störmer and K. W. Baldwin, Appl. Phys. Lett., **50**, 1826, (1987).

MINIBAND CONDUCTION IN SEMICONDUCTOR SUPERLATTICES

A. Sibille, J.F. Palmier and C. Minot
Centre National d' Etudes des Télécommunications
196 av. H. Ravera, BP 107, 92225 Bagneux cedex, FRANCE

ABSTRACT

We review experimental data on miniband conduction perpendicular to interfaces of semiconductor superlattices, and theoretical works on semiclassical miniband transport. The experiments adressed by this paper are based on electrical as well as optical techniques, and both without and with an externally applied perpendicular electric field. A comprehensive discussion of Esaki and Tsu's description of miniband conduction is carried out in the frame of Boltzmann equation, and shown to account for most SL non-linear transport properties. Finally the limits of this semiclassical approach, particularly concerning disorder or Wannier-Stark quantization, are discussed.

1 INTRODUCTION

The concept of superlattice (SL) was first introduced in 1970 by L. Esaki and R. Tsu as a new class of man-made artificial crystals characterized by tailorable electronic properties [1]. A superlattice is a periodic alternation of at least two different semiconductors obtained by sophisticated epitaxial growth techniques such as molecular beam epitaxy (MBE). The resulting superperiodicity breaks the conduction and valence bands into a series of minibands and minigaps when the individual layer thicknesses are small enough for quantum size effects to be significant. Additional phenomena may be related to periodic strain or internal electric fields in non-lattice matched or periodic doping systems [2,3].

Esaki and Tsu proposed that the nonparabolicity of the ground state miniband associated with Bragg reflection at the minizone boundary induced unusual nonlinear transport properties perpendicular to interfaces, like negative differential conductance (NDC) at the onset of Bloch oscillations [1]. Unfortunately the formidable technological effort required for the testing of these predictions prevented their successful validation in the seventies, because of the necessary control of interface quality on the atomic level. This challenge undoubtedly has very effectively stimulated the development of MBE, which has progressed steadily since then.

Structural disorder in SL of insufficient quality indeed destroys the coherence of inhomogeneously broadened electronic states to the point of invalidating the very concept of miniband. Many experimental works were then interpreted in terms of individual quantum well to quantum well tunneling phenomena [4-6]. Despite these problems, good indications of true miniband conduction in SL were found in a few cases, using both electrical and optical techniques. It was only recently nevertheless that Esaki and Tsu's original ideas were finally demonstrated, in state of the art GaAs/AlAs superlattices [7,8].

In this paper, we concentrate on both experimental and theoretical aspects of miniband conduction in composition SL and do not consider strain or internal electric fields related effects. Section 2 briefly recalls the notion of miniband and the calculation of a miniband spectrum, while section 3 reviews some experimental observations of miniband conduction in SL. In section 4 we discuss a series of experiments allowing to demonstrate and characterize negative differential velocity (NDV) and NDC in SL. Section 5 theoretically addresses semiclassical miniband conduction in the frame of Boltzmann equation, and highlights the role of interface roughness in dominating transport. The limits of this approach, particularly when either interface disorder or Wannier-Stark quantization prevail, are analysed in section 6.

2 MINIBAND STRUCTURE

In this section we derive the basic electronic properties of semiconductor composition SL from simple considerations of quantum mechanics. Our aim is to give the main features of their band structure, more fully developed in standard textbooks [9], as an introduction to the discussion of conduction properties.

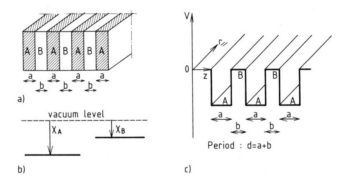

Fig. 1: a) A SL is a periodic stack of alternating layers of two epitaxially compatible materials. b) Due to the different electronic affinities of the two materials, the potential energies of conduction band electrons are offset. c) The electronic potential can be approximated by a periodic square well in the growth direction.

In the following, we specialize our presentation to composition superlattices formed as a succession of alternating layers of two distinct materials A and B (fig. 1a). In an ideal structure all layers of A and B with abrupt interfaces have identical thicknesses a and b respectively, so that a period d=a+b can be defined. This ideal SL can then be considered as a new intrinsic compound, whose properties can nevertheless be altered by extrinsic factors such as deviations from periodicity, interface roughness or doping. Assuming the concept of effective mass for electrons, a pseudo-free particle approximation is valid in bulk semiconductors in a moderate energy range around a band extremum, which amounts to having a flat electronic potential. When no electric field is present, the potential in vacuum is also flat, yielding a reference level which determines the absolute energy of a pseudo-free electron in either material. Figure 1b shows the position of a definite energy

level, for instance the Γ_6 conduction band minimum, as a function of a distance for two isolated layers of semiconductor. Once the B/A heterojunction has been formed, the electron experiences a potential step, the height of which is the difference between electronic affinities $W=|\chi_A-\chi_B|$. In practice W may range from less than 0.1eV up to ~ 1 eV. If $|\chi_A|>|\chi_B|$, material B acts as a potential barrier for an electron in A. The SL potential V can then be viewed as a periodic sequence of square barriers and wells along the z direction and isotropic in the (x,y) planes (fig. 1c). The variations of electronic potential in the underlying semiconductors are neglected (pseudo-free electron approximation), whereas only step variations of the heterojunction potential are retained (abrupt heterojunctions). Although this is a very simplified model for the heterojunction, it is quite successful in the multi-heterojunction structures we are interested in, because it accounts primarily for the dominant term of the potential relevant to weakly bound electrons. It belongs to a broader class of models, referred to as the envelope function scheme, in which the rapidly varying component of the SL wavefunctions is smoothed out on the atomic scale and yields parameters for the calculation of the slower component varying on the SL period scale[8]. The potential V only depends on z and the one-electron Hamiltonian:

$$H = T + V = \frac{1}{2} \vec{p} \frac{1}{m^*(z)} \vec{p} + V(z) \qquad (1)$$

can be solved exactly. Indeed an eigen function of the form:

$$\Psi(\vec{r}) = \phi(z) \exp(i\vec{k}_{//}\cdot\vec{r}_{//}) \qquad (2)$$

with $\vec{k}_{//}=(k_x,k_y)$ and $\vec{r}_{//}=(x,y)$ leads to the eigenvalue equation:

$$-\frac{\hbar^2}{2}\frac{\partial}{\partial z}\left(\frac{1}{m^*(z)}\frac{\partial \phi}{\partial z}\right) + \left(V(z)+\frac{\hbar^2\vec{k}_{//}^2}{2m^*(z)}\right)\phi(z) = \mathcal{E}\phi(z) \qquad (3)$$

where $\mathcal{E}$ is the eigenenergy of state Ψ. If the effective-mass is uniform, $\mathcal{E}$ is the sum of a free motion energy in the planes of the layers and a quantized energy ϵ due to the superlattice potential. As V(z) is piecewise constant, the solution in each individual layer is a linear combination of two plane waves with either real or imaginary wavevector depending wether $\mathcal{G}= \mathcal{E}-\hbar^2\vec{k}_{//}^2/2m^*(z) - V$ is negative or positive. An imaginary wavevector correponds to exponentially growing or decaying functions.

It is instructive to find approximate solutions of H under some simplifying assumptions. First, if b is large enough a sub-system can be isolated consisting of a quantum well between two thick barriers (fig. 2a). It is straightforward to obtain the eigenfunctions and eigenenergies by imposing adequate boundary conditions ($\phi(z)\to 0$ when $z \to \pm\infty$ and the continuity of $\phi(z)$ and $m^*(z)^{-1}\partial\phi/\partial z$ at both interfaces). There are two types of eigenstates: i) above the barrier energy, a quasi-continuum of states corresponding to a free-particle like unbound motion in which a plane wave impinging on the quantum well is partially reflected and partially transmitted ($\mathcal{G}_A>0$ and $\mathcal{G}_B>0$). ii) below the barrier energy, at least one or several discrete bound states with stationary wave behavior in the well reminiscent of quantum states in a box ($\mathcal{G}_A>0$) and evanescent propagation in the barriers ($\mathcal{G}_B<0$). In either case the wavefunctions are plane waves with a quadratic energy dispersion relation in the (x,y) planes. As a consequence the bound states are localized in the well

region in the z direction only, and divide into one (n=1) or several (n=1,2,...) two-dimensional subbands (fig. 2b). They have been extensively studied and clearly identified by a great deal of optical experiments, and provide a firm starting point to investigate structures with thinner and thinner barriers. This is not the case for states above the barriers, for there has been no comprehensive study of electronic conduction in that energy range.

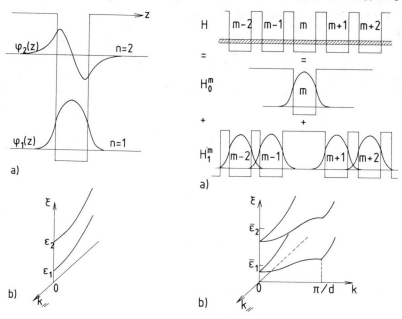

Fig. 2: a) Wavefunctions of an isolated quantum well in the growth direction. b) In-plane dispersion of the quantum well subbands.

Fig. 3: a) In a SL, the potential of adjacent wells can be considered as a perturbation to the isolated quantum well. b) Schematic view of the SL energy surfaces $\varepsilon(k,k_{//})$.

We are now in a position to derive the notion of miniband from first a tight-binding approach by considering the superlattice as a set of N (with N very large) identical isolated but weakly-coupled quantum wells. Then all n=1 bound states are N-fold degenerate in energy. As they are not stationary states of the SL Hamiltonian H, the degeneracy is in fact lifted by the SL potential, so that the N energy levels are regularly distributed over a small energy domain in a so-called miniband. More precisely, H can be split into two parts for each well: H_0^m for the mth isolated quantum well and a perturbation H_1^m (fig. 3a) which is just the potential of other wells. The superlattice ground state is constructed as a linear combination of the n=1 wavefunctions of all isolated wells:

$$\Phi_1(z) = \frac{1}{N} \sum_m c_m \, \phi_1(z - md) \qquad (4)$$

where:
$$H_0^m \phi_1(z - md) = \bar{\varepsilon}_1 \phi_1(z - md) \quad (5)$$

If the overlap of wavefunctions on different sites is neglected, consistently with the assumption of weak-coupling, Bloch's theorem implies that:

$$c_m = \exp(ikmd) \quad k = \frac{2\pi m}{Nd} \quad m = 0,1,...,N-1 \quad (6)$$

by virtue of translational invariance and Born Von-Karman cyclic boundary conditions. The full SL eigenvalue equation $(H_0+H_1)\Phi_1(z) = \varepsilon_1 \Phi_1(z)$ now becomes:

$$\sum_m \exp(ikmd) \{\bar{\varepsilon}_1 \phi_1(z-md) + H_1 \phi_1(z-md)\} = \varepsilon_1 \sum_m \exp(ikmd) \phi_1(z-md) \quad (7)$$

and using the weak-coupling approximation again to neglect overlap integrals:

$$\varepsilon_1 = \bar{\varepsilon}_1 + \sum_m \exp(ikmd) \, t_m \quad (8)$$

where:
$$t_m = \int_{-\infty}^{+\infty} \phi_1^*(z) H_1^0 \phi_1(z - md) dz \quad (9)$$

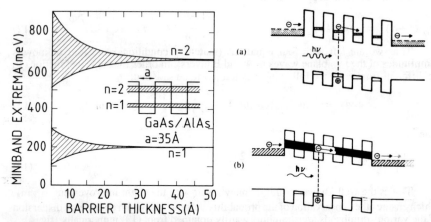

Fig. 4: n=1 and n=2 miniband extrema as a function of barrier thickness for GaAs/AlAs superlattices of fixed barrier thickness (a=35Å).

Fig. 5: effective mass filtering in a SL photoconductor in the case of electron hopping conduction a) or electron miniband conduction b) (from ref. [4])

The t_m integrals obey the relation $t_{-m}=t_m^*$ and are in fact real. In a nearest neighbor approximation, all integrals with m>1 are removed and the eigenenergies are given by:

$$\varepsilon_1 = \bar{\varepsilon}_1 + t_0 + 2t_1 \cos(kd) \quad k = \frac{2\pi m}{Nd} \quad m = 0,1,...,N-1 \quad (10)$$

t_0 and t_1 are called the shift and transfer integrals respectively. A similar approach could be applied to the n=2 bound state of the isolated quantum well if it exists.

The energy levels are thus labelled by k wavevectors and form "minibands" separated by energy gaps just as in bulk semiconductors (fig. 3b), the miniband width Δ being four times the transfer integral. It can be adjusted at will by modifying well and barrier thicknesses, and barrier height. Widely tailorable values ranging from 1 meV or less up to ~ 200 meV are at the origin of the interest for SL. Below 10meV Δ is of the order of the broadening of individual k states due to various mechanisms (see section 6); above 200meV usual composition superlattices become hardly distinguishable from random alloys.

The present approach clearly shows that miniband formation primarily originates from superperiodicity, i.e. Bloch's theorem applied to the SL period. The transfer integral only determines the magnitude of degeneracy lift. Although our derivation relies upon a weak-coupling approximation, this is not a necessary assumption and detailed tight-binding calculations give basically equivalent but more refined results.

We can now go back to the exact solution of Hamiltonian H (1) in the envelope function scheme. Proceeding along the same lines as for the isolated quantum well, we write matching conditions at the interfaces for $\Phi(z)$ and $m^*(z)^{-1}\partial\Phi/\partial z$. This may be done over one period only at both interfaces A/B and B/A$^+$ (A$^+$ is the A layer of the next period), since Bloch's theorem supplies the additional relations:

$$\Phi_{1A^+}(z+d) = \exp(ikd)\,\Phi_{1A}(z) \quad (11) \qquad \frac{\partial \Phi_{1A^+}}{\partial z}(z+d) = \exp(ikd)\frac{\partial \Phi_{1A}}{\partial z}(z) \quad (12)$$

Thus we are left with four equations (matching conditions) and four unknowns (amplitudes of the two plane waves in A and B layers). In order for the wavefunctions to be different from zero, the following relation has to be satisfied:

$$\cos(kd) = \cos(ak_A)\cos(bk_B) - \frac{1}{2}\left(\xi - \frac{1}{\xi}\right)\sin(ak_A)\sin(bk_B) \quad (13)$$

where:

$$k_A = \left(\frac{2m_A \mathcal{E}_A}{\hbar^2}\right)^{1/2}, \qquad k_B = \left(\frac{2m_B \mathcal{E}_B}{\hbar^2}\right)^{1/2} \quad \text{and} \quad \xi = \frac{k_A m_B}{k_B m_A} \quad (14)$$

This is the well-known Kronig-Penney result with the slight improvement $m_A \neq m_B$, which ensures the conservation of the probability current. The width and energy dispersion of the various minibands are straightforwardly obtained from (13) numerically (fig. 4).

This model is very commonly used to get insight into widths and spacings of minibands around a given band extremum in a superlattice of known parameters (a,b,W and effective masses). It is valid over large portions of the Brillouin zone and for most of the SL structures. Only far from band extrema and for short period SL in which the assumption of a slowly varying envelope function is unsatisfactory does it becomes questionable. Refinements are possible: band non-parabolicity of bulk materials can be accounted for through Kane's model, based on a coupling between several light particle bands. G. Bastard and J.A. Brum have shown that (13) remains valid provided the wavevectors k_A and k_B are derived from Kane's rather than parabolic (14) dispersion relations [9]. In-plane ($k_{//}$) dispersion in valence bands can also be treated more accurately (since the usual distinction between heavy and light holes is no longer legitimate in these directions). Finally the envelope function scheme is a reference model which other sophisticated calculations are

compared to. Indeed, tight-binding, pseudo-potential and local density functional methods have deepened our understanding of mixing between states from different valleys, of wavefunctions matching at interfaces or of the influence of strain on band line-ups. We refer the reader to a recent paper by L.J. Sham and Y.T. Lu for a review of these developments [10].

3 EXPERIMENTAL OBSERVATIONS OF LOW FIELD MINIBAND CONDUCTION

The experimental demonstration of miniband conduction in SL is not an easy task and is rarely direct. Many published works rely on measurements of carrier velocities which are expected to be high when a miniband can be defined. Let us first clarify this basic point: as discussed in section 2, the very existence of a miniband implies a significant wavefunction delocalization over several SL periods, so that successive quantum wells could interact constructively or destructively and provide dispersion. Conduction in this case is limited by scattering between Bloch extended states, and relatively high mobilities can be calculated. On the other hand severe scattering by intrinsic or extrinsic mechanisms combined with wide barriers (i.e. narrow minibands) can fully localize the electronic states (section 6). Conduction then proceeds by hopping in a manner similar to amorphous materials, and mobilities are much lower. In the following we concentrate on experimental results obtained in the absence of, or under low, electric field. Nonlinear transport is discussed in section 4.

F. Capasso [4] et al. first claimed miniband conduction in SL to be at the origin of high photoconduction gain in SL photoconductors,due to the widely differing electron and hole velocities ("effective mass filtering", fig. 5). However the same authors also reported [4] high gains in the hopping conduction regime of narrow miniband structures, so that miniband conduction is not crucial. In simple terms, the photoconduction gain is rougly equal to the ratio τ/t_e of recombination time τ to electron drift time t_e through the SL, and is thus proportional to the electron perpendicular velocity. However the large gains measured in SL in fact mainly result from long capture time constants due to residual traps and are therefore not an intrinsic effect.

Another purely optical technique was proposed by B. Deveaud et al [11], who performed subpicosecond luminescence experiments in graded-gap like SL samples consisting of a stack of successive SL's (~ 800 Å thick) with fixed well and barrier thicknesses but varying barrier heights. As a result the conduction band bottom goes down step by step, while the valence band goes up and the bandgap decreases. Photoexcited carriers flowing by diffusion within each SL step, also fall into the next step when they impinge on it. The ambipolar motion of electrons and holes is observed through the time variations of the luminescence spectrum, since each step has a characteristic emission energy (fig. 6). In the end the luminescence typical of an enlarged well (EW) located after the last SL step indicates that most carriers have finally reached the opposite end of the SL structure.

The modelling of the spectrum temporal evolution was performed by assuming usual diffusion equations with an appropriate ambipolar diffusion coefficient D and adequate boundary conditions. The experimentally observed linear variation of the carrier packet mean position as a function of time was well reproduced in three GaAs/GaAlAs SL (10/10, 20/20 and 30/30 for the well and barrier thicknesses in monolayers, the Al content in the barriers being changed from 35% to ~17% in steps of 2%) and a GaAlAs control alloy at 15K. The charge motion is in fact as fast in both 10/10 and 20/20 structures as in the control sample, from which it was concluded that the low-field charge transport in these SL was Bloch transport via extended miniband states. This was not the case however in the 30/30

structure for holes, whose mean free path was smaller than the period. Due to the ambipolar character of transport, D depends on carrier concentrations, and is equal to the hole or to the electron diffusion coefficient for high or low carrier densities respectively. Both diffusion coefficients D_e and D_h have thus been determined at 140K by varying the excitation conditions [12]. The results were in clear agreement with miniband transport for electrons since D_e varied as the inverse of the SL perpendicular effective mass m_e up to the value found for the GaAlAs alloy control sample, beyond which D_e remains constant. D_h behaved similarly but decreased much faster than $1/m_h$. Experiments performed by another group at a lower temperature (T=2K) on a wedge-shaped sample are not in agreement with this simple picture however, because they yield a somewhat too large diffusion coefficient [13]. It is likely that exciton states involved in the diffusion mechanism at very low temperatures complicate the analysis.

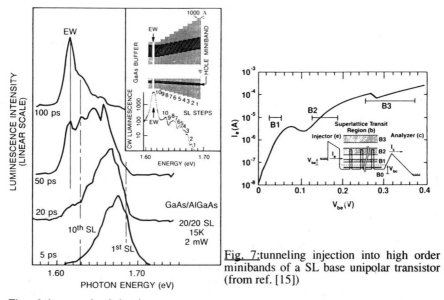

Fig. 6: time-resolved luminescence spectrum of a graded gap SL showing the ambipolar carrier sweepout. EW is an enlarged well (from. ref. [11])

Fig. 7: tunneling injection into high order minibands of a SL base unipolar transistor (from ref. [15])

The low-field properties of SL have also been investigated through time-resolved photocurrent measurements. In p-i-n structures changes of the electron drift velocity could be derived from changes of the peak photocurrent. In p-i-n GaAs/AlAs SL photodiodes under forward bias, the peak photocurrent exhibited a linear variation as a function of

applied voltage in a small region beyond flat band, which could be attributed to the linear low-field portion of a velocity-field V(F) relation (mobility regime) in spite of the small widths of the SL "minibands": 0.51meV and 1.53meV. Realistic momentum relaxation times of 130fs and 95fs were deduced. The non-linearity of the V(F) relation at higher fields (see 4.3) is also observed, and a miniband transport model taking into account non-zero temperature effects is found to fit the data properly from 50K to 130K [14].

In n-i-n structures the photocurrent response is more difficult to interpret due to carrier injection. Despite that, the mobilities of both types of carriers have also been obtained from time-resolved photocurrent measurements in GaAs/AlAs SL with the aid of careful numerical simulations, the details of which are given in section 4.3. In a series of samples having approximately the same well thickness a=36 Å but different barrier thicknesses (varying from 3 to 7 monolayers), electron and hole mobilities are shown to decrease according to the expected exponential variation of $1/m_e$ and $1/m_h$ at 300K. Electron mobilities are somewhat smaller than calculated for LO-phonon scattering in the n=1 miniband, so that additional scattering mechanisms have to be invoked. Contrarily, hole mobilities are larger than predicted for n=1 heavy holes and LO-phonon scattering. This is interpreted as resulting from a more complex transport regime involving light holes and/or n>1 heavy holes [15].

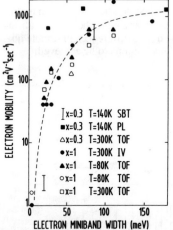

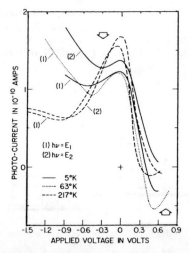

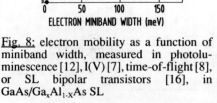

Fig. 8: electron mobility as a function of miniband width, measured in photoluminescence [12], I(V) [7], time-of-flight [8], or SL bipolar transistors [16], in GaAs/Ga$_x$Al$_{1-x}$As SL

Fig. 9: negative differential photocurrent in a reverse-biased SL Schottky diode, at different photon energies and temperatures (from ref. [19])

In another approach, a purely electrical technique was proposed by us in 1986 [16], relying on the measurement of the gain β of a SL base heterojunction bipolar transistor, β being approximately the ratio τ/t_d of the recombination time to the injected minority electron transit time through the base. Assuming a value for or measuring directly τ thus allows to

deduce the electron perpendicular diffusion coefficient from β. A typical electron mobility of 1000 cm^2/Vs at 300K was thus estimated for a GaAs/GaAlAs SL having $\Delta \sim 100$ meV, in approximate agreement with theoretical calculations based on LO-phonon scattering of Bloch states.

More recently the same kind of idea was developped by P. England et al. in unipolar transistors [17], for which quasi-ballistic electrons are injected in the second or third miniband of a SL base and collected by an analyser (fig. 7). Such a structure allows the spectroscopic determination of electron transit times and/or relaxation processes in the SL as a function of electron energy in the various minibands. Experimental transistor gain data on GaAs/AlAs SL were analysed by a simple model based on the calculation of the electron group velocity together with a relaxation time approximation involving LO-phonon emission. Comparison with the data yielded an extremely short scattering time ~ 20-50 fs, corresponding to only ~ 3 SL periods for the mean free path.

Finally G. Brozak et al. [17] claimed to have observed the thermal saturation predicted by a simple model of miniband conduction based on the relaxation time approximation (see section 5), through far infrared Drude conductivity measurements. The good agreement between this model and data is surprising, in view (see 6.1) of the narrow miniband involved (3 meV). One possible reason for this success is the exceptionally one-dimensional character of the SL (along the SL axis) enhanced by the large magnetic field required by the technique.

Some of these results are summarized in a plot of perpendicular electron mobilities measured by various techniques vs. Δ (fig. 8). As is apparent there is a substantial disagreement between them, which may not only be due to differences in sample quality. A reliable and precise determination of mobilities thus remains to be achieved.

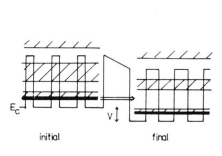

Fig. 10: conduction band diagram of a doped SL with an enlarged barrier (from ref. [23])

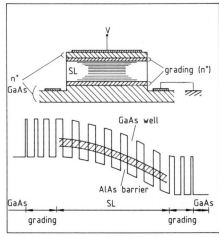

Fig. 11: typical structure of samples used for nonlinear transport (up), and band diagram of a biased n$^+$p$^-$(SL)n$^+$ device

4 NON-LINEAR MINIBAND TRANSPORT

As exemplified by Esaki and Tsu's pioneering proposal, it is clear that electronic conduction perpendicular to SL layers can be expected to exhibit interesting nonlinear transport properties under moderate or high electric fields related to the SL bandstructure. The first attempt in this respect was achieved by R. Tsu et al.[19] and later by F. Capasso et al. in another system [4], who found negative differential photoconductance in reverse-biased photodiodes (fig. 9). They interpreted their data in terms of electron localization by the electric field associated with a transition from miniband conduction to hopping conduction between localized states. The onset of photo NDC was found to agree satisfactorily with a criterion earlier derived by Tsu and Döhler for this transition [20] : e d $F_c \sim \Delta$. It was later realized that scattering profoundly affected F_c [21], particularly for wide minibands (see section 6).

Another approach consisted in studying tunneling processes between the minibands of two SL coupled by an enlarged barrier acting as a resistive quantum layer, over which a large fraction of the applied voltage was dropped [22,23]. NDC is also obtained in this case *in the dark*, when the top of the miniband on the anode side aligns with its bottom on the cathode side, thereby providing quantum reflection by the minigap for injected electrons (fig. 10). One interesting by-product of this phenomenon could be the measurement of Δ through the voltage at the current peak, if the elimination of extrinsic resistances was possible.

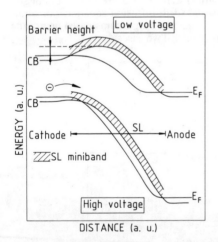 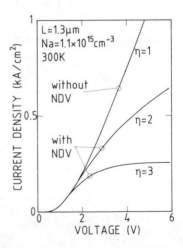

Fig. 12: band diagram of a biased $n^+p^-(SL)n^+$ device under low (up) or high (down) bias

Fig. 13: simulated I(V) characteristic of a $n^+p^-(SL)n^+$ device for various values of the parameter η

In the following we describe in some details a series of recent experiments performed in our laboratory demonstrating miniband NDV for GaAs/AlAs SL. Let us first recall that a SL can be qualified of anisotropic 3 dimensional semiconductor in the case of miniband

conduction. Such an anisotropy is characterized by a marked difference between the electron mobilities parallel and perpendicular to the layers, due to differing effective masses in the two directions, the perpendicular one being derived from the miniband along the growth axis. In an "effective medium" description of the SL, we forget its underlying well and barrier stucture to retain only its bulk material-like behavior, and interpret its transport properties in the manner of an ordinary anisotropic semiconductor. This allows to model transport using classical (drift-diffusion), or semiclassical (Boltzmann) transport equations in a miniband, as discussed in section 5. The three following paragraphs show results analysed in such terms and concerning SL conduction measured in either d.c., microwave, or time-resolved (after pulsed photoexcitation) regime .The sample structure is depicted in fig. 11 and, briefly stated, consists of a SL sandwiched between n$^+$ contact layers eventually including n$^+$ GaAlAs windows for optical access.

4.1 Static transport experiments

One serious problem in the extraction of significant information on, say velocity-field V(F) relations from experimental current-voltage I(V) characteristics, is the non-uniformity of the electric field along the SL growth axis. This difficulty is general for n$^+$-p$^-$n$^+$ as well as n$^+$-n$^-$n$^+$ structures, particularly in a medium exhibiting NDV and particularly under biasing, and arises from the interplay between charge density, electric field, and electron velocity. Fortunately such non-uniformities can be accurately modelled in the case of a effective medium approach by the numerical self-consistent resolution of Poisson and transport equations. The latter can be classical (drift-diffusion equations) or semiclassical (Boltzmann equation), using for instance standard algorithms based on a finite difference resolution schemes [24]. We will distinguish Two cases:

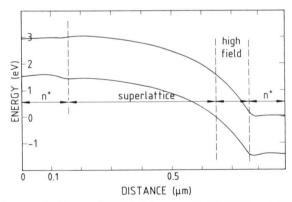

Fig. 14: band diagram of a biased n$^+$n$^-$(SL)n$^+$ device with NDV, biased slightly beyond F_c

i) *n+-p-(SL)-n+ structures:* in this case the band bending is marked under any bias, because of the negative charge of ionized acceptors in the SL (fig. 12). Electrons injected by the cathode have to surmount a macroscopic barrier before being collected by the anode, which is responsible for a temperature-activated rectification at low bias. In this regime the device conductance is exponentially sensitive to the barrier height, and only linearly sensitive to the electron mobility. At high bias in the punchthrough regime on the other

hand, the barrier nearly vanishes and the conductance is mainly determined by the electron velocity. Therefore determining the electron V(F) relation for such structures requires the quantitative analysis of the full I(V) characteristics in a wide voltage range, and is of limited accuracy at very low fields.

These ideas can be validated in a test case, by studying the sensitivity of calculated I(V) 's to the shape of the V(F) law. The following phenomenological function:

$$V(F) = \frac{\mu F}{1 + (F/F_0)^\eta} \tag{15}$$

covers all representative cases as regards NDV, i.e. no NDV at all ($\eta \leq 1$), saturating law for $\eta = 1$ (saturation velocity $V_s = \mu F_0$), or NDV for $\eta > 1$. of particular interest is $\eta = 2$ for which the peak velocity is $V_p = \mu F_0/2$ and the critical field is $F_c = F_0$ (Esaki-Tsu model, section 5). Fig. 13 very clearly shows the I(V) dependence on η for instance, *particularly the sublinearity of the high voltage I(V) for $\eta > 1$, which is therefore a characteristic feature of NDV*.

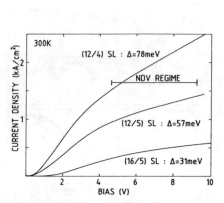 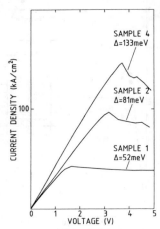

Fig. 15: I(V) characteristics for a series of $n^+p^-(SL)n^+$ devices of increasing miniband width (from ref. [7])

Fig. 16: I(V) characteristics for a series of $n^+n^-(SL)n^+$ devices of increasing miniband width (from ref. [27])

ii) *$n+$-n-(SL)-$n+$ structures:* in this case the electric field in the SL is relatively uniform under low bias particularly for a thick enough and highly enough doped SL, and the devices are ohmic. Nonlinear V(F) laws unfortunately destroys this ohmicity under large voltages, particularly in the presence of NDV which is responsible for the formation of a high field domain on the anode side (fig. 14). If the (doping level $N_d \times$ SL length L) product is high enough, true NDC occurs beyond a peak voltage $V_p \sim F_c L$. In this regime modelling is more difficult due to the natural instability of the numerical solution. All these features are actually reminiscent of the properties of Gunn effect devices [25,26], because they mostly derive from the existence of NDV rather than from its microscopic origin.

Experimental I(V) characteristics have been investigated [7] for a series of

GaAs/AlAs n⁺-SL-n⁺ devices whose miniband spectrum was calculated in the enveloppe function approximation (see section 2) using effective masses derived from k.p calculations.

As shown in fig. 15, all I(V) curves on p⁻ direct gap SL systematically showed sublinearity at high enough voltages, which is a signature of NDV as discussed above. The quantitative extraction of the whole V(F) relation required a careful fitting with numerical simulations, estimated accurate to ~ 30 % for Vp and Fc [7]. In n⁻ SL true NDC was also obtained as expected [27], which brought a definitive confirmation of the reality of SL NDV (fig. 16). The rather poor peak to valley ratio in these structures is a consequence of the anode high field domain. Much better ratios can fortunately be measured under resonant excitation at microwave frequencies (see next sub-section).

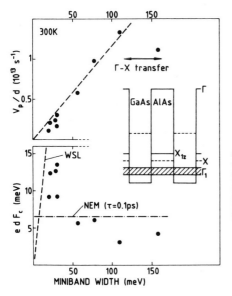

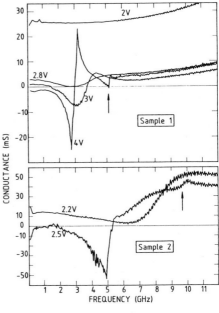

Fig. 17: dependence on Δ of the peak velocity divided by the period (up), and of the voltage drop per period at F_c (down), in GaAs/AlAs SL (from ref. 7)

Fig. 18: NDC microwave resonances in n⁺n-(SL)n⁺ devices, at the fundamental, and at the 1st harmonic of the transit time frequency V/L (from ref. [29])

All these results provide a striking agreement with Esaki and Tsu's premonition of SL NDC. Their simple calculation predicts NDV with a V(F) of the form as given by eq. (15), with η = 2, $edFc = \hbar/\tau$ and $Vp/d = \Delta/4\hbar$, τ being the scattering time. Plotted in fig. 17 are Vp/d and edFc as a function of Δ, for comparison. As can be seen, the nearly linear

variation of Vp/d with Δ is a very good indication of true miniband conduction in these samples, as predicted by Esaki and Tsu's theory. Nevertheless quantitative differences exist, particularly regarding Vp, and are discussed in section 5.

4.2 Microwave transport

Similarly to static transport, special features related to NDV are expected under a.c. electrical excitation in the microwave spectrum. This effect stems from the finite delay between the application of a voltage pulse on the cathode and the arrival of the corresponding charge packet at the anode. Resonances related to NDV in the microwave conductance spectra of samples biased beyond Fc have been predicted and observed in Gunn effect devices [28], at the fundamental and harmonics of the transit time frequency F_t=V/L, where V is the average electron velocity. These resonances are all the narrower as the electron velocity is well defined, i.e in structures with as uniform electric field as possible. Again the doping level by SL length ($N_d \times L$) product is the figure of merit, which should be as high as possible for narrow resonances. These effects have been perfectly confirmed [29] in the same kind of samples as described in the previous sub-section (fig. 18). Interestingly, the linear dependence of F_t with L^{-1} [30] proves that the SL behaves as a homogeneous entity, i.e. validates an effective medium description of transport. Similarly to Gunn effect devices, microwave oscillators taking advantage of SL NDC can potentially be developed, with possible advantages related to the differing microscopic mechanism [27].

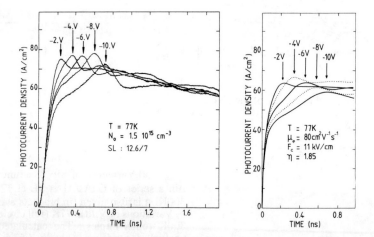

Fig. 19: left: Time evolution of the photocurrent in a Δ=22meV GaAs/AlAs SL as a function of applied voltage. right: Simulated responses.

4.3 Non-linear transport through time of flight experiments

Time-of-flight is a direct method to demonstrate NDV by measuring transit times over a known distance, under high-field conditions. For that purpose, excess carriers have to be injected in a superlattice either electrically, for instance by a sudden change in applied voltage, or optically by a short light pulse. Photoexcitation is easier to implement because high-frequency impedance matching problems are then minimized. Time-resolved photocurrent measurements have been performed in the n^+-SL(p^-)-n^+ structures, the static and microwave characteristics of which have been discussed above [15]. The experiment consists in illuminating the SL through the GaAlAs optical window by picosecond light pulses around 0.71µm, and measuring the induced photocurrent transient for various applied voltages.

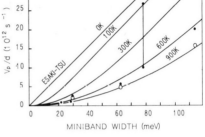

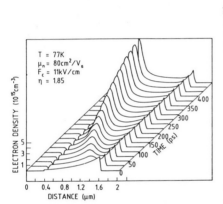

Fig. 20: Temporal evolution of the electron density in the simulation of the responses given in fig. 19

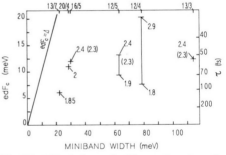

Fig. 21: a) Variations of edF_c as a function of Δ in a series of GaAs/AlAs SL at 77K and 300K. η is also given (in brackets at 300K). b) Variations of V_p/d at 77K (full circles) and 300K (open circles). The calculated temperature dependence is obtained from [32]

Fig. 19 shows the time-resolved photoresponse of a SL of miniband width Δ=22 meV. The wavelength is adjusted so that the photon energy is slightly above the SL bandgap, n=1 electrons only being created (the n=2 electron miniband is at least 200meV above the n=1 miniband). The photocurrent transient distinctly occuring at the outset is further delayed as the bias is increased, while the photocurrent amplitude simultaneously saturates. *This is a direct visual time domain evidence of NDV*, showing the longer and longer time

for photocreated electrons to drift throughout the SL as the internal field grows [8,31]. It is clear from the responses of different superlattices that the wider the miniband, the shorter the time scale of NDV, and the larger the bias marking the onset of the effect. The temperature dependence between 77K and 300K is smooth, as also found in static measurements.

In the same spirit as above (sec. 4.1), a better understanding of the transient photocurrent together with quantitative information can be extracted from raw data, through numerical simulations which carry out self-consistently the resolution of drift-diffusion and Poisson equations, now for both electrons and holes and including the time dependence. After instantaneous excitation, the photocreated electron population, which is predominantly located in the vicinity of the cathode, drifts towards the anode giving rise to a primary current. Its duration and shape reflect the primary electron drift throughout the SL, as visualized on the electron density temporal evolution (fig. 20) for a velocity law given by eq. (15) with $\eta=1.85$. Because of the internal field inhomogeneity, an electron bump forms and builds up as electrons coming from the cathode catch up with the slower electrons in the vicinity of the anode. The delayed current peak is associated with such a behavior (fig. 19), whereas it is absent in the simulated responses without NDV ($\eta \leq 1$).

Such effects have been observed in a series of superlattices having a miniband width between 22meV and 78meV, in agreement with static and microwave measurements. The experimental time responses have also been compared with the calculated ones, for various values of the adjustable parameters μ, F_c and η, the value of μ being derived from the low bias response. A good agreement was also found in a wide voltage range as far as the time variations are concerned [8], for instance from 0V to -10V in the 22 meV miniband sample (fig. 19).

Again all these results are in favor of a band conduction mechanism in the first miniband (Γ_1) at 77K and 300K. The Esaki-Tsu derivation supplies the basic concepts in this interpretation. There is a rough agreement with their model ($\eta=2$) for each individual sample, for η remains close to 2. The potential drop per period at the critical field F_c then yields a straightforward estimate of the momentum relaxation time (τ) in the SL, according to the formula $edF_c=\hbar/\tau$. These time of flight investigations indicate relaxation times ranging from 100fs down to 30fs, with an apparent tendency to diminish as the miniband width increases (fig. 21). They are much shorter than usual phonon relaxation times but not irrealistic if other interaction processes take place such as interface disorder or interface roughness (see section 5). There is an appreciable discrepancy with static data (fig. 17), which is not yet understood. The peak velocity V_p is also always smaller than predicted by Esaki-Tsu model. The discrepancy can be accounted for by invoking carrier heating and a slightly improved ($T \neq 0$) model [32] (see section 5) as shown in figure 21.

5 THEORY AND MODELS OF SEMI-CLASSICAL MINIBAND CONDUCTION

The recent development of SL physics has been initiated by the miniband NDV predicted by Esaki and Tsu [1], as explained above. One main characteristic attached to such an effect is the rough proportionality to the miniband width of Vp in the SL growth direction. The deep-lying physics is Bragg scattering for electrons having a wavevector near $\pm \pi/d$. The existence of a maximum drift velocity in the miniband simply stems from the maximum group velocity in the miniband itself, at $k_z = \pm \pi/2d$, and to Bragg reflection. In the present section, we review the different models which have been used to describe these effects, in the frame of the rigid band Bloch conduction model. In other words the

SL is considered as a perfect crystal in which both the electric field and the lattice imperfections, phonons, etc.., act as weak perturbations. We start from the oldest models best represented by Esaki and Tsu's in 1970; then we present the general problem of solving the Boltzmann Transport Equation (BTE) which is their natural extension to treat of scattering realistically, and discuss the different solutions proposed in the past. In the following sub-section we specialize to the pertinent scattering processes for real SL, and analyse the salient features exhibited by numerical solutions like lattice temperature or SL parameters dependences of V(F) laws. In the last part, magneto-transport effects in crossed electric and magnetic field will complete this short review of semiclassical miniband transport theories.

5.1 Esaki-Tsu model

The original model published in [1] predicts a variation of the drift velocity as:

$$V = <V_z> = V_{MAX} \frac{F^*}{1+F^{*2}} \tag{16}$$

in which $V_{MAX} = 2 V_p$ is the maximum group velocity in the miniband attained precisely at the inflexion point of the dispersion relation, and F^* the electric field in reduced units of $\hbar/e\tau d$, τ being a constant collision time. This result is trivially derived from a simplified Chambers expression of the drift velocity:

$$<V_z> = eF\hbar^{-2} \int_0^\infty (\partial^2\varepsilon_z/\partial k_z^2)\exp(-t/\tau)dt \tag{17}$$

with the tight binding expression (see section 2) for the miniband dispersion law : $\varepsilon(k_z) = \Delta(1 - \cos(k_z d))/2$. Obviously other expressions of $\varepsilon_z(k_z)$ would lead to V(F) relations deviating from this simple function, as discussed in the original paper [1]. The tight binding expression of the miniband energy dispersion is indeed invalid for wide minibands. Lebwohl and Tsu [33] have shown that the shape of the V(F) law is not considerably modified if the position of the inflexion point is shifted from its medium position, as obtained from a cosine dispersion law, as far as the edge of the minizone: the NDV is indeed maintained, but Fc increases by about 50 %. Similarly, it is easily shown that annihilating the superperiodicity by blocking $\varepsilon(k_z)$ to a constant value beyond π/d has the same effect. These remarks mean that NDV is governed not only by Bragg reflection but also by the negative effective mass of the dispersion relation which slows down accelerated electrons beyond the inflexion point. Obviously in a real SL for which the dispersion is periodic, both effects are intimately mixed.

It has been conclusively established in the previous section that Vp steadily follows the miniband width Δ in GaAs/AlAs SL [7]. This is in *qualitative* but not *quantitative* agreement with Esaki-tsu's model which predicts Vp/d to be equal to $\Delta/4\hbar$, i.e. several times in excess of the experimentally measured values depending on the samples. Such discrepancies are certainly not surprising in an oversimplified model which does not even include thermal broadening for instance. In the following sub-section, more exact approaches to miniband transport in the frame of BTE will provide a substantial improvement of Esaki-Tsu model, although not in stark conflict with its essential features, i.e. :

    * V_p approximately scales with Δ * Fc approximately scales with τ^{-1}

5.2 Solution of BTE for miniband transport.

Several approaches to the resolution of BTE in the case of miniband transport limited by various scattering mechanisms have been proposed in the literature. We start from the simplest analytical theories dealing with a constant collision time, and proceed with more sophisticated calculations. The general BTE for electron transport in a miniband, notwithstanding Pauli exclusion terms or spatial dependences, can be written as:

$$\frac{\partial f}{\partial t} + \frac{eF}{\hbar}\frac{\partial f}{\partial k_z} = S(f) - \frac{f}{\tau} \tag{18}$$

in which S and τ are, respectively, the scattering-in and the scattering-out collision terms, defined through scattering transition probabilities W per unit time:

$$S(f) = \Sigma \quad W(\vec{k'} \to \vec{k})f(\vec{k'}) \qquad \text{summed on} \quad \vec{k'} \tag{19}$$

$$\frac{1}{\tau} = \Sigma \quad W(\vec{k} \to \vec{k'}) \qquad \text{summed on} \quad \vec{k'} \tag{20}$$

In order to have an analytical expression of the distribution function f, Lebwohl and Tsu [33], Shik [34], Bass and Rubinshtein [32], for example, have simplified the collision term simply by writing:

$$S(f) = \frac{f_0}{\tau} \tag{21}$$

in which f_0 is the equilibrium distribution function and the collision time τ is constant. Different methods can be used to solve (18) whith the approximation (21). Let us present a classical one which we shall extend later to the magneto-transport problem. Remarking that (18) is linear in the partial derivatives, it possesses the characteristic solution parametrized by the trajectory time s:

$$\frac{dt}{ds} = 1 \quad (22) \quad ; \quad \frac{dk_z}{ds} = \frac{eF}{\hbar} \quad (23) \quad ; \quad \frac{df}{ds} = \frac{f_0(k_z(s), t(s))}{\tau} \tag{24}$$

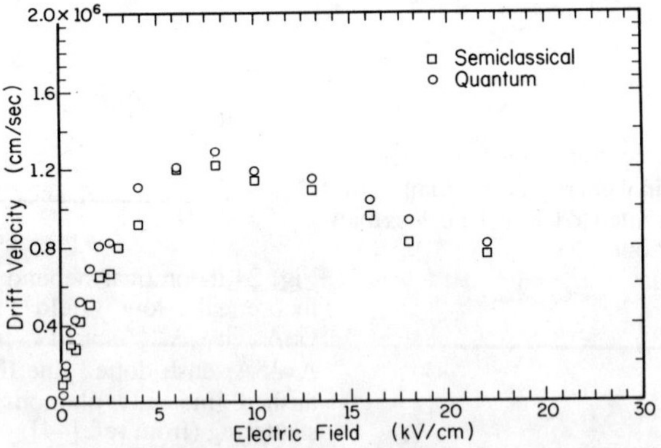

Fig. 22: V(F) relation calculated by Monte-Carlo techniques without (semiclassical) or with (quantum) inclusion of collisional bradening (from ref. [37])

Concentrating on the steady-state distribution the solution can be written as:

$$f(\vec{k}) = f_0(k_x, k_y) \int_0^\infty e f_0(k_z^*(t)) \exp(-t/\tau)\tau^{-1} dt \quad (25) \qquad \text{with} \qquad k_z^*(t) = k_z - \frac{eFt}{\hbar} \quad (26)$$

It immediately appears that such a simplification eliminates electronic heating, i.e. the deviation of the k_x and k_y dependences of the distribution function from what it is at equilibrium. This is clearly quite unphysical, since collisions actually redistribute the energy supplied by the electric field among all directions, *including kx and ky*. Fortunately, the distribution function expressed in (25) can be used to derive an improved expression of V(F). Starting from the Boltzmann approximation for f_0 a result quite similar to (16) can be easily derived:

$$V = <V_z> = V_{MAX} \frac{F^* \ I_1(\Delta/2k_BT)}{1 + F^{*2} I_0(\Delta/2k_BT)} \quad (27)$$

in which I_n is a modified Bessel function. The ratio I_1/I_0 asymptotically tends to 0 at infinite temperatures ("thermal saturation", ref. 18) and to 1 at 0 K. Very simply stated, this result expresses the equiprobability at high temperatures of all k_z states because of thermal agitation, which is responsible for a vanishing average velocity.

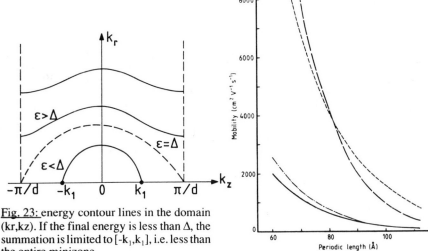

Fig. 23: energy contour lines in the domain (kr,kz). If the final energy is less than Δ, the summation is limited to $[-k_1,k_1]$, i.e. less than the entire minizone.

Fig. 24: theoretical dependence on d of the theoretical low field mobility in a GaAs/Ga$_{0.7}$Al$_{0.3}$As SL. dashed line:IF, Λ=35Å; dash-dotted line:IF, Λ=200Å; long dashed line: LO phonons; full line: total scattering (from ref. [44])

It is remarkable that the exact solution of the simplified BTE only scales the Esaki-Tsu formula by a temperature dependent band filling factor. However this model is still too simple, and the experimental results of section 4 require a realistic approach of the diffusion processes, which is done by solving (18) with appropriate collision terms. One of the widely

used methods for this is the Monte Carlo stochastic technique, as applied by Price [35], Andersen and Haas [36] for example to SL, with no major changes of the technique with respect to bulk materials. Artaki and Hess [37] have introduced an interesting improvement in modifying the Fermi golden rule to take into account collisional broadening in the final density of states, but this effect does not seem to change calculated drift velocities significantly (fig. 22).

The Monte Carlo method is not the only way to solve (18) nevertheless. Two deterministic solutions of the homogeneous BTE are currently used in the authors' laboratory, the first one having been developed in collaboration with INRIA (Rocquencourt, France). It consists in using an upwind finite difference approximation of the first member of (18) together with a deterministic conservative calculation of the collision integrals. The resulting algebraic system is subsequently solved by a least squares method in the form of a conjugate gradient algorithm [38]. This technique is very robust but requires preliminary explicit determination of all transition probabilities in order to calculate the adjoint of operator S. Relaxing the total energy conservation constraint by using a finite width function rather than the Dirac delta function circumvents this drawback.

One of the authors has implemented a different method, based on Budd's free trajectory iterative algorithm [38]. It rests upon the remark that (18) can be replaced by an inhomogeneous Fredholm equation using characteristics (free) electron trajectories and is therefore equivalent to the widely used particle simulation methods [35-37]. In this case (25) is replaced by the homogeneous integral equation:

$$f(\vec{k}) = \int_0^\infty S(f(k^*(t))) \exp u(t) dt \quad \text{with} \quad u(t) = -\int_0^t \tau^{-1}(k^*(t')) dt' \quad (28)$$

As proposed by Budd, (28) can be transformed into an inhomogeneous integral equation in order to safely apply iterative methods [39]. In this case the seed for the solution may be given by the numerical evaluation of the second member of (28) with $S(f) = f_0/\tau$. Prior to get accurate BTE solutions, we have to adress scattering terms in detail.

5.3 Scattering mechanisms; role of interface roughness

The collision terms in the second member of (18) are written as the sum of all relevant scattering-out and scattering-in probabilities:

$$\frac{1}{\tau} = \frac{1}{\tau_{PO}} + \frac{1}{\tau_{AC}} + \frac{1}{\tau_{IF}} \qquad S(f) = (S_{PO} + S_{AC} + S_{IF})(f) \qquad (29)$$

in which PO means for polar optical phonon scattering, AC for acoustic phonon scattering and IF for interface fluctuation scattering.

The particular details of the integration scheme concerning PO and AC mechanisms have been described in [40], we only summarize them here in order to give an example of derivation which could be extended to any other scattering mechanism. The consideration of constant energy surfaces in (k_r, k_z) plane where $k_r = \sqrt{k_x^2 + k_y^2}$ as drawn in fig (23), constitutes a firm basis for the derivation of collision terms [41]. Scattering events imply transitions from one constant energy curve to to the same or to another. If the final energy is less than 2Δ there is a restriction in the k_z' summation (fig. 23). The summation over k' states for all transitions probabilities is performed in the axisymmetric coordinate system with k_z as symmetry axis, instead of the polar system used for an isotropic band. Using the undimensionned quantities u_r, u_z and u_z', the scattering time limited by PO scattering is given by:

$$\frac{1}{\tau_{PO}} = \gamma((N_0+1)J^+ + N_0 J^-) \quad (30) \qquad\qquad with:$$

$$J^{\pm} = \int_\Gamma \frac{du'_z \Theta(K^{\pm})}{\sqrt{(1-K^{\pm})^2 + 2(1+K^{\pm})(u_z - u'_z)^2 + (u_z - u'_z)^4}} \qquad K^{\pm} = u_r^2 \mp \Omega + \varepsilon(u_z) - \varepsilon(u'_z)$$

in which γ is a constant factor, Θ the Heaviside step function and N_0 the phonon distribution.

The above formula can be generalized for occurrence of Umklapp processes. We point out that there is no restriction to the PO scattering even for a miniband width lower than the phonon energy, in contrast with a pure 1D transport in the z direction. Elastic collisions such as electron-acoustical phonon scattering or electron-impurity scattering can be treated along the same line. The neglect of phonon confinement effects in this model is questionable [42], but may be justified by recent theoretical works claiming electron-phonon in SL to be marginally different from that in the equivalent alloy of same average composition [43].

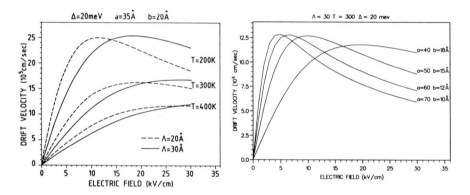

Fig. 25: Effect of IF scattering on V(F) characteristics. Solid lines: $\Lambda = 30$ Å; dotted lines : $\Lambda = 20$ Å.

Fig. 26: Illustration of the importance of IF scattering on V(F) curves in GaAs/AlAs SL, for a given miniband width, but for a varying well thickness (from ref. [46])

Considering now random potential fluctuations at interfaces (IF scattering), we note that Dharssi and Butcher have proposed a model [44] which is an extension to perpendicular transport in SL of the work of Sakaki et al [45] on parallel transport in quantum wells. In this theory, interfaces are randomly modulated due to growth imperfection and a Gaussian autocorrelation function parametrized by Λ of these random fluctuations is assumed. These

authors showed that for realistic values of Λ IF scattering can dominate over PO scattering at 300K (fig. 24). The following τ_{IF} can be derived:

$$\frac{1}{\tau_{IF}} = \frac{1}{\tau_{IF}^0} \exp\left(-\left(\frac{\Lambda^2(\varepsilon(u_z) + 2u_r^2)}{4}\right)\right) \int_{-u_1}^{u_1} \exp\left(\frac{\Lambda^2 \varepsilon(u_z')}{4}\right) I_0\left(\frac{\Lambda^2 u_z' u_1}{2}\right) du_z' \qquad (31)$$

with τ_{IF}^0 given by:

$$\tau_{IF}^0 = \frac{\pi}{d}\left(\frac{(a + 2\varepsilon_1/\chi W)(W - \varepsilon_1)}{\varepsilon_1 A \Lambda W}\right)^2 \qquad (32)$$

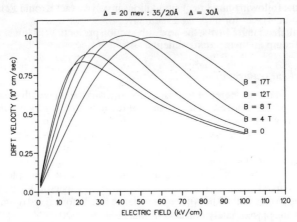

Fig. 27: V(F) characteristics for an increasing transverse magnetic field (from ref. 49)

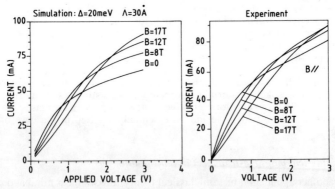

Fig. 28: comparison between experiment and theory (ref. [49]) of magnetotransport: a) result of a numerical simulation at 300K assuming Λ = 30 Å, with the SL sample parameters of ref [50]; b) experimental I(V) characteristics of ref [50].

in which u_1 is a variable boundary determined by the total energy (see fig (1)), u'_1 a final state u_r component. All quantities in (32) are undimensionned, i.e. a means for $k_0 a$, W for W/u_T etc... ϵ_1 is the offset miniband origin with respect to the well origin, W the barrier energy, χ the logarithmic derivative of the wavefunction at the interface, A the layer thickness fluctuation amplitude, and a the well thickness. It is worthwhile pointing out that $\tau_{IF} \sim a^6$ when $W \to \infty$, i.e. we recover the infinite quantum well case. However although already complicated formulae (31-32) may not be convenient enough for wide minibands since the wavefunctions may be quite different from those of simple quantum wells. For that reason, the following numerical simulations involve exact Kronig Penney probability density at the interfaces for improved accuracy.

Using all these ingredients, the next sub-section presents V(F) laws calculated in this manner, and compare them to experiments.

5.4 Some salient features of numerical results.

In spite of the high degree of sophistication entering the BTE with the above scattering probabilities, we first notice that a few general features are entirely predicted by simpler models such as yielding eq. (27). For instance, fig. 25 shows the changes in V(F) with Λ for A = 2.83 Å in a moderately wide miniband (25 meV) GaAs/AlAs SL(a = 38 Å, b = 20 Å). We observe that not only is V_p independent of Λ, but also do its lattice temperature variations agree with the simple model of eq. (27). This justifies the interpretations of section 4 in terms of Esaki-Tsu transport on the basis of such simple models. On the contrary, Fc markedly depends on Λ, and even worse on specific SL parameters a (well thickness) and b (barrier thickness); see for instance fig. 26 for a fixed roughness parameter Λ= 30 Å and an approximately constant miniband width ~ 20 meV. Finally in the case of wide minibands we clearly found a smaller dependence on lattice temperature than given by eq. (27) [46].

There is no doubt that transverse heating is responsible for this effect, due to the much less effective thermal randomization of k_z for wide minibands. It has been considered by Lei, Horing and Cui specifically in the case of carrier-impurity scattering [47], who have applied to SL miniband conduction the approximate solution of BTE proposed by Lei and Ting [48]. The results reported in [47] are in surprisingly remarkable agreement with the experimental data of section 4.1 [7]. However this agreement should be appreciated cautiously since the assumed impurity content was unrealistic, IF scattering limited transport being more likely in the (undoped) samples in question. As a tentative explanation, we suggest that V(F) laws are more sensitive to the *magnitude* of the elastic scattering than to its *nature*, a feature which we indeed essentially verified in the case of IF vs. PO scattering.

5.5 Semiclassical magneto-transport in a miniband.

As another application of semiclassical theory to non-linear miniband conduction, we now discuss magneto-transport in a crossed fields configuration, the electric field F being along z and the magnetic field B along x, i.e. parallel to the layers. In this particular geometry we neglect Landau quantization which requires $\mu_{eff} B > 1$, μ_{eff} being a low field mobility averaged along x and z. In a typical SL for which mobilities are small, significant quantization is repelled to very high values of B.

Miniband transport in crossed fields has first been theoretically investigated by A. Shik [34], but restricted to small magnetic fields in order to expand the distribution function f in powers of B. Here, the full BTE is solved under any field with the realistic

scattering mechanisms considered above, i.e. PO and IF scattering [49]. The technique again consists in calculating a characteristic solution of the electron dynamics along the free trajectories associated with an iterative convergent algorithm. The main effect of the transverse magnetic field is semiclassically interpreted in terms of a modification of the free trajectories under the action of the Lorentz force, which repels the onset of Bragg scattering to higher electric fields. A direct comparison with experimental I(V) is performed by taking into account the electric field non-uniformity earlier discussed in 4.1.

Since B lies along x, the free trajectories are deviated along y. Assuming therefore $F_y=0$ in a typical sample geometry suited to perpendicular transport studies, f now obeys the BTE

$$\frac{e}{\hbar}\left[F_z\frac{\partial f}{\partial k_z}+v_zB\frac{\partial f}{\partial k_y}-v_yB\frac{\partial f}{\partial k_z}\right]=S(f)-\frac{f}{\tau(\vec{k})} \qquad (33)$$

The method of characteristics consists in transforming (33) into an ordinary differential the solution of which is given by eq. (28). Using variables $u_z = k_z d$ and $u_y = k_y d$, the free trajectories are given by the following autonomous system:

$$\frac{du_z^*}{ds} = F^* - \lambda_1 u_y^* \qquad (34) \qquad\qquad \frac{du_y^*}{ds} = \lambda_2 \sin(u_z^*) \qquad (35)$$

in which $\lambda_1 = \omega_c \tau$ and $\lambda_2 = \Omega_c \tau$. ω_c and Ω_c are the cyclotron frequencies associated, respectively, with the (k_y,k_z) effective mass, assumed to be close to the bulk effective mass, and the effective mass in the k_z direction ($2\hbar^2/\Delta d^2$). τ is the time unit for s. The resulting pendulum equation is obtained for u_z^*, as also stated by Shik [34]:

$$\frac{d^2 u_z^*}{ds^2} + \lambda^2 \sin(u_z^*) = 0 \qquad (36)$$

in which $\lambda^2 = \lambda_1 \lambda_2$. Equation (36) has solutions in terms of tabulated functions, the reduced time s being explicitely given as a function of first kind elliptic integrals; the major drawback for the use of these functions is they are multiform and not really easy to handle with. It is therefore more efficient to directly integrate eqs. (34-35) by numerical techniques [49]. The mean drift velocity has a maximum for $F^*=1$, which corresponds to a free flight limited in the Brillouin zone. When a magnetic field is applied in the x direction, the free trajectories are twisted in the u_y direction. This distorsion occurs at any order in B, the effect being scaled to the F/λ_1 ratio. A detailed discussion of the Poincaré stability analysis of the autonomous system (33-36) involving both open and closed trajectories is given in [49]. Let us summarize the general features: the low field mobility decreases with B, F_c is shifted to higher values and the magnetoresistance becomes negative beyond F_c. All these conclusions remain valid with the realistic scattering processes considered above. The magnetic field dependence of the full V(F) relation at 300 K is shown in fig. 27, for SL parameters corresponding to an experimental work [50]. The only unknown parameter related to scattering is Λ, which has been adjusted to its best-fit value without magnetic field, i.e. $\Lambda = 30$ Å. The shift of the peak velocity to larger F_z values for B>0 is qualitatively in agreement with the effect on free trajectories devised above. We note that a large enough magnetic field somewhat enhances V_p in comparison with its zero field value. We conjecture that the distorsion of free trajectories is mainly responsible for that rather than scattering.

The calculated V(F,B) laws have been injected into the computer code allowing to take into account non-uniformity effects (section 4.1) in the n$^+$nn$^+$ structures used for

magneto-transport experiments of ref. 50. The important improvement obtained by solving the BTE rather than simply extending Esaki-Tsu model [50] now allows a satisfactory agreement between theoretical and experimental I(V) characteristics (fig. 28).

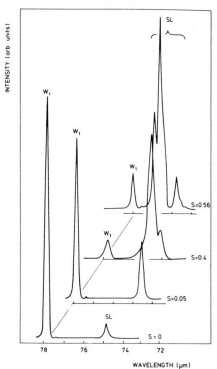

Fig. 29: luminescence spectra of SL of disorder parameter S, showing the enhanced contribution of SL luminescence compared to the enlarged well W_1 when carriers are localized by disorder (from ref. [54])

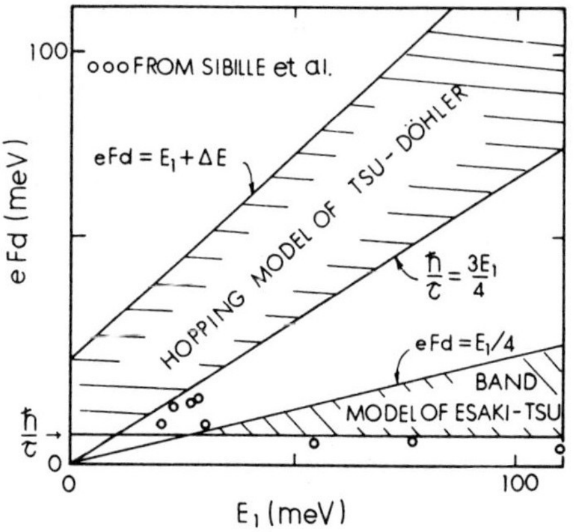

Fig. 30: existence domain of hopping conduction vs. miniband conduction in SL, depending on the electric field and the miniband width, here noted E_1 (from ref. [21]). Miniband conduction is actually valid also below $\hbar/\tau$ for eFd.

6 LIMITS OF SEMICLASSICAL MINIBAND TRANSPORT

In the previous sections, we have shown that semiclassical miniband conduction adequately described electron transport perpendicular to interfaces of SL for minibands wider than ~ 20 meV. We now adress the limits of validity of this approach by considering localization by disorder on the one hand, and Wannier-Stark localization by the electric field on the other.

6.1 Localization by disorder

Disorder is responsible for destroying the long distance coherence of the electron states, and may even provoke a phase transition resulting in electron localization. This a vast field of solid-state physics, which will only be briefly tackled here in the context of SL. Obviously the perturbational approach of the last section is inadequate. We discuss two classes of disorder:

6.1.1 homogeneous broadening:phonons

Phonons disturb the strict periodicity of the lattice by random fluctuations of the atom positions, resulting in a homogeneous broadening of electron states by $\hbar/\tau$, where τ is a typical phonon scattering time. The strong temperature dependence of τ for phonon emission and absorption processes imply a greater coherence length at low temperatures, and a possible transition from miniband conduction at low temperatures to hopping conduction between disorder localized states at high temperatures. One possible signature of such a transition would be a minimum in mobility vs. temperature plots, since scattering inhibits band conduction whereas it enhances hopping. Possible observations of this effect have been reported for narrow minibands [51,52].

6.1.2 inhomogeneous broadening: structural disorder

Microscopic fluctuations of local composition of wells or barriers in the case of alloys, as well as interface roughness are responsible for a spatially dependent structural disorder, at the origin of inhomogeneous energy broadening δE. A typical localization length δl is given by:

$$\delta l = d \frac{\Delta}{\delta E} \quad (37)$$

. Miniband conduction obviously ceases to be meaningful when $\delta l < d$, since the electron does not feel more than a single SL period [53]. We therefore adopt the following criterion for miniband conduction to be valid:

$$\delta E \ll \Delta \quad (38)$$

. This criterion directly relates the validity of miniband conduction to SL structural quality, and stresses the importance of interface perfection, particularly for narrow minibands.

A nice experiment [54] has demonstrated localization driven by random fluctuations of the *position of interfaces*, purposely controlled during MBE growth (which is strictly speaking not equivalent to interface roughness). An optical technique was used, consisting in monitoring the photoluminescence of an enlarged well situated at the far end of the SL. Capture of e^--h^+ pairs by this well required their prior diffusion perpendicular to many interfaces, so that the ratio I_{WW}/I_{SL} of wide well luminescence intensity to that of the SL provided a qualitative measure of their diffusion coefficient. As shown in fig. 29, the capture efficiency and thus the ambipolar diffusion coefficient is rapidly quenched when the disorder parameter is increased. The onset of localization is in satisfactory agreement with its estimation from a simple tight-binding model using a criterion equivalent to eq. (38).

6.2 Wannier-Stark localization by the electric field

An old prediction of solid-state physics was the presumed ability for an electric field to localize Bloch electrons in a periodic lattice [55]. This phenomenon, intimately connected to Bragg reflection at the Brillouin zone boundaries has been the subject of a considerable debate among theoreticians in the last 20 years. Only recently has the associated quantization of electron energies into the so-called Wannier-Stark ladder of

localized levels been unambiguously demonstrated in superlattices. It must be precised that more properly should one speak of resonances rather than truly localized states, because of their inherently delocalized character associated with scattering as well as with tunneling via higher lying minibands. This pseudo-localization is nevertheless appreciable as soon as the rates corresponding to these two effects are smaller than the Bloch oscillation pulsation $eFd/\hbar$, or equivalently as soon as the energy spacing eFd between two resonances of the Wannier-Stark ladder is larger than its lifetime broadening. This writes $eFd > \hbar/\tau$, where τ incorporates all scattering processes. It is educating that this criterion very exactly corresponds to the onset of NDV in Esaki-Tsu model [56,21]. Such a correspondance is not surprising since NDV requires an appreciable fraction of the electrons to be Bragg reflected between scattering events. The deep link between Wannier-Stark localization and NDV thus underlined has actually been experimentally demonstrated by their correlated presence *in the same sample*, for nearly identical critical fields [57].

The observation of Wannier-Stark resonances has first been reported by optical absorption or reflectance measurements in SL subjected to a perpendicular electric field [58,59]. This technique yields a qualitative estimation of the electron wavefunction coherence length δl from the enumeration of the resonances. Values up to ~ 1000 Å at low temperatures for a miniband width of 100 meV were thus suggested [60], i.e. much more than a typical SL period (~50-100 Å). An analogous technique now using a magnetic field parallel to the layers has yielded a value of ~ 400 Å for a 110 meV wide miniband [61]. The relatively large value of this coherence length explains why true superlattice effects are now commonly observed. Again we emphasize that both wide minibands together with good SL quality characterized by τ are crucial features for a maximized coherence length.

In the presence of Wannier-Stark quantization, a semiclassical description of transport probably fails. As a rule of thumb, we propose the departure from miniband conduction to be significant when the inter-Wannier-Stark resonance spacing eFd is an appreciable fraction of Δ, say 50 %. In this case, the electron wavefunction extends over only two periods or less, and conduction is more appropriately described in terms of hopping between localized states, in the spirit of Tsu and Döhler's theory [20]. Therefore by combining the latter criterion to the Esaki-Tsu critical field $Fc = \hbar/e\tau d$ one can expect semiclassical miniband conduction to describe satisfactorily Esaki-Tsu NDV if Δ is greater than about $2\hbar/\tau$, which should be compared to (38) with $\delta E = \hbar/\tau$. Finally even when this criterion is verified, it is nevertheless clear that a semiclassical picture is inadequate when the applied electric field is such that $eFd > \Delta$ since the Wannier-Stark energy is greater than the miniband width and the corresponding localization length is ≤ one period. More appropriate then seem calculations of hopping rates between Wannier-Stark levels [20], on condition scattering is properly accounted for [21]. The approximate domain of applicability of these two limiting cases is shown in the F vs. Δ diagram of fig. 30.

The apparent simplicity of the above experimentalist's view should not wrongly suggest the reader that the physics of biased SL is simple and definitely understood. The available theories are basically insufficient since they either neglect quantization or underestimate scattering and disorder, and a both complete and consistent theory of superlattice transport is still missing.

7 CONCLUSIONS

We have attempted in this paper to devise a topical but necessarily partial view of miniband conduction in superlattices, based on the available experimental data as well as on existing theoretical models of transport. An impressive progress has been done in the last years in this field, with in particular:

   i) the conclusive observation of tailorable perpendicular miniband conduction by adjusting SL parameters

   ii) the experimental verification of the pertinence of Wannier-Stark quantization related to Bloch oscillations in biased superlattices

   iii) the experimental validation of the 20 years old ideas of Esaki and Tsu on nonlinear miniband transport, including a quantitative agreement with experiments through sophisticated semiclassical calculations

   iv) the clarification of the relation between the two latter effects

Nevertheless the story of SL is far from finished, and the authors would feel uneasy for eventually giving this impression unvoluntarily. This paper was restricted to transport properties of electrons in a *single electron miniband of the simplest case of SL*. But SL have the fantastic advantage of versatility, and certainly much remains to be understood and even imagined. Let us only modestly suggest some possibly promising routes, to be investigated in the near future: miniband conduction in the valence band, with expected effects related to light and heavy hole bands and resonance effects under strain or applied fields; multi- electron miniband systems with similar effects and/or inter-miniband tunneling; complex miniband structures involving more than two semiconductors for the unit cell; miniband conduction in lateral SL... In addition to these exciting but academic perspectives, practical applications of SL transport properties in e.g. micro-opto-electronic devices will probably emerge, as the result of improved SL structural quality and improved understanding of SL physics.

8 ACKNOWLEDGEMENTS

The authors take the opportunity of this paper in order to express their sincere gratitude to all contributors to superlattice transport work conducted at Bagneux. In particular no experimental result could have been obtained without the tight collaboration with F. Mollot and R. Planel of CNRS-L2M (Bagneux) on GaAs/AlAs SL. The participation of H. Le Person, H. Wang and M. Hadjazi on time of flight investigations and microwave measurements have been invaluable, as well as that of G. Etemadi to Boltzmann transport modelling. Magnetotransport experiments carried out by A. Celeste and J.C. Portal at INSA (Toulouse) provided the powerful confirmation of miniband conduction in SL presented in section 5. Finally we heartily acknowledge all the other contributions, including help in SL characterization, device processing, and discussions on SL physics with other workers in the field.

9 REFERENCES

[1]   L. Esaki and R. Tsu IBM J. Res. Develop. 14, 61 (1970)

[2]   C. Mailhiot and D.L. Smith, Phys. Rev. B 35, 1242 (1987)

[3]   G.H. Döhler, H. Künzel, D. Olego, K. Ploog, P. Ruden, H.J. Stolz and G. Abstreiter, Phys. Rev. Lett. 47, 864 (1981)

[4] F. Capasso, K. Mohammed and A.Y. Cho, IEEE J. Quantum Electron. QE-22, n° 9, 1853 (1986)

[5] K.K. Choi, B.F. Levine, R.J. Malik, J. Walker and C.G. Bethea, Phys. Rev. B 35, 4172 (1987)

[6] L. Esaki and L.L. Chang, Phys. Rev. Lett. 33, 495 (1974)

[7] A. Sibille, J.F. Palmier, H. Wang and F. Mollot, Phys. Rev. Lett. 64, 52 (1990)

[8] C. Minot, H. Le Person, F. Mollot and J.F. Palmier, Proceedings Series, SPIE, vol. 1362, 301, (1991).

[9] G. Bastard and J.A. Brum, IEEE J. of Quantum Electron., QE-22, 1625, (1986). see also:
G. Bastard, "Wavemechanics applied to Semiconductor Heterostructures", Les Editions de Physique, Les Ulis, France

[10] L.J. Sham and Y.T. Lu, J. of Luminescence, 44, 207, (1989).

[11] B. Deveaud, J. Shah, and T.C. Damen and B. Lambert and A. Regreny, Phys. Rev. Lett., 58, 2582, (1987).

[12] B. Lambert, F. Clerot, B. Deveaud, A. Chomette, G. Talalaeff, A. Regreny and B. Sermage, J. of Luminescence, 44, 277, (1989).

[13] J. Benhlal, P. Lavallard, C. Gourdon, R. Grousson, M.L. Roblin, A.M. Pougnet and R. Planel, Journal de Physique (Paris), 48, suppl. n°11, C5-471, (1987).

[14] H.T. Grahn, K. von Klitzing, and K. Ploog, G.H. Döhler, Phys. Rev. B, 43, 12094, (1991).

[15] C. Minot, H. Le Person and J.F. Palmier and R. Planel, Superlattices and Microstructures, 6, 309, (1989).

[16] J.F. Palmier, C. Minot, J.L. Lievin, F. Alexandre, J.C. Harmand, C. Dubon-Chevallier and D. Ankri, Appl. Phys. Lett. 49, 1260 (1986)

[17] P. England, J.R. Hayes, E. Colas and M. Helm, Phys. Rev. Lett. 63, 1708 (1989)

[18] G. Brozak, M. Helm, F. DeRosa, C.H. Perry, M. Koza, R. Bhat, and S.J. Allen, Jr., Phys. Rev. Lett. 64, 3163 (1990)

[19] R. Tsu, L.L. Chang, G.A. Sai-Halasz, and L. Esaki, Phys. Rev. Lett. 34, 1509 (1975)

[20] R. Tsu and G. Döhler, Phys. Rev. B 12, 680 (1975)

[21] R. Tsu and L. Esaki, Phys. Rev. B 43, 5204 (1991)

[22] R.A. Davies, M.J. Kelly and T.M. Kerr, Phys. Rev. Lett. 55, 1114 (1985)

[23] R.A. Davies, M.J. Kelly, and T.M. Kerr, Appl. Phys. Lett. 53, 2641 (1988)

[24] J.F. Palmier, J. Dangla, E. Caquot and M. Campana, in *proceedings of the 3rd int. Conf. on numerical analysis of Semiconductor Devices and Integrated Circuits*, ed. J.J.H. Miller (Boole Press, Dublin, 1985)

[25] see for instance: S.M. Sze, "Physics of semiconductor devices", 2nd ed., Wiley, New York (1981), chap. 11

[26] P. Gueret, Phys. Rev. Lett. 27, 256 (1971)

[27] M. Hadjazi, A. Sibille, J.F. Palmier and F. Mollot, Electr. Lett. 27, 1101 (1991)

[28] B.W. Hakki, J. Appl. Phys. 38,808 (1967)

[29] A. Sibille, J.F. Palmier, H. Wang, J.C. Esnault and F. Mollot, Appl. Phys. Lett. 56, 256 (1990)

[30] A. Sibille, J.F. Palmier, F. Mollot, H. Wang and J.C. Esnault, Phys. Rev. B 39, 6272 (1989)

[31] H. Le Person, J.F. Palmier, C. Minot, J.C. Esnault and F. Mollot, Surface Science, 228, 441, (1990).

[32] F.G.Bass and E.A.Rubinshtein, Soviet Phys. Solid State, 19, 800 (1977)

[33] P.A.Lebwohl and R.Tsu, J. Appl. Phys. 41,2664 (1970)

[34] Y. Ya Shik, Soviet Physics Semicond. 7, 187 (1973)

[35] P.J. Price, IBM J; Res. Develop. 17,39 (1973)

[36] D.L.Andersen and E.J. Aas, J. Appl. Phys. 44, 3721 (1973)

[37] M. Artaki and K. Hess, Superlattices and Microstructures, 1, 489 (1985)

[38] J.F.Bourgat, R.Glowinski, P. Le Tallec and J.F.Palmier, Computing Methods in applied Science and Engineering,p. 325, edited by SIAM(1990), Philadelphia

[39] H.F. Budd, Phys. Rev. 127, 4 (1962); see also: H.F. Budd, J. Phys. Soc. Japan, 18, 142 (1963)

[40] J.F.Palmier and A. Chomette, J. de Physique (Paris) 43, 381 (1982).

[41] A.G. Cheban, B.V.Prepelitsa and Sh D. Tiron, Soviet Phys. Semicond. 11, 30 (1977)

[42] I. Dharssi and P.N. Butcher, J. Phys. Condens. Matter 2,119 (1990).

[43] T. Tsuchiya and T. Ando, 7th int. Conf. on hot carriers in Semicond., to appear in Semiconductor Science and Technology

[44] I.Dharssi and P.N.Butcher, J. Phys. Condens. Matter 2, 4629 (1990)

[45] H.Sakaki, T.Noda, K.Hirakawa, M.Tanaka and T.Matsusue, Appl. Phys. Lett. 51, 1934(1987).

[46] J.F.Palmier, G.Etemadi, A. Sibille, M.Hadjazi, F. Mollot and R. Planel, to appear in Surface Science; see also: G. Etemadi, "modelling of perpendicular electronic transport in semiconductor superlattices", PhD thesis, Université Paris 7, Jussieu, january 1992.

[47] X.L.Lei, N.J.Horing and H.L.Cui, Phys. Rev. Lett. 66, 3277 (1991)

[48] X.L. Lei and C.S. Ting Phys. Rev. B 32, 1112 (1985)

[49] J.F.Palmier, A.Sibille, G.Etemadi, A.Celeste and J.C.Portal, 7th int. Conf. on hot carriers in Semicond., to appear in Semiconductor Science and Technology

[50] A. Sibille, J.F. Palmier, A. Celeste and J.C. Portal, Europhysics Letters 13, 279 (1990).

[51] A. Sibille, J.F. Palmier, C. Minot, J.C. Harmand and C. Dubon-Chevallier, Superlattices and Microstructures 3, 553 (1987)

[52] T.H.H. Vuong, D.C. Tsui and W.T. Tsang, J. Appl. Phys. 66, 3688 (1989)

[53] R. Lang and K. Nishi, Appl. Phys. Lett. 45, 98 (1984)

[54] A. Chomette, B. Deveaud, A. Regreny and G. Bastard, Phys. Rev. Lett. 57, 1464 (1986)

[55] G.H. Wannier, Phys. Rev. 117, 432 (1960)

[56] F. Beltram, F. Capasso, D.L. Sivco, A.L. Hutchinson, Sung-Nee G. Chu and A.Y. Cho, Phys. Rev. Lett. 64, 3167 (1990)

[57] A. Sibille, J.F. Palmier and F. Mollot, to be published

[58] E.E. Mendez, F. Agullo-Rueda and J.M. Hong, Phys. Rev Lett. 60, 2426 (1988)

[59] P. Voisin, J. Bleuse, C. Bouche, S. Gaillard, C. Alibert and A. Regreny, Phys. Rev. Lett. 61, 1639 (1988)

[60] E.E. Mendez, F. Agullo-Rueda and J.M. Hong, Appl. Phys. Lett. 56, 2545 (1990)

[61] G. Belle, J.C. Maan and G. Weimann, Sol. Stat. Com. 56, 65 (1985)

Barrier Width Dependence of Optical Properties in Semiconductor Superlattices

J.J. Song, J.F. Zhou, and J.M. Jacob
Department of Physics and Center for Laser Research
Oklahoma State University, Stillwater OK. 74078-0444, USA

Abstract

In recent years there have been many developments in the field of semiconductor quantum well physics. In this review article, we will discuss the various optical properties of superlattices, emphasizing those which depend on the barrier widths, L_b's. The well-to-well coupling of electronic wave functions present in thinner L_b superlattices leads to subband dispersion which is associated with many interesting optical properties. These include exciton lineshape changes due to subband-dispersion and subband-mixing, as well as line narrowing effects. We will discuss the L_b dependence of optical properties in III-V superlattices which are mostly observed in the excitonic spectra associated with the confined n=1 and n=2 transitions, and the first unconfined transitions.

1. Introduction

The vast area of scientific research concentrated on semiconductor heterostructure materials has been widely investigated for many years.[1,2] Rapid advancements in epitaxial growth techniques [2], such as molecular beam epitaxy (MBE), have made possible the high quality fabrication of many important heterostructures. Of these, the most widely studied are the simple quantum well (QW) or multiple quantum well (MQW) structures. However, in recent years there has been a surge of interest in the multiple coupled-quantum well structure commonly referred to as the superlattice (SL) structure.[3] The various studies conducted on all these structures exposed many new and interesting physical phenomena.

One of the most fundamental properties of QW structures is the spatial confinement of particles and quasiparticles (e.g. electrons, holes, excitons, phonons). In MQW structures, the sample parameters which influence confinement the most are the QW width (L_z) and the barrier height. However, when the barrier width (L_b) is small a coupling between particles (quasiparticles) of the QW's exists. Therefore, in SL structures L_b is also an important parameter in the confinement of particles. The electronic confinement leads to a varying degree of energy band structures in superlattices. The ability to engineer electronic band structures in solids is an important issue in device technology, thus making superlattice structures good candidates for many device applications.[2,4,5]

The main differences between MQW and SL structures arise because of well-to-

well coupling. To understand the importance of coupling consider the following simple example of electrons in QW's. The growth of a QW structure spatially confines the electrons in one of three dimensions, usually taken to be the z-direction. If the confinement scale (L_z) is less than the electron mean free path, size quantization will occur. For QW's with infinite barriers the quantization results in two dimensional properties of the electrons. This may be seen in the following equations for the energy of an electron in a semiconductor quantum well.[6]

$$E = E_n + \frac{\hbar^2(k_x^2 + k_y^2)}{2m^*}$$

$$E_n = \frac{\hbar^2 \pi^2 n^2}{2m^* L_z^2} \quad (1)$$

E_n represents the discrete energy levels originating from size quantization along the z-direction. The parabolic terms, k_x and k_y, are similar to those in bulk semiconductors, and reveal the two dimensional (2D) freedom of electrons in QW's. The corresponding QW density of states is shown in Fig.1 as the 2D step-like function. The parabolic density of states symbolizes the 3D freedom of electrons in bulk semiconductors.

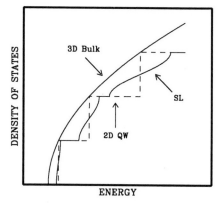

FIG. 1. Schematic representation of the density of states in bulk, QW, and SL semiconductor structures. The effect of subband dispersion in the SL smooths out the sharp step-like features in the QW density of states as shown by the dashed lines.

In practice, QW structures have potential barriers of finite height which influence the degree of electronic confinement. One effect is to limit the number of quantized energy levels bound to the quantum well.[7] The finite barrier height also allows for penetration of electronic wave functions into the barrier region (see Fig. 2a), which lowers the quantized energy levels E_n. In a SL, the L_b's are thin enough to allow electronic wave functions confined in adjacent wells to penetrate through the barriers and couple together (see Fig. 2b). The most pronounced effects of well-to-well coupling may be seen in the origins of subband dispersion as well

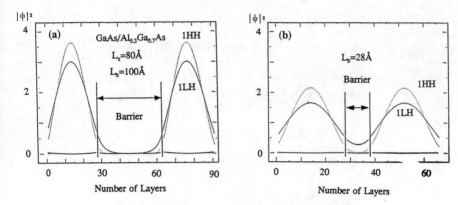

FIG. 2. Calculated wave function amplitudes for n=1 heavy- and light-holes in GaAs/Al$_{0.3}$Ga$_{0.7}$As superlattices. The L_z's are fixed at 80Å while the L_b's are either 28Å or 100Å. As L_b decreases the larger spatial overlap between wave functions gives rise to stronger coupling.

as tunneling phenomena along the z direction. Using a Kronig-Penney model in the envelope function approximation or a tight binding model [2,7,8], the subband dispersion along the z direction may be expressed as;

$$E_n(q) = E_n^{QW} + S_n - \frac{W_n}{2}\cos(qd) \quad (2)$$

E_n^{QW} is the quantized energy level for the uncoupled QW, S_n represents the shift of the center of the subband gravity with respect to E_n^{QW}, and W_n denotes the subband width. The period of the SL is given by $d = L_z + L_b$. This SL subband dispersion is plotted in Fig. 3. The modified 2D density of states for a SL is shown in Fig. 1 as the dashed lines. An effect of well-to-well coupling in the SL is smoothing out the sharp step-like features of the QW density of states. This results in many new and interesting subband dispersion related phenomena in SL's.

Superlattice structures fabricated from the semiconductors GaAs and Al$_x$Ga$_{1-x}$As have received much experimental and theoretical attention.[2,7,8-10] Much of the studies were concentrated on size quantization related phenomena. To a first order approximation the well widths do not influence wave function penetration, but only size quantization. However, the barrier height, i.e. x value, significantly influences both size quantization and wave function penetration.[6,7] In addition, it is easier to control layer widths than alloy concentrations during the growth of these structures. Therefore, most superlattice

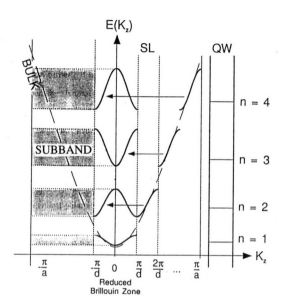

FIG. 3. Conduction subband structure of an electron in a superlattice with period d.

samples used for size quantization related experiments are designed with constant L_b and x, while the L_z's varied.[2,7,9] As we have discussed, SL structures have the interesting property of well-to-well coupling. The influence of this unique property on SL phenomena has not received much attention. In SL structures the parameter L_b is very important in determining the degree of well-to-well coupling. The effects of this coupling on SL phenomena may be systematically investigated in a series of samples where the parameters L_z and x remain constant while only L_b varies. This is because the L_b's mainly influence only the spatial overlap of the wave functions. To a first order approximation, neither the wave function penetration or size quantization depends on L_b. Therefore, the constant parameters L_z and x insure that electronic wave function penetration depths are almost constant. As L_b decreases the spatial overlap of these wave functions becomes larger so that well-to-well coupling is more significant. In this review article, we will mainly discuss well-to-well coupling related properties of SL's. We will devote most of our attention to the barrier width dependent phenomena (i.e. fixed L_z and x). One of the most distinguishing features of well-to-well coupling is the subband dispersion. In the next sections we will discuss several SL phenomena which are associated with subband dispersion. These are in the following order, energy shifts in excitonic transitions, subband associated excitonic resonances, unconfined states, exciton line narrowing effects, and the effects of subband dispersion on subband mixing

phenomena. We will also briefly discuss electronic tunneling, electric field effects, and the coupling of LO phonons in SL's. This article is limited to discussions on nominally undoped SL's.

2. Exciton Peak Shift

Photoluminescence (PL) and Photoluminescence excitation (PLE) spectroscopies have been used extensively to investigate optical properties associated with band structures in superlattices. The most pronounced features in these optical spectra are those due to fundamental exciton states. In QW structures, the energies of individual exciton states are given by;

$$E_n^{ex} = E_n^{QW} - E_{bd} \qquad (3)$$

E_n^{ex} is the optical transition energy of an exciton associated with the nth subband. E_n^{QW} is the transition energy between the nth conduction and valence subbands in an isolated QW. E_{bd} denotes the exciton binding energy associated with the nth subbands. An effect of subband dispersion in a SL is to shift the energies of excitonic spectra relative to that in the isolated QW. This shift results from a redistribution in the joint density of states arising from the well-to-well coupling, and may be seen in the energies of excitonic states in SL's given by Eq. 4.

$$E_n^{ex} = E_n^{QW} + S_n - \frac{W_n}{2} - E_{bd} \qquad (4)$$

E_n^{ex}, E_n^{QW} and E_{bd} take the same meaning as in Eq. 2. W_n is the combined width of the nth subband for electrons and holes. The parameter S_n describes the overall shift in the center of gravity for the nth electron and hole bands. Unlike QW's, the excitonic peak positions given by Eq. 4 depend on L_b as well as L_z and the barrier height. In the following three subsections we will discuss the individual terms of Eq. 4 and how they affect the excitonic spectra. In the fourth subsection we will discuss the experimental energy seperation between the n=1 heavy- and light-holes in GaAs/Al$_x$Ga$_{1-x}$As SL's. The L_b dependence of this seperation will be disussed by examining Eq. 4.

2.1 Binding Energy

The exciton binding energy has always been an important issue in the optical spectroscopies of semiconductor materials. In a semiconductor superlattice, the spatial confinement of excitonic wave functions may strongly influence the exciton binding energies away from the corresponding bulk values. In an ideal QW structure, the exciton

wave functions are perfectly confined within the well, i.e. an infinite square well. Therefore, the confined two-dimensional (2D) exciton has a binding energy four times larger than that of the corresponding bulk material.[11] This implies that the binding energy of a 2D exciton (E_{bd}(2D)) in an ideal GaAs QW is approximately 16meV, since E_{bd}(3D) is ~4meV in bulk GaAs (for additional discussion see References [12,13]). However, QW's used in practice have non-infinitesimal well widths and finite barrier heigths so that the quantum confined excitons are of quasi-two-dimensional nature. The finite barrier height implies that exciton wave functions are not perfectly confined, which leads to penetration of the wave function into the barrier regions. Thus the wave function extends along the z-direction covering the quantum well width (L_z) and the exciton penetration depth λ. The effect of this wave function distribution on exciton binding energies in GaAs can be seen in Fig. 4.[14] When L_z is much larger than

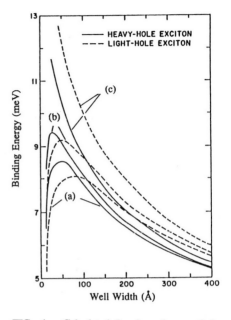

FIG. 4. Calculated L_z dependence of the ground state binding energies for the n=1 heavy- and light-hole excitons in GaAs/Al$_x$Ga$_{1-x}$As QW's. The barrier heigths correspond to Al concentrations of (a) x=0.15 and (b) x=0.3, and (c) for an infinite well. (Figure copied from reference 14)

the exciton diameter in bulk GaAs (~300Å) [15], the exciton in the GaAs QW is essentially bulk-like and has a binding energy ≈4meV. With a reduction in L_z, the exciton binding energy will increase to a maximum value somewhere between E_{bd}(2D) and E_{bd}(3D). This maximum binding energy corresponds to the situation of strongest quantum confinement. When reducing L_z the penetration of excitonic wave functions into the barrier regions is enhanced since it lifts the confined energy levels toward the barrier height. This effect tends to reduce the strength of the confinement resulting in lower binding energies. Therefore, in the limit of $L_z \rightarrow 0$, the exciton binding energy approaches the 3D bulk value of the barrier material.[14] Since the barrier heigths strongly influence the wave function penetration depth, the binding energy will also depend on x (see Fig. 4).[14] For a constant L_z the exciton binding energies decrease with smaller barrier heigths. This arises from larger wave function penetration depths with decreasing barrier heigths. In the case of a SL, the exciton wave functions in adjacent wells spatially overlap. Therefore, the spatial extent of the wave function along the z-direction is significantly larger than in QW's. This implies that the binding energies in SL's should

be less than that in similar QW's, i.e. similar L_z and x.

There have been many experimental results published concerning exciton binding energies in QW's.[16-22] In several cases, the binding energies are estimated using photoluminescence excitation [16-18] and magneto-optical [23-26] spectroscopy. In magneto-optical spectroscopy, the Landau levels associated with excitonic and band-to-band transitions have different characteristics, thus providing an estimate to the exciton binding energies. PLE spectroscopy is capable of resolving the energy difference between 1s and 2s exciton states. This also provides a good estimate of exciton binding energies in QW's.[16-18] An example of this is seen in the PLE spectra of high quality GaAs/Al$_x$Ga$_{1-x}$As QW structures (see Fig. 5). [17] The exciton binding energies in SL's may be more difficult to experimentally estimate using the aforementioned techniques. This is because subband dispersion plays an important role in optical spectra of SL's. Therefore, it is necessary to understand the effects of subband dispersion on the optical spectra of SL's.

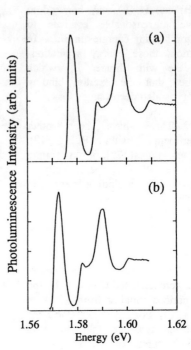

FIG. 5. Low temperature PLE spectra from GaAs/Al$_x$Ga$_{1-x}$As MQW samples with (a) L_z=82Å, x=1 and (b) L_z=75Å, x=0.35. (Figure copied from reference 17)

The theoretical and experimental studies of Chu and Chang [27], and Song et. al. [28] have clearly identified the effects of subband dispersion on the n=1 excitonic spectra in GaAs/Al$_x$Ga$_{1-x}$As SL's. Experimental (PLE) [28], and theoretical (photoabsorption) [27] lineshapes of specific SL's are shown in Fig. 6. The details of these lineshapes will be discussed fully in section 3, our focus now is the exciton binding energy in SL's. The peaks labeled 1HH and 1LH originate from fundamental excitonic transitions occuring at the M_0 critical point, and are also observed in QW structures.[16-18] Those exciton peaks labeled H and L correspond to a mixture between 2s excited states and excitonic resonances with the subband continuum. In thick barrier SL's, these peaks are mainly 2s exciton states and are very similar to those observed in isolated QW's. However, when L_b is thin enough such that significant well-to-well coupling exists, these peaks (H and L) are unique only to the admixture of states originating from subband dispersion in SL's. The origins of these spectra are described well by the theory of Chu and Chang.[27]

According to this model, the energy seperations between the peaks 1HH (1LH) and H (L) may be used to derive a reasonable estimate to exciton binding energies in SL's. In addition, these energy seperations decrease with thinner L_b, which implies that the exciton binding energy is also decreasing.

There have been other experiments using PLE [19], magneto-optical [29], or temperature dependent PL [19] spectroscopy to estimate exciton binding energies in SL's.

The experimental studies of Chomette et. al. [19], have estimated exciton binding energies in several superlattices with different periods. The GaAs/Al$_x$Ga$_{1-x}$As superlattices used were designed with $L_b = L_z$ and SL periods ranging from ~25Å to 200Å.[19] For large values of d, the band structures resemble those in QW structures. In this instance, the peak H in PLE spectra is most likely 2s excited states. For the short period SL's, the excitonic peaks are more difficult to observe in the PLE data. This is caused by several effects of strong coupling in short

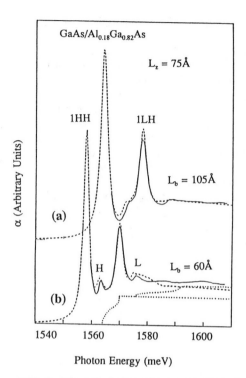

FIG. 6. Theoretical absorption spectra (dashed curves) and experimental PLE data (solid curves) for GaAs/Al$_{0.18}$Ga$_{0.82}$As SL's with L_z=75Å and L_b's equal to 100Å and 60Å. The continuum states of 1HH and 1LH for L_b=60Å are shown by the dotted lines. (Figure copied from reference 28)

period SL's. For example, stronger coupling results in larger subband dispersion which tends to reduce the binding energy. Also, interface imperfections become more important in short period SL, thus affecting sample quality. However, temperature dependent PL was used to determine exciton binding energies in short period SL's.[19] For superlattice periods which were not too thin, yet allowed for significant well-to-well coupling, the peaks 1HH and H were still resolved in the PLE spectra. It was suggested that the energy seperation between the 1HH and H peaks could be used to estimate the heavy-hole exciton binding energy, since it was assumed that peak H was related to the onset of the subband continuum.[19]

There have been several theoretical studies related to exciton binding energy in SL's.[19,27,30-34] Several of these have used the data of Chomette et. al. to test the

validity of the theoretical models.[19,30,34] An example of this is shown in Fig. 7.[34] The data were fitted with the theoretical model of Signam and Sipe.[34] This model used an exciton envelope function in terms of localized exciton Wannier functions in the basis of "two-well" exciton states. As seen in Fig. 7, the maximum binding energy for $GaAs/Al_{0.3}Ga_{0.7}As$ SL's is approximately 9meV, and occurs at the SL period around 140Å ($L_z=L_b=70$Å). A comparison of Fig's. 4 and 7 shows that the maximum exciton binding energy in $GaAs/Al_{0.3}Ga_{0.7}As$ QW's occurs at a much narrower L_z (~30Å) [13,14], or L_z (~40Å).[34] This is an indication of the influence that well-to-well coupling has on the exciton binding energies in SL's. If the effects of coupling were not significnt the maximum binding energy would be expected to occur near L_z ~ 30Å to 40Å for short period SL's.

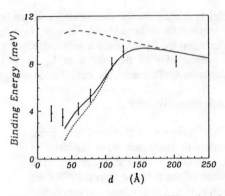

FIG. 7. Binding energies for the 1HH exciton in $GaAs/Al_{0.3}Ga_{0.7}As$ superlattices as a function of d. The solid curve represents the theory [34] which agrees well with the experimental data.[19] (Figure copied from reference 34)

2.2 The Shift S_n

The presence of subband dispersion in a SL shifts the subband center of gravity. The sum of the overall shift of the nth subband center of gravity for both electrons and

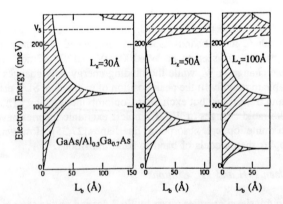

FIG. 8. Barrier width dependence of the band structure for electrons in $GaAs/Al_{0.3}Ga_{0.7}As$ superlattices with fixed well widths. The hatched area denotes the subband. (Figure copied from reference 8)

holes is denoted by the parameter S_n. It has been proven that this shift is always small in comparison to the subband width W_n, especially when L_b is not very thin.[8] A model calculation by Bastard of the subband edges as a function of L_b is shown in Fig. 8.[8] A general idea of the effects of L_b on W_n and S_n may be ascertained from these calculations. Therefore, the contribution of S_n to the peak shift may often be omitted.[8]

2.3 Subband Width W_n

As we have discussed, the spatial overlap of electronic wave functions in SL's results in subband dispersion. The widths of the subband dispersion for the combined n=1 heavy-hole and conduction bands in GaAs/Al$_{0.3}$Ga$_{0.7}$As SL's are shown in Fig. 9.[8] It can be seen that the width of this dispersion depends exponentially on the barrier thickness. In thick barrier SL's, W_n may be nearly 0 meV, whereas it can be larger than 200meV in thin barrier SL's. This change in the subband width as a function of L_b is much faster than the corresponding change in exciton binding energies. This is seen in the comparison of Fig. 7 [34] to Fig. 9.[8] The binding energies for GaAs/Al$_{0.3}$Ga$_{0.7}$As SL's with $L_z = L_b$ are shown in Fig. 7. It can be seen that the maximum binding energy of $\approx$9meV

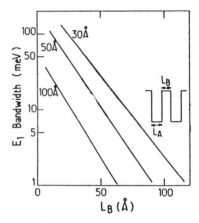

FIG. 9. The combined bandwidth for the n=1 conduction and heavy-hole subbands in GaAs/Al$_{0.3}$Ga$_{0.7}$As superlattices versus the L_b's for several fixed L_z values. (Figure copied from reference 8)

occurs when $L_b=L_z\approx$70Å, while the minimum binding energy of $\approx$4meV occurs when $L_b=L_z\approx$30Å. The width of the subband dispersions associated with the minimum and maximum binding energies are seen to be $W_1\approx$100meV and $W_1\approx$2meV, respectively. This is a considerable change in W_n while the binding energy only changes by $\approx$5meV. Therefore, it is possible to roughly fit the peak position of excitons in SL's using a simple model of band-to-band transitions, but excluding Coulomb interactions. This is because E_{bd} varies slowly with rapid changes in W_n. Complete excitonic theories have been used successfully to determine optical absorption lineshapes.[27,35] However, these are generally more complex than models of band-to-band transitions.

2.4 Seperation between 11H and 11L Excitons

The excitonic transition energies given by Eq.4 depend on the sample parameters, (L_z, L_b, x) to various degrees. The influence that L_b's have on the transition energies may be studied by properly designing L_b dependent measurements. PLE data taken from two

GaAs/Al$_{0.18}$Ga$_{0.72}$As SL's with L$_z$ ~75Å and L$_b$ = 60 or 100Å are shown in Fig. 6. For additional data from a different SL series, see Fig. 17. It is observed that as the L$_b$'s decrease, the n=1 heavy- (11H) and light-hole (11L) excitons red shift and their energy seperations also decrease. This is understood by examining the energy difference in the excitonic transitions given by;

$$\Delta_{11L-11H} = E_{11L}^{ex} - E_{11H}^{ex}$$

$$= E_{11L}^{QW} - E_{11H}^{QW} - \frac{W_{11L} - W_{11H}}{2} + E_{bd}(11H) - E_{bd}(11L) + S_{11L} - S_{11H} \quad (5)$$

When L$_z$ and x are constant parameters in a series of GaAs/Al$_x$Ga$_{1-x}$As SL's, some of the terms in Eq. 5 will not depend strongly on L$_b$. To a first order approximation E$_n^{QW}$ is independent of L$_b$, therefore the term E$_{11L}^{QW}$ - E$_{11H}^{QW}$ may be considered constant with changing L$_b$. The quantity S$_n$ is influenced by L$_b$, however, both S$_{11L}$ and S$_{11H}$ have nearly the same dependence on L$_b$.[8] Therefore the term S$_{11L}$ - S$_{11H}$ may also be considered constant with changing L$_b$. If the sample parameters are properly chosen, the binding energies of both 11L and 11H excitons can have a very similar dependence on L$_b$. In this case E$_{bd}$(11L) - E$_{bd}$(11H) may also be considered constant with changing L$_b$. Therefore, $\Delta_{11L-11H}$ will depend mostly on the difference in the 11H and 11L subband widths. As L$_b$ decreases, the subband dispersion becomes larger and the difference between the 11L and 11H subbands becomes smaller. This leads to a decrease in the energy seperation between the 11L and 11H exciton transitions.

3. Effects of Subband Dispersion and Saddle Points on Excitonic Spectra

In superlattices, the zone-folding effects of the band structure along k$_z$, the wave vector along the sample growth direction, can lead to the energy subband dispersion and the formation of M$_1$-type critical points (saddle points).[36] The effects of subband dispersion and saddle point singularities may be evident in the excitation (absorption) spectra of excitons in SL's. In fact, features associated with these subband properties have been identified in the excitonic spectra of SL's.[28,36,37] The most prominent structure in the excitonic spectra of QW's appears near the M$_0$ critical point. However, in a SL with reasonable subband dispersion, additional peaks near the M$_0$ critical point, as well as small peaks near the M$_1$ saddle point appear in the excitonic spectra.

The subband dispersion in a SL reflects the amount of well-to-well coupling, and is altered upon a change in L$_b$. For a fixed L$_z$ and x, the subband dispersion will increase as the L$_b$'s decrease. The effects of increasing subband dispersion on the n=1 excitonic spectra may be seen in Fig.6 and Fig. 10.[28] The n=1 light-hole (1LH) spectrum taken from the SL with L$_b$=105Å is shown in Fig. 10(a). In contrast to the 1LH spectra from

ordinary QW structures, additional peaks (H and L) appear in the excitonic lineshape. Additional peaks H' and L' also appear in the 1LH spectra taken from the thinner barrier SL's.[28] These features are associated with saddle point excitonic resonances and will be discussed in a following paragraph.

A theoretical model for the absorption spectra in SL's has been used to compare with the experimental excitation spectra. This model takes into account the mixing of valence subbands, and excitonic effects of different subbands as well as the M_1-type singularities.[27,28,35] A comparison of theory and experiment can be seen in Fig. 6. According to the theoretical lineshape, the peaks H and L are attributed to a mixture of 2s exciton states and the band edge resonance arising from a redistribution of the oscillator strengths caused by the Coulomb interaction. In a QW or a SL with thick L_b, Fig. 10(a), the peaks H and L are most likely 2s excited exciton states merged into a step-like band continuum. However, in a SL with thinner L_b's a significant portion of H and

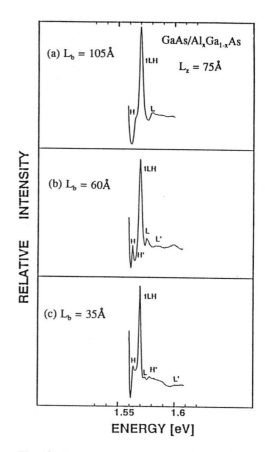

Fig. 10. Low temperature PLE spectra taken from a series of GaAs/Al$_{0.18}$Ga$_{0.82}$As superlattices with $L_z \approx 75$Å. These samples have barrier widths of (a) 105Å, (b) 60Å, and (c) 35Å. (Figure copied from ref. 28)

L arise from the lower subband edge resonances.[27,28] In contrast to the nearly symmetric lineshape of the fundamental exciton peaks, the peaks H and L have an asymmetric lineshape. The sharp rise of these peaks may be attributed to the 2s excited states while the slowly decreasing tail is caused by subband edge resonances.[27,28]

As mentioned earlier, the curvature of the subband dispersion in SL's leads to the formation of M_o- and M_1-type singularities. The most pronounced features in optical spectra of superlattices most often originate near the M_o singularities. This corresponds,

for example, to peaks 11H (11L) and H (L). However, weak excitonic resonances associated with M_1 saddle points have also been clearly observed in the excitation and magneto-optical spectra of superlattice structures.[28,36,38-40] These resonances have been observed with unconfined states [36,38] as well as the n=1 heavy- and light-hole excitons.[28,39,40] The weak structures associated with M_1 saddle point excitons are shown in Fig. 10. The peaks H' and L' refer to the 11H and 11L exciton peaks appearing at the Brillouin-zone edge, respectively. The assignment of these peaks to excitonic resonances at the M_1 critical point is based on the consideration of the corresponding transition energies.[28] The occurance of these resonances are not limited to only $Al_xGa_{1-x}As/GaAs$ SL's. They have also been observed in $In_xGa_{1-x}As/GaAs$ strained layer SL's.[41-43]

It was theoretically suggested that saddle point excitons in solids should be observed.[44] This was later verified in alkali halides, [45] and bulk semiconductors.[46] However, the issue of saddle point excitons in semiconductors is still somewhat controversial.[44,46,47] The nature of saddle-point associated excitonic resonances in SL's is more complicated than the M_0 exciton problem. Recent studies of excitonic band structures in thin L_b SL's calculate a negative binding energy at the M_1 critical point in thin barrier SL's.[34] In addition, theoretical and experimental studies have shown that M_1 associated excitonic resonances involve all of the subband states unlike M_0 resonances, which involve only those states in the near vicinity of the critical point.[27,28,40] Therefore, it is more difficult to discuss the excitonic lineshapes occuring at the saddle point. In contrast, at the M_0 singularity the fundamental and subband dispersion associated excitonic resonances have symmetric and asymmetric lineshapes, respectively. The origins of these lineshapes are also well understood.[27,28,40]

Since the properties of SL's often depend on subband dispersion, it is useful to simply determine the amount of dispersion for a given SL sample. A convenient way to estimate the subband dispersion in a SL is from excitation or absorption spectra. Since excitonic resonances can occur at both M_0 and M_1 critical points, the energy seperation of the corresponding spectra is a good estimate to the amount of subband dispersion.

4. Unconfined States

In semiconductor superlattices such as $GaAs/Al_xGa_{1-x}As$, there are quantized electronic energy states confined mostly within the GaAs well layers with energies below the $Al_xGa_{1-x}As$ potential barriers. Quantized energy subbands can also be formed above (below) the conduction- (valence-) band barriers.[38] These subbands, often called unconfined energy states, are more difficult to detect than confined states since they are superimposed on top of the large continua states.

By means of resonance Raman spectroscopy performed with LO phonons, Zucker et. al. detected optical transitions between the first unconfined electron and hole states in $GaAs/Al_xGa_{1-x}As$ superlattices with $L_b \sim 100$Å and 200Å.[48] Unconfined states were

also observed by Song et. al. in the PLE spectra of a series of GaAs/Al$_x$Ga$_{1-x}$As superlattices with varying L_b's.[38] The strengths of the transitions involving the first unconfined heavy-hole and conduction bands were found to depend strongly on the L_b. For GaAs/Al$_{0.2}$Ga$_{0.8}$As SL's with L_z=150Å, the transition is very strong when L_b=150Å, but is very weak for L_b's of 70Å and 30Å.[38] This was the first work that showed the barrier width dependence of unconfined transitions. Since then, the unconfined states have been observed by modulation photoreflectance [43,49-52], and photocurrent experiments [53], from a variety of QWs' and SLs'.

The L_b dependence of the unconfined transition strength may be understood by examining the wave functions of these states. Figure 11 shows the wave functions of these states in GaAs/Al$_{0.2}$Ga$_{0.8}$As SL's with L_z = 150Å and for three different L_b's.[38] The wave functions are calculated with a tight-binding model.[54] In this series, the first unconfined valence and conduction states are HH6 and CB4, respectively. In unconfined transitions, the Δn=0 selection rules break down. An examination of Fig.11 shows that there is a large overlap between the heavy hole and conduction states in the barrier layers for all three cases. The overlap in the well layers is small due to the different number of valence and conduction-state nodes present. The total overlap is large for thick L_b's because the wave function amplitudes are much larger in the barrier layers than in the well layers. For this case the wave function is considered to be resonant in the barrier region. In general, the first unconfined transitions in most of superlattices with large L_b's are strong, although the strengths also depend on L_z and x.

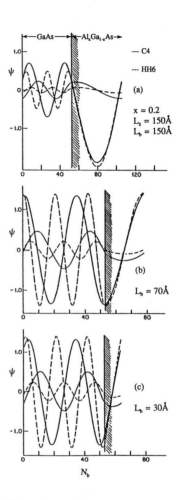

FIG. 11. The wave functions representing the first unconfined valence (HH6), dashed line, and conduction (C4), solid line, states at k_z=0 in GaAs/Al$_{0.22}$Ga$_{0.78}$As superlattices. The wave functions are composed of s- and p-type orbitals. These are shown for HH6 as the smaller amplitude dashed and solid lines respectively. N_b is the number of GaAs lattice constants (~5.7Å). (Figure copied from ref. 38)

As L_b gets larger the transition energies decrease and approach the energy gap of the bulk material. This trend can be seen in Fig.12[36] which shows PL spectra taken from a series of GaAs/Al$_{0.22}$Ga$_{0.78}$As SL's. The inset shows the PLE spectrum for the n=1,2 confined levels, whereas the unconfined transitions are indicated by a thick arrow. All the samples exhibit essentially the same spectra for the confined regions. For the unconfined transitions, however, striking variations are noticed with L_b changes. In Fig.12, two peaks are found for each spectrum between 1.8 and 1.85eV, except for the one with L_b=70Å. Both of the doublet peaks arise from the first unconfined subband transitions. The peaked structure in the PLE spectra discussed above implies that there is some kind of excitonic enhancement occurring for the transitions of concern here. The lower energy peak is attributed to the Brillouin zone (BZ) edge transition, whereas the higher one to the BZ center

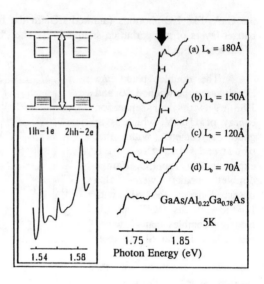

FIG. 12. PLE spectra taken from GaAs/Al$_{0.22}$Ga$_{0.78}$As superlattices with L_z = 150Å. The L_bs range from 180 to 70Å as shown in (a)-(d). The PLE doublets are indicated by the thick arrow in the spectra, and the corresponding transitions are depicted on the upper left corner. The seperations of the doublets are indicated by the horizontal bars. The inset shows the low energy side of the PLE spectrum taken from the SL sample with L_b = 150Å. Similar spectra were obtained in this region from the other SL's. (Figure copied from reference 36)

transition. For the states involved in this transition, the BZ edge and center correspond to the M_0 and M_1 critical points, respectively. A distinct difference in the lineshapes between n=1 confined states (see section 3) and unconfined transitions is noticed. In particular, with those states associated with the saddle-points. The energy seperation ΔE, between the peaks associated with transitions at $k_z = 0$ and π/d respectively, increases with increasing subband dispersion. This is seen in the L_b dependent PLE spectra in Fig. 12. It was found that the experimental values of ΔE agreed well with the theoretical calculations, which assumed that the binding energies were approximately equal at the two critical points. In addition, the calculations show that ΔE is not very sensitive to the aluminum concentration. Therefore the measurement of ΔE for unconfined transitions can be a very useful way to determine L_b, in much the same way that the confined transition energies relate to the well width. The unconfined transition doublet peaks can also be observed by resonance Raman spectroscopy (RRS). Fig.13 shows the Raman profile of

the GaAs-like LO-phonons (marked LO_1 in the figure) in the 180Å-thick $Al_xGa_{1-x}As$ barrier layers of a superlattice sample.[55] The PLE spectrum of this sample is shown at the top of Fig.12.

The doublet peaks are clearly seen between 1.80 and 1.83 eV with their separation energy of 11meV, which is in good agreement with that observed in the PLE experiments. The LO_1 Raman doublet peaks were also observed from the SL with L_b=150Å. Its RRS profile closely resembles that shown in Fig. 13. In the SL samples studied here the electron and hole wave functions of the confined states exhibit damping in the barrier layers, leading to small transition strengths in the barrier regions. Hence, no significant enhancement of barrier phonons will be expected at confined transition energies. Therefore, the observation of clear resonant enhancement of the barrier phonons at the PLE doublet peak positions confirms that those doublets indeed come from unconfined transitions. This is also consistent with the idea that unconfined states in thick barrier SL samples have large amplitudes in the barrier region as we discussed in the begining of this section.

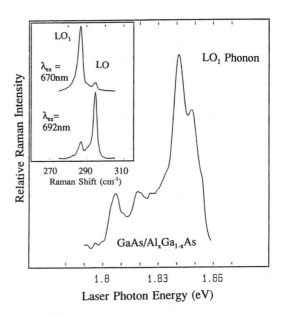

FIG. 13. The resonant Raman profile of the LO phonons in the barriers of a $GaAs/Al_{0.22}Ga_{0.78}As$ SL with L_b = 180Å. The doublet peaks are detected just above 1.80eV. The corresponding outgoing resonance peaks appear superimposed on top of other unconfined transitions. The inset shows the phonon spectra taken at two different wavelengths; below and above the barrier gap. LO and LO_1 denote the longitudinal optic phonons from the well and the barrier layers, respectively. (Figure copied from ref. 55)

5. Line Narrowing

The optical linewidth in semiconductor superlattices is an important issue among physicists and engineers. The inhomogeneous broadening of this optical linewidth originates from both structural and compositional imperfections. These imperfections naturally occur during the growth of a semiconductor heterostructure and are usually

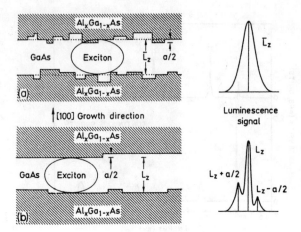

FIG. 14. Illustrated effects of layer imperfections on the optical spectra in superlattices. The resulting lineshape depends on the differences between confined exciton diameters and the lateral size of these imperfections. (Figure copied from ref. 2)

appreciable at the interfaces. It is well known that structural imperfections may appear as formations of islands or terraces during growth. Fig. 14 [2], depicts this interface roughness as shown in the diagram of a QW structure. The effects that this imperfection has on the optical spectra of QW's are illustrated in this figure. In a simplified model, if the lateral size (L_{xy}) of these islands are smaller than the quantum confined exciton diameter, a line broadening exists (Fig. 14a). However, if L_{xy} is larger than the confined exciton diameter, multiple peaks appear in the optical lineshape (Fig. 14b). The presence of these islands in a QW gives rise to a distribution of L_z's in the x-y plane. When L_{xy} is small, the exciton diameter can cover many islands. Thus, a spatial average in the distribution of L_z's results in line broadening. However, when L_{xy} is large, the exciton diameter will cover only one island resulting in multiple peaks in the optical lineshape.

The interface quality of MBE grown superlattices can be greatly improved using interrupted growth methods.[56] Photoluminescence spectra of single QW samples grown by MBE with and without growth interruption are shown in Fig. 15.[56] The spectra taken from the interrupted growth samples reveal multiple peaks associated with layer width fluctuations. These have been identified with monolayer fluctuations, however it has been suggested that these fluctuations may be different than 1 monolayer.[57] The multiple peaks from the interrupted growth samples have been attributed to the atomically smooth interfaces within the islands.[58] In contrast, the spectra taken from the uninterrupted growth samples display only broadened peaks. Recently, several experiments using the chemical lattice image [59], PL spectroscopy [60,61], and Raman scattering [61,62], have shown that both small and large scale atomic layer fluctuations exist even in high quality superlattice samples.

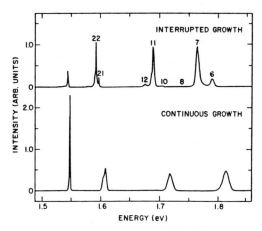

FIG. 15. Low temperature PL spectra of single QW structures grown with and without interruption. The structure seen in the top spectra are attributed to L_z's of 6-8,10-12,21, and 22 monolayers. (Figure copied from ref. 56)

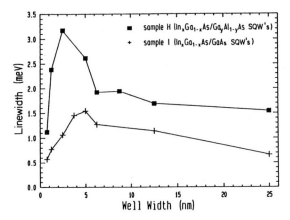

FIG. 16. PL linewidths are plotted against L_z. The samples are $In_xGa_{1-x}As/Al_yGa_{1-y}As$ single QW structures with L_b and x,y fixed. (Figure copied from ref. 64)

There have been several reports of experimental and theoretical investigations on the effects of superlattice layer widths on this inhomogeneous broadening.[63-66] The effects of L_z on exciton linewidths has been reported by Reithmaier et. al. in single QW structures.[64] The PL lineshapes are shown for the various L_z values in Fig. 16.[64] As L_z decreases from approximately 250Å to 30Å, the PL linewidth increases to a maximum. Further reduction in L_z results in a rapid decrease in the PL linewidth. For L_z's

significantly larger than 30Å, the spatial confinement of the excitons in QW's is relaxed and the interaction between the confined exciton and interfaces is reduced. As L_z decreases below 30Å, the confinement energy increases thus enhancing the penetration of excitonic wave functions into the barrier region. This effect tends to reduce the inhomogeneous broadening associated with interface roughness.

In a SL, the subband dispersion influences this inhomogeneous broadening. This has been investigated in a series of GaAs/Al$_x$Ga$_{1-x}$As superlattices designed for L_b dependent studies.[65] It was found that both heavy- and light-hole exciton linewidths decrease with an increase in subband dispersion. The PLE spectra taken from a series of MBE grown Al$_{0.14}$Ga$_{0.86}$As SL's (L_z=59Å, L_b=31-130Å) are shown in Fig. 17. [39,65] As L_b decreases from 130 to 31Å the exciton linewidth for both heavy- and light-holes decreases by nearly a factor of four. A theory using the coherent potential approximation to model layer-width fluctuations in the SL agrees well with the experimental results. The heavy- and light-hole exciton linewidths from both experiment and theory are shown in Fig. 18.[39,65] If the layer width fluctuations are considered as perturbations to the exciton states, a broadening in the optical

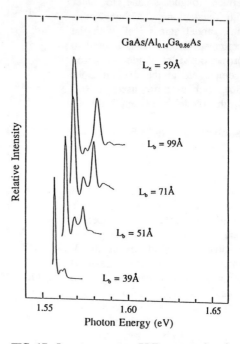

FIG. 17. Low temperature PLE spectra taken from a series of GaAs/Al$_{0.14}$Ga$_{0.86}$As superlattices. The spectra changes dramatically as the L_b's decrease. In particular, a distinct line narrowing exists. (Figure copied from ref. 65)

transitions should occur. In a weak scattering limit, this broadening is proportional to the exciton density of states. In a SL, decreasing L_b results in an increase in subband dispersion which is associated with stronger coupling between adjacent quantum wells. This results in a smaller density of states causing an excitonic line narrowing.

Exciton line narrowing in PL spectra of GaAs/Al$_x$Ga$_{1-x}$As short period SL's has also been observed.[66] The PL linewidths were shown to decrease with increasing subband dispersion. The experimental results were explained using a simple statistical model. Averaging the interface effects over a constant length (exciton diameter), shorter

period SL's result in a smaller standard deviation. This is because there are more SL periods contained within an exciton wave packet reducing the average effect of interface roughness due to layer width fluctuations. Although this simple model agrees well with the experimental results, it is limited to short period SL's. If the SL period is comparable to the exciton Bohr radius, such as those used in Ref. [65], this model is less applicable.

6. Subband mixing in SL's

In semiconductor superlattices, a series of subbands are formed by the confinement of electronic wave functions. This confinement also lifts a degeneracy which exists between heavy and light-hole valence bands at the Γ point. It was first realized theoretically that in-plane dispersion curves for valence subbands could mix together in III-V semiconductor superlattices.[54,67] This was later verified experimentally in GaAs/Al$_x$Ga$_{1-x}$As QW structures by several authors.[37,39,68-70] In

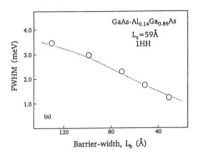

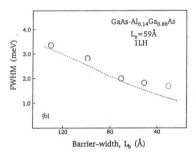

FIG. 18. The experimental (circles) and theoretical (dashed lines) of the FWHM for 1HH and 1LH excitons plotted against L_b. In this superlattice series the linewidth decreases by nearly a factor of 4 as the L_b varies from thick to thin. (Figure copied from ref. 65)

III–V semiconductor superlattices the valence subband dispersion curves along the z directon are heavy- and light-hole like. However, along the in-plane direction the heavy hole has a light mass and the light hole has a heavy mass dispersion. This is known as the mass reversal effect.[71] If no mixing exists among these hole subbands, the dispersion curves for heavy- and light-holes would intersect at some non-zero in-plane wave vector, k_{xy}. This is because the heavy-hole has a higher energy at the Γ point and a larger in-plane curvature than the light-hole, as seen in Fig. 19.[71] The presence of subband dispersion along the z direction in a SL may induce further valence subband mixing.[37,72] The crossing or anti-crossing of these valence subbands depends upon the symmetry properties of the SL. A SL which is grown on [111] oriented substrates has a higher symmetry along the growth direction than those grown on [001] oriented substrates. This results in allowed subband crossing for [111] orientations and an anti-crossing of subbands for [001] orientations.[73]

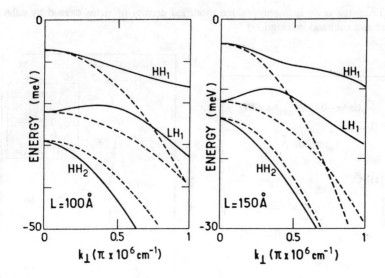

FIG. 19. In plane dispersion relations for the n=1 valence subbands in GaAs/Al$_{0.3}$Ga$_{0.7}$As QW's. The solid lines illustrate band-mixing effects, whereas they are excluded for the dashed lines. (Figure copied from ref. 71)

One of the most pronounced effects of subband mixing is the appearance of the forbidden transition from the n=2 conduction band to the n=1 light hole band (2CB - 1LH). This occurs because of the admixture between the 2HH and 1LH subbands.[54,74] In PLE or optical absorption in QW's, this effect appears as a doublet-like peak.[69,70] This is evident in the spectra taken from QW's using PLE shown in Fig. 20.[27] The peak labelled E_{21L} arises from the forbidden transition between the n=2 conduction subband to the n=1 light-hole subband (2CB - 1LH). Similarly, peak E_{2h} originates from 2CB - 2HH. In the following text we will refer to these peaks as 21L and 22H.

Becuase of the subband dispersion in SL's, the n=2 region may have some additional features other than those associated with subband mixing in QW's. The doublet-like structures originate from the transitions 21L and 22H near the M_o saddle point. This feature may be seen in both QW and thick barrier SL structures. A good example of the effects of subband dispersion and subband mixing on the optical spectra in SL's is shown in Fig. 21.[37] An unusual lineshape appears near the n=2 exciton region. Along with the typical doublet-like features, an additional broad peak (H) exists. The origins of this peak are the same as those labelled H in the n=1 spectra (see section 3). In Fig. 21, PLE data (a) and a theoretical SL lineshape (b) are shown. The theoretical model includes subband mixing, subband dispersion, Coulomb interactions and the coupling of excitonic states.[27,35] A theoretical plot of the lineshape without including Coulomb interactions, i.e. no exciton formation, is shown in Fig. 21(c). This

peculiar lineshape is an indication of the modified density of states caused by subband dispersion and subband-mixing.[39]

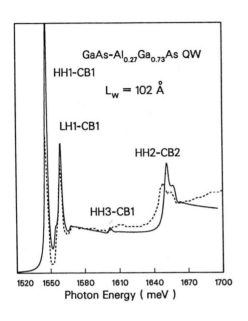 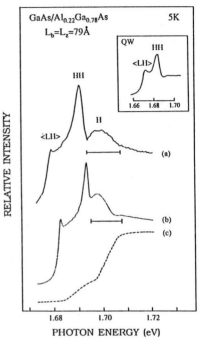

FIG. 20. A comparison between an experimental excitation spectrum and a theoretical absorption spectrum in GaAs/Al$_x$Ga$_{1-x}$As MQW structures is shown. The model includes band-mixing effects and illustrates the corresponding features in the absorption lineshape. (Figure copied from ref. 27)

FIG. 21. The effects of subband-dispersion and -mixing is shown experimentally (a) and theoretically (b) for the n=2 exciton region. These effects on the joint density of states may be ascertained from (c). (Figure copied from ref. 37)

In Fig. 22 experimental PLE results which display the effects of subband dispersion and subband mixing in SL's with different barrier widths are shown. In both Al$_{0.18}$Ga$_{0.82}$As/GaAs SL's the well widths are approximately equal to 75Å, but the barrier widths are either 150Å or 60Å. Calculated subband dispersion ranges are also shown on the figure as the solid bars.[39,75] In the thick barrier sample the well-to-well coupling is weak and the subband dispersion is small. In this case, the subband dispersion contributions to the excitonic resonances forming the peak H are small, and the peak H becomes hard to distinguish. The n=2 lineshape from this sample mainly displays the features of an isolated QW. In the thinner barrier sample, the well-to-well coupling is

strong and the subband dispersion becomes large, in this case the peak H looks like a flat plateau. A comparison of the spectra obtained from the two SL samples points to an enhancement of the peak H in the presence of significant subband dispersion.

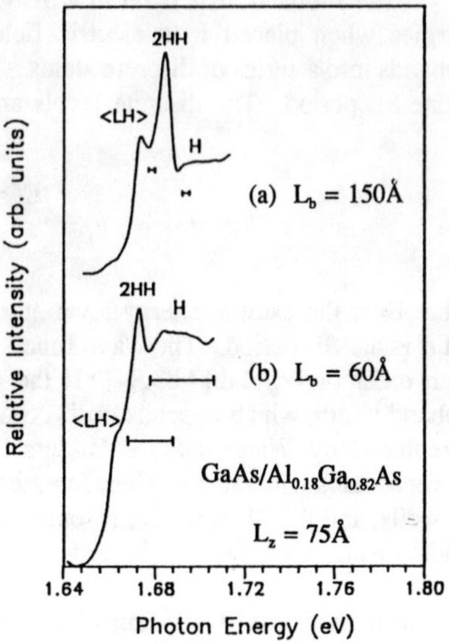

7. Tunneling Phenomena on Optical Properties of Superlattices

Tunneling phenomena in superlattice structures have recieved much attention due to the potential for device applications. [4] For more discussion on tunneling phenomena, see reference [2,76]. The effects of tunneling phenomena on optical properties have been studied in SL's [77-80], and coupled-double-QW [2,81] structures. In addition, much attention has been given to the effects of resonant tunneling phenomena on optical properties in these structures.[76,80] In practice, for thin L_b tunneling can lead to trapping of electrons at the surface of the SL structure. This escape mechanism allows the electrons to tunnel out of the wells through the barriers and be trapped by surface states. The effects of this have been seen in the quenching of PL signals in thin barrier SL's.[77] It was shown that quenching could be avoided if the thin barrier SL's were between thick cladding layers of $Al_xGa_{1-x}As$ with band gaps larger than the barrier height.

FIG. 22. Experimental PLE data near the 2HH exciton in $GaAs/Al_{0.18}Ga_{0.82}As$ SL's with $L_z \sim 75$Å. In (a) $L_b = 150$Å, and there is very little subband dispersion. This lineshape is similar to those observed in isolated QW structures. In (b) $L_b = 60$Å, and the lineshape displays the large resonances (H) arising from subband dispersion. The amount of subband dispersion is indicated by the solid bars. (Figure copied from ref. 39)

8. Stark Effect

If an electric field is applied along the z direction in a semiconductor QW, a red shift in the ground state exciton energy results. This field polarizes the exciton which leads to an interaction between the induced dipole and the electric field. This results in a red shift which depends almost quadratically on the field.[7,82] This effect cannot be observed in bulk semiconductors because the field ionizes the exciton. However, the quantum confinement of the excitons prevents them from ionization caused by the electric field.

The subband dispersion in a SL gives rise to a blue shift in the exciton transition energies when placed in a electric field along the z direction. The field splits the subbands into a series of discrete states. The energy splitting is equal to the electric work of one SL period. The discrete levels are given by;

$$E_m = E_0 \pm meFd \qquad (6)$$

where E_0 is the exciton energy level at zero bias, m is an integer, F is the electric field, and d is the SL period. The wave functions in the SL are localized along the z-direction to an order of W_n/eFd.[7,83,84] If the potential associated with eFd is larger than the subband width, which depends on L_b, Stark localization occurs and the subband structure is replaced by Wannier-Stark ladders. When this occurs, the well-to-well coupling between states will vanish. Therefore, the wave functions are completely localized within the wells, and the SL structure resembles that of MQW's. This causes a blue shift in the band-to-band transitions of the order $W_u/2$.

9. Confinement and Coupling of LO phonons in Superlattices

It is well known that GaAs LO phonons are almost perfectly confined to the GaAs layers in GaAs/AlAs superlattices. This is simply understood by examining the bulk phonon dispersion curves of GaAs and AlAs. As seen in Fig. 23, [85] GaAs phonon energies fall between the energies of AlAs optical and acoustic phonons. Therefore, the energy of the GaAs LO phonon will not be able to propagate through the AlAs material. However, the LO phonon wave functions should penetrate a small distance into the AlAs material. When the barrier material is $Al_xGa_{1-x}As$, the LO phonon penetration should be larger than that in pure AlAs.

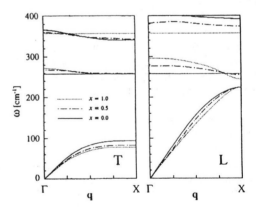

FIG. 23. Phonon dispersion curves in $Al_xGa_{1-x}As$ alloys for 3 different Al concentrations, x. The dispersion is along the ΓX direction. (Figure copied from ref. 85)

This is partly due to the smaller energy seperations between the LO phonons in the constituent materials of the SL, i.e well and barrier phonons. Therefore, as the barrier width (L_b) between adjacent GaAs wells becomes small, a coupling between the GaAs LO phonon wave functions may occur. The penetration of GaAs LO phonons into the alloy

barrier regions is not well understood, and the problem is further complicated by the two phonon mode behavior of the alloy barrier.[86]

From cw Raman measurements, it has been shown that GaAs LO phonons are confined by AlAs layers as well as AlAs phonons being confined by GaAs layers.[87-90] The spatial confinement of the phonon results in quantizing the wave vector as $q \approx m\pi/L$, where m is an integer and L is the scale of the confinement. The scale of the confinement is considered to be the well (barrier) width plus the phonon penetration depth. The cw Raman spectra of confined phonons contain multiple peaks associated with the different wave vectors. These peaks are shifted down in energy from the bulk LO phonon energies of the constituent material. This is because the dispersion of the phonons yields smaller energy for larger wave vector. Theoretical studies have shown that the penetration depths of GaAs (AlAs) phonons into the AlAs (GaAs) layers are as little as 1 monolayer.[91] There have been other similar Raman studies which reveal that GaAs LO phonons are confined by alloy $Al_xGa_{1-x}As$ layers, [92,93] and that the GaAs-like LO phonons in the alloy layers are confined by AlAs layers.[94] However, these did not treat the penetration of GaAs (AlAs) LO phonons into the alloy material or vice versa.

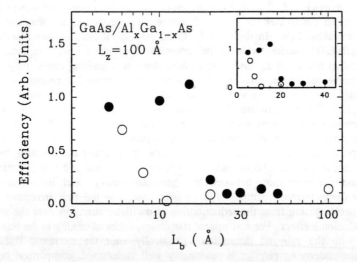

FIG. 24. The hot phonon generation efficiency in GaAs/Al_xGa_{1-x}As SL's with L_z = 100Å. The data represent two series where x = 1 (closed circles) and x = 0.4 (open circles). The abrupt rise in the generation efficiency with decreasing L_b's leads to the phonon penetration depth. (Figure copied from ref. 95)

Recently, a picosecond Raman scattering experiment has revealed the penetration depth for GaAs LO phonons into $Al_xGa_{1-x}As$ layers.[95] It was found that GaAs LO

phonon penetration into the $Al_xGa_{1-x}As$ layers changes significantly with x. This was observed by examining the dependence of hot phonon generation rates on the barrier widths in series of GaAs/$Al_xGa_{1-x}As$ SL's. In Fig. 24, the hot phonon generation efficiency of the GaAs LO phonons is shown for various barrier widths.[95] The phonon penetration depth is determined by the abrupt rise in the generation efficiency. The efficiency depends upon the Fröhlich interaction, which governs much of the intra-valley scattering of hot LO phonons.[95] This interction is inversely proportional to the phonon wave vectors. The confined LO phonons in GaAs have a wave vector given by π/L_z, and is generally large compared to bulk phonon wave vectors. This results in a generation efficiency much less than that for bulk GaAs. However, when the L_b's become thin enough so that significant overlap of the GaAs LO phonon wave function exist, the generation efficiency rises abruptly. This is because the coupled phonon wave vectors contain bulk-like features.[95] The penetration depth is simply taken to be one half of the L_b where the abrupt rise in the efficiency occurs. However, the area of phonon coupling in a SL needs more theoretical and experimental work.

Concluding Remarks

Optical properties of superlattices which are associated with the subband dispersion were discussed. The subband dispersion results from electronic wave function coupling through the barrier layers, and therefore these properties are sensitive to the barrier layer widths, L_b's. In the large L_b limit, the superlattices behave like isolated quantum wells (2D behavior), whereas their properties approach the bulk properties of the well material in the small L_b limit. Inbetween these two limiting cases, they exhibit quasi-2 dimensional characteristics. In practice, these quasi-2D properties can be systematically investigated in a series of superlattices with fixed L_z and barrier heights but varying L_b's. The L_z's are usually designed to be of substantial thickness to minimize any complications arising from layer width irregularities that may occur during the sample growth. We have discussed diferences in optical properties which can be observed in this type of superlattice when L_b's are varied as a tool of changing the coupling between electronic wave functions. The subband dispersion effects appear in the optical transition energies and the seperation between the quantized heavy- and light-hole states. Particularly distinct from quantum well or bulk cases are the rich structures in the excitonic spectra arising from the redistribution of oscillator strengths over the subband due to the Coulomb effect. For n>1 optical transitions, subband mixing in the hole states combined with the subband dispersion drastically alter the excitonic lineshapes. Although the underlying physics is reasonably well understood, comparison between theory and experiments is lacking for n>1 superlattice excitonic transitions. Such comparison is also in need for superlattice binding energies. The subband dispersion also leads to line narrowing and additional peaks or steps in the appearance of unconfined transition spectra. For the latter, the lineshapes have not been theoretically examined yet. Also, highly in need is the theoretical modeling of phonons in superlattices with substantial well widths and relatively thin barriers. The issue of phonon coupling in superlattices is recent and its physical nature is not completely understood.

Most of the experimental data which reveal the electronic well-to-well coupling effects discussed in this article, were from GaAs/Al$_x$Ga$_{1-x}$As superlattices. With the fast advance in thin film growth techniques, fabrication of other high quality superlattice samples with well-controlled sample parameters will become more easily available. The authors hope that this article has provided a basis for understanding optical effects that will be observed in those superlattices. The references sited here cannot be exhaustive and any omission is not intentional. The authors would like to thank Professor Y.C. Chang and Dr. J.N. Schulman for their theoretical collaboration in the investigations performed in our laboratory. This work was supported in part by ONR.

REFERENCES

1. L. Esaki, R. Tsu, IBM J. Res. Develop. **14**, 61 (1970).
2. E.O. Göbel, K. Ploog, Prog. Quant. Electr. **14**, 289 (1990).
3. The term superlattice has a general meaning which includes both MQW and coupled-MQW structures. We will use the abbreviation SL only when we are referring to coupled-MQW structures.
4. F. Capasso, K. Mohammed, A.Y. Cho, IEEE J. Quantum Electron. **QE22**, 1853 (1986).
5. D.S. Chemla, D.A.B. Miller, P.W. Smith, Opt. Engng. **24**, 556 (1985).
6. R. Dingle, W. Wiegmann, C.H. Henry, Phys. Rev. Lett. **33**, 827 (1974).
7. G. Bastard, J.A. Brum, R. Ferreira, Solid State Physics **44**, p.229, Eds. H. Ehrenreich, D. Turnbull, Academic Press (1991).
8. G. Bastard, *Wave Mechanics Applied to Semiconductor Heterostructures*, Les Editions de Physique, Les Ulis, Halsted Press, New York, 1988.
9. L. Esaki, IEEE J. Quant. Electr. **QE22**, 1611 (1986).
10. D.L. Smith, C. Mailhiot, Rev. Mod. Phys. **62**, No.1, 173 (1990).
11. H.I. Ralph, Sol. Stat. Commun. **3**, 303 (1965).
12. The upper limit for the exciton binding energy of ~16meV for an ideal GaAs QW is not certain. In a recent study, Andreani and Pasquarello [13] used a model including band-mixing effects, Coulomb coupling between different subbands, and band non-parabolicity to show that the exciton binding energy in GaAs/Al$_x$Ga$_{1-x}$As SL's could be larger than $E_{bd}(2D) \approx 16$meV.
13. L.C. Andreani, A. Pasquarello, Phys. Rev. **B42**, 8928 (1990).
14. R.L. Greene, K.K. Bajaj, D.E. Phelps, Phys. Rev. **B29**, 1807 (1984).
15. A. Baldereschi, and N.C. LiPari Phys. Rev. **B3**, 439 (1971).
16. R.C. Miller, D.A. Kleinman, W.T. Tsang, A.C. Gossard, Phys. Rev. **B24**, 1134 (1981).
17. P. Dawson, K.J. Moore, G. Duggan, H.I. Ralph, C.T.B. Foxon, Phys. Rev. **B34**, 6007 (1986).
18. D.C. Reynolds, K.K. Bajaj, C. Leak, G. Peters, W. Theis, P.W. Yu, K. Alavi, C. Colvard, I. Shidlovsky, Phys. Rev. **B37**, 3117 (1988).
19. A. Chomette, B. Lambert, B. Deveaud, F. Clerot, A. Regreny, G. Bastard,

Europhys. Lett. **4**, 461 (1987).
20. E.S. Koteles, J.Y. Chi, Phys. Rev. **B37**, 6332 (1988).
21. L.W. Molenkamp, G.E.W. Bauer, R. Eppenga, C.T. Foxon, Phys. Rev. **B38**, 6147 (1988).
22. K.J. Moore, G. Duggan, K. Woodbridge, C. Roberts, Phys. Rev. **B41**, 1090 (1990).
23. J.C. Maan, G. Belle, A. Fasolino, M. Altarelli, K. Ploog, Phys. Rev. **B30**, 2253 (1984).
24. D.C. Rogers, J. Singleton, R.J. Nicholas, C.T. Foxon, K. Woodbridge, Phys. Rev. **B34**, 4002 (1986).
25. A.S. Plaut, J.Singleton, R.J. Nicholas, R.T. Harley, S.R. Andrews, C.T.B. Foxon, Phys. Rev. **B38**, 1323 (1988).
26. D.D. Smith, M. Dutta, X.C. Liu, A.F. Terzis, A. Petrou, M.W. Cole, P.G. Newman, Phys. Rev. **B40**, 1407 (1989).
27. H. Chu, Y.C. Chang, Phys. Rev. **B39**, 10861 (1989).
28. J.J. Song, P.S. Jung, Y.S. Yoon, H. Chu, Y.C. Chang, C.W. Tu, Phys. Rev. **B39**, 5562 (1989).
29. N.J. Pulsford, M.J. Lawless, R.J. Nicholas, G. Duggan, K.J. Moore, C. Roberts, K. Woodbridge, Superlat. Microstr. **8**, 151 (1990).
30. N.F. Johnson, J. Phys. Condens. Matter **2**, 2099 (1990).
31. M.F. Pereira Jr., I. Galbraith, S.W. Koch, G. Duggan, Phys. Rev. **B42**, 7084 (1990).
32. D.S. Chuu, Y.C. Lou, Phys. Rev. **B43**, 14504 (1991).
33. D.M. Whittaker, Phys. Rev. **B41**, 3238 (1990).
34. M.M. Dignam, J.E. Sipe, Phys. Rev. **B41**, 2865 (1990).
35. H. Chu, Y.C. Chang, Phys. Rev. **B36**, 2946 (1987).
36. J.J. Song, Y.S. Yoon, P.S. Jung, A. Fedotowsky, J.N. Schulman, C.W. Tu, J.M. Brown, D. Huang, H. Morkoç, Appl. Phys. Lett. **50**, 1269 (1987).
37. J.F. Zhou, P.S. Jung, J.J. Song, C.W. Tu, Appl. Phys. Lett. **56**, 1880 (1990).
38. J.J. Song, Y.S. Yoon, A. Fedotowsky, Y.B. Kim, J.N. Schulman, C.W. Tu, D. Huang, H. Morkoç, Phys. Rev. **B34**, 8958 (1986).
39. J.J. Song, J.M. Jacob, J.F. Zhou, S.J. Hwang, P.S. Jung, Y.C. Chang, C.W. Tu, SPIE Vol. **1283**, p.344 (1989).
40. B. Deveaud, A. Chomette, F. Clérot, A. Regreny, J.C. Maan, R. Romestain, G. Bastard, H. Chu, Y.C. Chang, Phys. Rev.**B40**, 5802 (1989); Superlat. Microstr. **6**, 183 (1989).
41. K.J. Moore, G. Duggan, K. Woodbridge, C. Roberts, N.J. Pulsford, R.J. Nicholas, Phys. Rev. **B42**, 3024 (1990); K.J. Moore, G. Duggan, A. Raukema, K. Woodbridge, Phys. Rev. **B42**, 1326 (1990); Superlat. Microstr. **7**, 303 (1990).
42. D. Morris, C. Lacelle, A.P. Roth, P. Maigne, J.L. Brebner, Surf. Sci. **228**, 347 (1990).
43. S.H. Pan, H. Shen, Z. Hang, F.H. Pollak, W. Zhuang, Q. Zu, A.P. Roth, R.A. Masut, C. Lacelle, D. Morris, Phys. Rev. **B38**, 3375 (1988).
44. J.C. Phillips, Solid State Physics **18**, Eds: F. Seitz, D. Turnbull, p.55, Academic

Press, New York 1966.
45. J.C. Phillips, Phys. Rev. **136**, A1705 (1964).
46. E.O. Kane, Phys. Rev. **180**, 852 (1969).
47. C.B. Duke, B. Segall, Phys. Rev. Lett. **17**, 19 (1966).
48. J.E. Zucker, A. Pinczuk, D.S. Chemla, A. Gossard, and W. Wiegmann, Phys. Rev. B **29**, 7065 (1984).
49. U.K. Reddy, G. Ji, T. Henderson, H. Morkoç, J. Appl. Phys. **62**, 145 (1987).
50. H. Shen, S.H. Pan, F.H. Pollak, M. Dutta, T.R AuCoin, Phys. Rev. **B36**, 9384 (1987); H. Shen, S.H. Pan, Z. Hang, F.H. Pollak, R.N. Sacks, Sol. Stat. Commun. **65**, 929 (1988).
51. G. Ji, W. Dobbelaere, D. Huang, H. Morkoç, Phys. Rev. **B39**, 3216 (1989).
52. M.W. Peterson, J.A. Turner, C.A. Parsons, A.J. Nozik, D.J. Arent, C. Von Hoof, G. Borghs, R. Houdré, H. Morkoç, Appl. Phys. Lett. **53**, 2666 (1988).
53. R.H. Yan, R.J. Simes, H. Ribot, L.A. Coldren, A.C. Gossad, Appl. Phys. Lett. **54**, 1549 (1989).
54. Y.C. Chang, J.N. Schulman, Appl. Phys. Lett. **43**, 536 (1983).
55. P.S. Jung, Y.S. Yoon, A. Fedotowsky, J.J. Song, J.N. Schulman, C.W. Tu, R.F. Kopf, and H. Morkoç, J. of Superlattices and Microstructures, **4**, 581 (1988).
56. C.W. Tu, R.C. Miller, B.A. Wilson, P.M. Petroff, T.D. Harris, R.F. Kopf, S.K. Sputz, M.G. Lamont, J. Cryst. Growth **81**, 159 (1987).
57. C.A. Warwick, W.Y. Jan, A. Ourmazd, K. Leo and J. Shah, Bull. Am. Phys. Soc. **35**, 519 (1990).
58. For example, see R.F. Kopf, E.F. Schubert, T.D. Harris, R.S. Becher, Appl. Phys. Lett. **58**, 631 (1991).
59. A. Ourmazd, D.W. Taylor, J. Cunningham, C.W. Tu, Phys. Rev. Lett. **62**, 933 (1989).
60. C.A. Warwick, W.Y. Jan, A. Ourmazd, T.D. Harris, Appl. Phys. Lett. **56**, 2666 (1990).
61. D. Gammon, B.V. Shanabrook, D.S. Katzer, Phys. Rev. Lett. **67**, 1547 (1991).
62. B. Jusserand, F. Mollot, J.M. Moison, G.L. Roux, Appl. Phys. Lett. **57**, 560 (1990).
63. D.C. Bertolet, J.K. Hsu, S.H. Jones, K.M. Lau, Appl. Phys. Lett. **52**, 293 (1988); D.C. Bertolet, J.K. Hsu, K.M. Lau, E.S. Koteles, D. Owens, J. Appl. Phys. **64**, 6562 (1988).
64. J.P. Reithmaier, R. Höger, H. Riechert, Phys. Rev. B **43**, 4933 (1991).
65. P.S. Jung, J.M. Jacob, J.J. Song, Y.C. Chang, C.W. Tu, Phys. Rev. B **40**, 6454 (1989).
66. M. Krahl, J. Christen, D. Bimberg, G. Weimann, W. Schlapp, Appl. Phys. Lett. **52**, 798 (1988).
67. M. Altarelli, Phys. Rev. B **28**, 842 (1983).
68. R. Sooryakumar, D.S. Chemla, A. Pinczuk, A. Gossard, W. Weigmann, L.J. Sham, J. Vac. Sci. Technol. B **2**, 349 (1984).
69. R.C. Miller, A.C. Gossard, G.D. Sanders, Y.C. Chang, J.N. Schulman, Phys. Rev. B **32**, 8452 (1985).

70. W.T. Masselink, P.J. Pearah, J. Klem, C.K. Peng, H. Morkoç, G.D. Sanders, Y.C. Chang, Phys. Rev. B **32**, 8027 (1985).
71. G. Bastard, J.A. Brum, IEEE J. Quant. Elect. **QE22**, 1625 (1986)
72. J.F. Zhou, S.J. Hwang, G.D. Sanders, J.J. Song, C.W. Tu, J.F. Klem, Bull. Am. Phys. Soc. **36**, 452 (1991).
73. J.N. Schulman, Y.C. Chang, Phys. Rev. B **33**, 2594 (1986).
74. J.N. Schulmang, Y.C. Chang, Phys. Rev. **B31**, 2056 (1985); Y.C. Chang, J.N. Schulman, Phys. Rev. **B31**, 2069 (1985).
75. J.N. Schulman, Y.C. Chang, Phys. Rev. **B24**, 4445 (1981).
76. E.E. Mendez, *Physics and applications of Quantum Wells and Superlattices*, Edited by E.E. Mendez and K. von Klitzing, Plenum Press, New York, 1987.
77. A. Chomette, B. Lambert, B. Clerjaud, F. Clérot, H.W. Liu, A. Regreny, Semicond. Sci. Technol. **3**, 351 (1988).
78. B. Lambert, F. Clérot, B. Deveaud, A. Chommette, G. Talalaeff, A. Regreny, B. Sermage, J. Lumin. **44**, 277 (1989); B. Lambert, B. Deveaud, A. Chommette, F. Clérot, A. Regreny, B. Sermage, Surf. Sci. **228**, 446 (1990).
79. K. Fujiwara, N. Tsukada, T. Nakayama, A. Nakamura, Phys. Rev. B **40**, 1096 (1989).
80. P. Tsu, L. Esaki, Appl. Phys. Lett. **22**, 562 (1973); L. Esaki, L.L. Chang, Phys. Rev. Lett. **33**, 495 (1974).
81. G. Bastard, C. Delalande, R. Ferreira, H.W. Liu, J. Lumin. **66**, 247 (1989).
82. D.A.B. Miller, D.S. Chemla, t.C. Damen, A.C. Gossard, W. Wiegmann, T.H. Wood, C.A. Burrus, Phys. Rev. Lett. **53**, 2173 (1984); Phys. Rev. B **32**, 1043 (1985).
83. E.E. Mendez, f. Agulló-Rueda, J.M. Hong, Phys. Rev. Lett. **60**, 2426 (1988).
84. J. Bleuse, G. Bastard, P. Voisin, Phys. Rev. Lett. **60**, 220 (1988).
85. S. Baroni, S. de Gironcoli, P. Giannozzi, Phys. Rev. Lett. **65**, 84 (1990).
86. B. Zhu, K.A. Chao, Phys. Rev. B **36**, 4906 (1987).
87. A.K. Sood, J. Menéndez, M. Cardona, K. Ploog, Phys. Rev. Lett. **54**, 2111 (1985); Phys. Rev. B **32**, 1412 (1985).
88. D.J. Mowbray, M. Cardona, K. Ploog, Phys. Rev. B **43**, 1598 (1991).
89. F. Calle, D.J. Mowbray, D.W. Niles, M. Cardona, J.M. Calleja, K. Ploog, Phys. Rev. B **43**, 9152 (1991).
90. B. Jusserand, D. Paquet, Phys. Rev. Lett. **56**, 1752 (1986).
91. T. Tsuchiya, H. Akera, T. Ando, Phys. Rev. B **39**, 6025 (1989).
92. A.K. Arora, E.K. Suh, A.K. Ramdas, F.A. Chambers, A.L. Moretti, Phys. Rev. B **36**, 6142 (1987).
93. B. Jusserand, D. Paquet, A. Regreny, Phys. Rev. B **30**, 6245 (1984).
94. B. Jusserand, D. Paquet, F. Mollot, Phys. Rev. Lett. **63**, 2397 (1989).
95. D.S. Kim, A. Bouchalkha, J.M. Jacob, J.F. Zhou, J.J. Song, J.F. Klem, Phys. Rev. Lett., accepted.

RADIATIVE PROCESSES IN GaAs/AlGaAs HETEROSTRUCTURES

P.O. Holtz, B. Monemar
Department of Physics and Measurement Technology, Linköping University, S-581 83 Linköping, SWEDEN

and

J.L. Merz
Center for Quantized Electronic Structures (QUEST), University of California at Santa Barbara, Santa Barbara, CA 93106, USA

ABSTRACT

The radiative processes related to the two dimensional electron gas (2DEG) confined in the notch potential of n-channel GaAs/AlGaAs heterostructures have been studied by means of different photoluminescence (PL) related techniques. Two emission bands, the so-called H-bands, involving the 2DEG are observed in PL. A distinctive feature of these H-bands is the striking dependence of the PL peak energy on the excitation intensity level. This characteristic behavior is explained in terms of varying band bending across the active GaAs layer. The spatial separation between the electron and hole, which are involved in the H-band recombination, is strongly dependent on this potential as is directly demonstrated in PL decay measurements. The proposed band bending model can be simulated by PL experiments performed in an electric field. Not only the band bending, but also the Fermi level can be deliberately altered by application of an electric field, which makes the system suitable for investigations of Fermi edge singularities. Finally, important information about the electronic structure of the 2D carriers has been provided from PL excitation spectroscopy.

1. INTRODUCTION

By the introduction of modulation doping in semiconductor structures almost 15 years ago [1, 2], major progress was made in semiconductor technology: A high concentration of charge carrriers could be achieved in a high-purity and high-mobility semiconductor, without the presence of mobility limiting impurities. Since then impressive developments have resulted in many high-performance devices such as the High Electron Mobility Transistor (HEMT) and the selectively doped heterostructure transistor (SDHT).

However, modulation doped structures are also of great interest from a fundamental physical point of view: The properties of the 2D electron gas (2DEG) or the 2D hole gas (2DHG), the quantum Hall effect, Fermi edge phenomena etc. From different optical experiments, when the radiative processes related to the heterostucture interface are monitored, we can extract information on e.g. the band bending, the electronic energy levels, band offsets, electron-hole interaction and effective masses of the involved electronic particles.

Doping the barrier on one side of the heterojunction will cause a band bending across the structure, while the chemical potential, i.e. the Fermi level, remains constant in equilibrium throughout the whole structure. For the case of n-type modulation doping on the AlGaAs side of a GaAs/AlGaAs heterostructure, electrons will transfer from the donors on the wide bandgap AlGaAs side to a

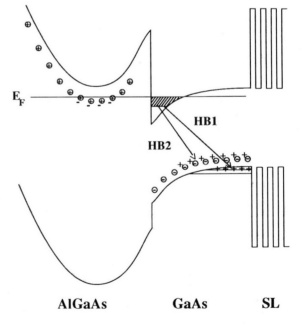

Fig. 1 The energy-band diagram of a n-type modulation doped heterostructure. The electrons are filling up the notch from the E_0 ground state to the Fermi energy, E_F, forming the 2DEG. Electrons from this 2DEG recombine with either free holes from the GaAs valence band (HB1) or acceptor bound holes (HB2). Also the hole states experience the confinement and split into the hh- and lh-states.

quasi-triangular "notch" potential on the GaAs side until the Fermi energy is raised to the same level as the Fermi energy on the AlGaAs side. The donors close to the interface thus become ionized. This fact results in a depletion layer next to the interface on the AlGaAs side of the heterostucture. This, in turn, will result in a characteristic bandstructure (Fig. 1) for a n-type modulation doped GaAs/AlGaAs heterostucture, in which the charge carriers (electrons) are spatially separated from the mobility-limiting donor impurities. To further separate, and accordingly further reduce the Coulomb interaction between the ionized donors and the channel electrons, an undoped spacer AlGaAs layer is usually introduced. This charge transfer effect, which results in conducting electrons in a high-mobility semiconductor, has attracted great interest both from a fundamental physical point of view and from its applicability as high-performance devices as mentioned above. We will in this article concentrate on the former aspect, the physical properties of the GaAs/AlGaAs heterostucture, which can be extracted from the radiative recombination processes related to the heterojunction interface.

The unusual characteristics of this radiative recombination were first reported in 1984 by Yuan *et al* [3, 4]. A broad asymmetric line close to the GaAs bandgap energy was observed in the low temperature photoluminescence (PL) measured on a GaAs layer grown on a AlGaAs buffer layer (Fig. 2). This band, usually denoted the H-band, appears between the shallow impurity bound exciton lines and the donor-acceptor pair (DAP) transitions in GaAs. However, the properties of this luminescence, such as the spectral shape and energy position, were found to be strongly dependent on the experimental conditions, e.g. the sample design, temperature, excitation intensity and energy. The H-band emission is interpreted as due to the recombination between 2D electrons in the notch potential and free holes in the valence band of the active GaAs layer.

Similar PL results have been obtained for p-type GaAs/AlGaAs heterojunctions [5]. In this case, the corresponding H-band luminescence is due to the recombination between a free electron from the conduction band in the GaAs layer and a 2D hole confined in a quantized hole subband near the interface.

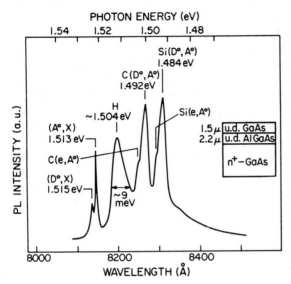

Fig. 2 The PL spectrum measured at 1.4K of a GaAs/AlGaAs heterostructure showing the HB1 recombination (denoted H in the figure). The sample is excited above the AlGaAs bandgap (4880Å) at an intensity level of 5 mW/cm^{-2}. The sample used is grown by LPE and its design is shown in the insert.

Analogous PL spectra related to 2D carriers at a semiconductor surface or interface have been obtained for the case of a Si - SiO$_2$ [6] MOS structure or a semiconductor-electrolyte interface [7]. However, these systems will not be discussed in this work.

2. THEORETICAL ASPECTS

Due to the confinement in the notch potential on the GaAs side, the electrons will exhibit 2D properties and the energy levels become quantized. These electronic energy levels can be calculated by solving the Schrödinger and Poisson equations self-consistently by using approximations of various degree of sophistication [8].

The electronic wave function, $\psi(r)$, consists of the periodic Bloch function, $u(r) = \exp(ikR)$, and a slowly varying envelope function, $f(r)$:

$$\psi(r) = u(r) \cdot f(r) \qquad (1)$$

The envelope function, $f(r)$, is determined from the Schrödinger equation:

$$[(p_x^2 + p_y^2)/2m^*(z) + p_z^2/2m^*(z) + V(z) - e\phi_{sc}(z)] f(r) = E \cdot f(r) \quad (2)$$

where the z-direction is chosen to be perpendicular to the heterostructure interface. The quantity, $m^*(z)$, is the z-dependent effective mass of the electron. $V(z)$ is the static potential energy, while the $\phi_{sc}(z)$ term in Eq. 2 includes many-particle effects such as the electron-electron and electron-impurity interactions. Calculations on the many-body effects have been carried out for the electronic energy levels in several papers [9 - 11], but will not be taken into account here. Instead the potential, $e\phi_{sc}(z)$, which takes the many body effects into account, is replaced by an average potential, $V_e(z)$. The prerequisite to fulfill the boundary conditions of the Schrödinger equation is continuity of $[m^*(z)]^{-1}df(z)/dz$ and $f(r)$ across each interface. The average potential, $V_e(z)$, can be derived by solving the Poisson equation:

$$\frac{d^2V_e(z)}{dz^2} = \frac{e\,\rho(z)}{\kappa(z)} \quad (3)$$

where $\kappa(z)$ is the dielectric constant of the GaAs and AlGaAs material, respectively. $\rho(z)$ is the space charge density of states, given by the electron density and ionized impurity density:

$$\rho(z) = e\,(N_d^+(z) - \sum_n N_n |X_n(z)|^2 + N_a^-(z)) \quad (4)$$

where $X_n(z)$ is the envelope function, which is derived by solving the Schrödinger equation. N_d^+ and N_a^- are the concentrations of the ionized donors and acceptors, respectively. N_n is the charge carrier concentration in the filled state, n, which is derived from the 2D density of states, $g_n(E)$, and the Fermi-Dirac distribution, $f(E)$

$$N_n = \int g_i(E) \cdot f(E)\,dE_i \quad (5)$$

By the introduction of the boundary conditions together with the prerequisite that the Fermi energy is at the same level throughout the entire structure, the equations given above can be solved selfconsistently and the potential and eigenenergies of the

electrons can be derived. The resulting bandstructure is shown in Fig. 1. The electrons are filling up the notch from the E_o ground state energy up to the Fermi energy, E_F, corresponding to a certain value of the electron density, N_s, according to

$$E_F = E_o + N_s/\rho_{2D} = E_o + N_s \pi h^2/m^* \qquad (6)$$

The formation of a 2DEG, spatially separated from the mobility limiting impurities (Fig. 1), is the basic mechanism for many devices like the HEMT. The calculated mobility in a 2DEG is found to be strongly dependent on the electron density, N_s, and the unintentional impurity concentration in the active GaAs layer, but fairly insensitive to the impurity concentration in the AlGaAs layer [12]. However, the 2DEG in the notch potential is also applicable for fundamental studies of the 2D properties of electrons and has attracted great interest ever since its first observation [13]. In this article, we will focus on the optical properties of the 2DEG as they can be derived from PL studies.

In a manner analogous to the electronic case described above, the eigensolutions for the holes in the valence band can be derived. Due to the limited width of the active GaAs layer in many of the n-channel heterostructures used here, the hole states are also quantized. The lowest hole states are no longer degenerate, but split into the heavy hole (hh) and light hole (lh) states due to the confinement, as illustrated in Fig. 1. This fact has recently been experimentally demonstrated in luminescence excitation spectra for the first time and will be described more fully in this article.

3. SAMPLES

Radiative recombination processes related to the GaAs/AlGaAs heterostructures have been observed in a wide variety of samples. The first observation of this interface-related recombination was performed on liquid phase epitaxy (LPE) material [3, 4], but later investigations on the similar recombination processes have been performed on metalorganic chemical vapor deposition (MOCVD) [14, 15] and molecular beam epitaxy (MBE) [16 - 20] material. In most cases, the 2DEG has been investigated in n-channel GaAs/AlGaAs heterostructures, but the properties of the 2D hole gas (2DHG) in p-

channel heterostructures have also been studied in PL experiments [16, 21].

In the pioneering studies of heterostructure recombinations made on LPE material [3, 4], the H-band emission (cf Fig. 2) was found to originate from the interface between an undoped 1 - 8 µm thick GaAs layer and a 1 µm thick p-type or undoped AlGaAs layer. Carefully controlled step-etching experiments confirmed that the emission originated from the interface. Further, this feature in the PL spectrum was observed only when care was taken to grow the most abrupt interfaces attainable by LPE, suggesting that the effect should be stronger when using either MOCVD or MBE. That was indeed found to be the case. In later studies [14 - 21], specially designed structures grown using both of these techniques have been investigated. On top of either a semi-insulating (SI) or a conducting substrate and an epitaxial buffer layer consisting of GaAs or a GaAs/AlGaAs superlattice (SL), the active GaAs layer was grown with a thickness varying in a wide range, 400 Å - 2 µm. The AlGaAs layer has been either left with the background doping ($p \approx 3 \times 10^{16}$ cm^{-3}) [15] or intentionally p- or n-type doped over a wide range of doping levels (10^{16} - 10^{18} cm^{-3}), and was spatially separated from the active GaAs layer by an undoped AlGaAs spacer layer [16 - 21].

4. PHOTOLUMINESCENCE MEASUREMENTS ON GaAs/AlGaAs HETEROSTRUCTURES

The properties of the interface-related recombination processes observed in PL, the so called H-bands, are found to be strongly dependent on the sample design and experimental conditions. For example, a distinctive feature of the H-band emission is the excitation-intensity dependence of the PL peak energy position. The H-bands shift towards higher energy with increasing excitation intensity. The emission band, that was first observed, hereafter referred to as H-band 1 (HB1), is found to shift roughly from the region of the bound exciton (BE) recombination down to the free-to-bound (FB), C(e, A°) transition in GaAs [3, 4].

Different recombination processes have been suggested to explain the properties of the H-band luminescence: In the first study of the radiative heterostructure recombination, Yuan *et. al.* [3,

4] explained the observed H-band emission (Figs. 1 and 2) in terms of recombination between 3D like holes and 2D electrons, which tunnel away from the notch potential at the interface to find a sufficient concentration of holes in the flat region of the valence band. Balslev interpreted the H-band recombination as being weakly bound interface excitons in analogy with excitons in quantum wells (QWs) in an external electric field [22].

Kukushkin *et. al.* [17] observed two different H-band recombination channels (denoted A and B in their work). The 2D electrons can recombine with either the free holes in the valence band (band A) or with acceptor bound holes (band B). Broad emission bands are observed in PL spectra with a width corresponding to the energy separation from the E_0 electron ground state up to the Fermi energy, E_F (cf. Eq. 6 and Fig. 1) When the concentration of 2D electrons is increased, the first excited 2D electron subband is populated, which is directly observed in the PL spectrum as another (n=2) band at higher energy.

In a later study, Kukushkin *et. al.* [23] investigated the dependence of the intensity and spectral position of the B-line as a function of the distance, z, between the interface and a δ-doped layer of acceptors. The recombination probability depends primarily on two effects: (1) the hole concentration, which decreases with decreasing z, and (2) the overlap between the wavefunctions of the 2D electrons and the acceptor holes. For the case of a 500Å wide GaAs layer, they estimated this distance to be 180Å for the most efficient recombination with 2D electrons.

More recently, a dynamic study of specially designed undoped heterostructures was reported by Gilliland *et. al.* [15]. The sample design was different from that used by other authors in several important respects: The structure used was symmetric (double heterostructure), the active GaAs layer was wide (0.1 - 20 μm), and, probably most importantly, the AlGaAs layer was *not* intentionally doped, which means that no 2D gas was formed at equilibrium. The observed PL emission was explained as being of excitonic origin with long lifetimes (from ns up to μs range), i.e. a system, which differs significantly from the H-band emission described above. The many-particle problem of the H-band recombination related to the 2DEG is, in the excitonic case, reduced to a two-particle system.

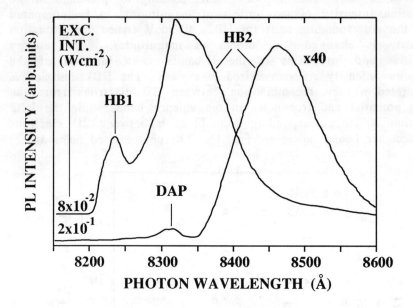

Fig. 3 PL spectra for a modulation doped heterostructure with a 500 Å wide active GaAs layer measured at 2K with excitation above the AlGaAs bandgap at two different intensity levels. The two dominating emission bands related to the interface, HB1 and HB2, shift significantly with the excitation intensity. Also the intensity ratio, I(HB1)/I(HB2), changes with the intensity level. The peak denoted DAP is due to the donor-acceptor pair recombination in bulk GaAs.

A distinctive feature of the H-band emission is the striking dependence of the PL peak energy on the excitation intensity, as mentioned above. This fact is demonstrated in Fig. 3, which shows two PL spectra measured with cw above bandgap excitation at two different intensity levels, differing by a factor of 40. At the higher excitation intensity level, two different H-band emission lines appear (denoted HB1 and HB2 in Fig. 3). With decreasing excitation intensity level, the PL intensity ratio, I(HB1)/I(HB2), decreases and below a critical value of excitation intensity, only HB2 is observed in the PL spectrum, as is obvious in Fig. 3. In fact, the HB1 peak is

observed to shift about 15 meV (corresponding to a change in the excitation intensity of two orders of magnitude), to be compared with the corresponding shift for HB2, 40 meV (when the excitation intensity is changed four orders of magnitude). The energy positions and intensities of the H-bands as a function of the excitation intensity are summarized in Fig. 4. The HB1 emission is interpreted as the recombination between 2D electrons from the notch potential and free holes in the valence band, while the HB2 emission is the corresponding recombination between 2D electrons and acceptor bound holes (see Fig. 1). The photoexcited holes are

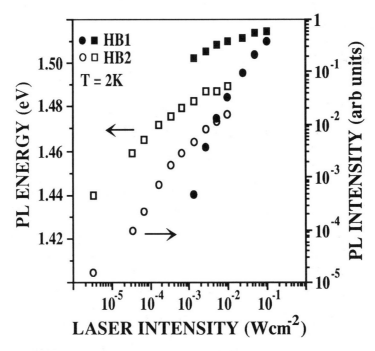

Fig. 4 The dependence of the PL peak position and the PL intensity of HB1 and HB2 on the excitation intensity level. The squares refer to the peak energy with the scale on the lefthand side, while the circles correspond to the PL intensity with the scale on the righthand side. The filled squares/circles refer to HB1, while the unfilled ones refer to HB2.

rapidly captured by the ionized acceptors, until all available acceptors outside the notch region are neutralized. Thus, the HB2 emission will dominate the PL spectrum, until the acceptors are neutralized and free holes in the valence band are available for the HB1 emission. This hypothesis is consistent with the results of Fig. 3.

Both H-bands shift towards lower energy with an approximately constant energy separation between HB1 and HB2 with decreasing excitation intensity level, as illustrated separately in Fig. 4. A similar behavior has been observed for samples with varying dopant concentration in the AlGaAs layer [18]. The striking intensity dependence is mainly due to the fact that the potential across the active GaAs layer is affected by the excitation conditions. The photoexcited holes may neutralize the ionized acceptors starting at the SL/GaAs interface (opposite to the notch), which results in a reduced band bending in the same region and a decreasing region with depleted acceptors. The reduced band bending will in turn give rise to a shift towards higher energy of the H-bands with increasing excitation intensity [18]. These effects are, of course, dependent on the background doping level in the GaAs layer, and may be small for high purity GaAs ($\approx 10^{14}$ cm^{-3}). Also the photoexcited electrons will contribute to the upward shift of the H-bands, but to a significantly smaller extent. The Fermi level, E_F, will raise according to Eq. 6, since the total electron concentration, N_s, will increase as a result of the photoexcitation

$$E_F = E_o + N_s/\rho_{2D} = E_o + (N_{ep} + N_{eq})/\rho_{2D} \qquad (7)$$

where N_{eq} is the equilibrium electron concentration and N_{ep} is the electron concentration caused by the photoexcitation. On the other hand, as the total electron concentration, N_s, increases, the bottom of the conduction band is also raising, approximately proportional to $(N_{ep} + N_{eq})^{2/3}$ [8]. However, due to the fact that normally $N_{ep} \ll N_{eq}$, the resulting effect on the H-band shift due to the photoexcited electrons is small, at least one order of magnitude smaller than the band bending effects caused by the photoexcited holes [19].

As mentioned above, both HB1 and HB2 shift towards lower energy with an approximately constant energy separation with decreasing excitation intensity. The energy separation observed

between HB1 and HB2 is about 20 meV, whereas the binding energy is 26.3 meV for the C-acceptor in bulk GaAs [24], which is the most common residual acceptor in MBE material. This deviation has been explained by the Coulomb energy originating from the fact that the acceptor is ionized in the final state [17], i.e. an analogy to the energy shift of the donor acceptor pair emission. The spatial distribution of the bound holes is also expected to give rise to broadening effects for the HB2, in agreement with the experimental observations (e.g. cf. Fig. 3). Another source for the deviation can be found in the band bending of the active layer as described above. While the free holes in the valence band will be confined close to the GaAs/SL interface due to the built-in potential, the photoexcited holes will be captured at acceptors distributed over a certain region up to a quasi-Fermi level in the valence band. When the excitation intensity is increased, more acceptors are photo-neutralized, the quasi-Fermi level is raised and the region with depleted acceptors is reduced. The recombination probability is strongly dependent on the spatial separation between the wave functions of the 2D electron, $\Psi^e(z)$, and the hole, $\Psi^h(z)$. The overlap $<\Psi^e(z)|\Psi^h(z)>$ increases with decreasing spatial separation between the electron and hole. This means that the 2D electron recombines with either a bound or a free hole, which are at a different spatial separation from the electron, in the HB1 or HB2 recombination, respectively. Accordingly, the energy separation between HB1 and HB2 will differ from the acceptor binding energy due to the band bending.

The PL intensity ratio, I(HB1)/I(HB2) is strongly temperature dependent. With increasing temperature, the intensity of HB2 decreases much faster than HB1 and the HB2 emission is completely quenched at a temperature of about 45K [18]. This fact is consistent with the proposed identities of the two H-bands: The HB2 intensity will decrease as increasing number of acceptors are thermally ionized.

5. DYNAMICAL STUDIES

The measured PL decay times of the H-bands are found to be strongly dependent on the photon energy, increasing with decreasing photon energy, in the range 1 - 100 ns for HB1 and 100

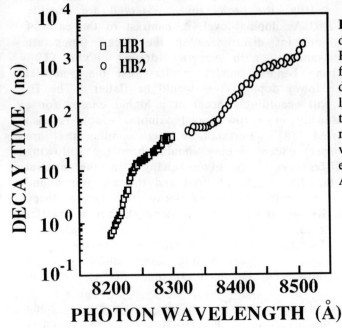

Fig. 5 The PL decay times for HB1 and HB2 as a function of the detected wavelength. The decay times have been measured at 2K with the excitation energy above the AlGaAs band gap.

ns - 10 μs for HB2, as summarized in Fig. 5 [19]. This behavior reflects the varying spatial separation between the recombining electron and hole. Our heterostructure system has in this respect similarities with a quantum well (QW) in an applied electric field. A red shift of the photon energy with increasing field has been attributed to the quantum confined Stark effect. Accordingly, the recombination lifetime increases as a consequence of the reduced electron-hole overlap [26]. In our heterostructure system, the recombination energy decreases with increasing spatial separation between the hole and the 2D electron in the notch potential, due to the band bending across the active GaAs layer. Accordingly, the recombination probability decreases with increasing spatial separation, as described above, and the decay time can be expressed as

$$\tau = \tau_0 / |\int \Psi^e(z) \Psi^h(z)|^2 dz \tag{8}$$

where τ_o corresponds to the decay time for a vertical transition.

For the case of HB1, the decay times observed are roughly independent of the AlGaAs doping level, in contrast to the case of HB2, for which significant differences in the decay times are observed between samples with varying doping level [19]. Consequently, the band bending model predicts that the potential for a sample with a lower doping level would be flatter. The H-band recombination will accordingly occur at a higher energy for a given excitation intensity in a low doped sample, which is also experimentally verified [18]. Furthermore, two samples that are similar in every respect except doping should show the following similarities and differences. A given change in the spatial separation between the 2D "notch" electron and the acceptor bound hole should produce a similar change of decay time (since decay time depends primarily on separation, not doping), but a smaller change in the HB2 recombination energy for the sample with the lower doping level. Therefore, in a plot of decay time versus photon energy, the sample with the lower doping should show a larger slope. This fact has also been confirmed experimentally as shown in Fig. 6 [19].

Optical studies of the dynamics of the intrinsic PL band observed in an undoped *double* heterostructure was recently reported by G.D. Gilliland *et. al.* [15]. The observed PL band exhibited the intensity dependent behavior characteristic for the H-band emission as described above. Also the measured decay times of the observed PL band were found to be strongly dependent on the detection photon energy and varied in the range from ns up to >50µs. However, the dependence on temperature was stronger than that observed for HB1 in modulation doped heterostructures [18]. The observed PL band vanished at a temperature as low as ≈15K, consistent with earlier results on undoped heterostructures [4]. An activation energy of 0.75 meV was determined, and interpreted as the binding energy of a quasi-2D exciton, formed by the Coulomb interaction between an electron and a hole bound intrinsically at the interface. The authors proposed that such exciton binding is effective for electron-hole distances up to several thousand Å [15]. In the modulation doped heterostructures, on the other hand, the situation is different [18]. Due to the modulation doping, a 2D electron gas is formed in the notch potential (from E_0 up to E_F in Fig.

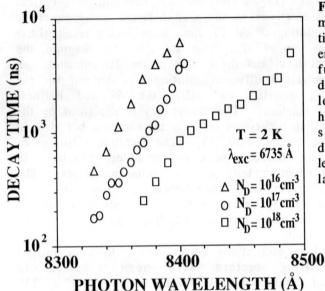

Fig. 6 The measured PL decay times for the HB1 emission at 2K as a function of the detected wavelength for three heterostructure samples with differrent doping level in the AlGaAs layer.

1), which makes the formation of the conventional exciton impossible. At high carrier densities the conventional two particle exciton ceases to exist. For the case of modulation doped QWs (with L_z = 200Å), it has been demonstrated that excitonic features are quenched for electron densities exceeding 3 x 10^{11} cm^{-2} [25]. In this case, the electron-hole recombination probabilities are also much larger than in the undoped case, a fact confirmed by the significantly shorter decay times measured in the modulation doped heterostructures [19]. Many body interactions have to be taken into account, and instead edge singularities are formed, as will be further discussed below (Sec. 8). It is believed by the authors of this review that the differences between the results of Gilliland *et. al.* and other authors reviewed herein originate primarily from the differences in sample structure and doping, rather than in some lack of quality of all other samples, as suggested by Gilliland *et. al.* [15].

6. MAGNETO-OPTICAL SPECTROSCOPY

Kukushkin *et. al.* [17] performed the first magneto-optical investigation on 2D electrons in the GaAs/AlGaAs heterostructure. They observed the splitting of the PL lines into Landau levels (LLs) at magnetic fields as low as 2T. From a Landau fan diagram, the position of the Fermi level and the bottom of the 2D subbands can be estimated. However, a different splitting was derived for the bands corresponding to HB1 and HB2 (the A- and B-lines, respectively, in their notation). The discrepancy is explained by the fact that the LL splitting occurs not only for the electron, but also, in the case of HB1, for the free hole [17]. The cyclotron splittings of the electron and hole therefore have to be added, which is in agreement with the experimentally derived cyclotron masses. The observed discrepancy in the LL splitting between HB1 and HB2 thus supports the proposed interpretation of the two bands as being the recombination between 2D electrons and free holes or acceptor bound holes, respectively.

Shubnikov-de Haas oscillations corresponding to both integer and fractional filling factors have been reported in magnetotransport and magneto-optical investigations [17, 27 - 29]. The agreement between experimental results and model calculations is poor, however [29]. It turns out that factors like g-factor enhancements are crucial in the modeling.

By detecting the PL in a magnetic field, it is possible to monitor all occupied LLs up to the Fermi level, in contrast to, for example, cyclotron resonance experiments, in which the splitting between an empty and occupied LL is measured. This fact makes it possible to directly measure the separation between occupied LLs. In this way, Plaut *et. al.* [28] estimated the dependence of the electron effective mass on the electron concentration by comparing different samples with varying concentration. The effective mass, m^*, of the electron was found to increase by as much as 30% with decreasing electron concentration due to exchange and correlation effects.

The 2DEG in n-channel GaAs/AlGaAs heterostructures has also been studied by means of Optical Detection of Cyclotron Resonances (ODCR) [30, 31]. The intensity of the H-band is monitored as a function of a microwave field to which the sample is exposed. The ODCR signal obtained in this way results from cyclotron resonant

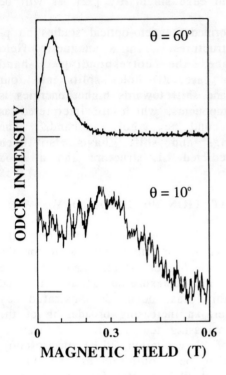

Fig. 7 Optically detected cyclotron resonance (ODCR) spectra detecting the HB1 emission in two different magnetic directions with respect to the interface of the heterostructure. The observed anisotropy in the ODCR response proves the 2D character of the involved carriers.

heating of either the 2D electrons or the 3D free carriers. In order to determine which process is responsible, an angular dependence of the ODCR spectrum is performed. Such measurements exhibit a pronounced anisotropy (Fig. 7), which is a clear indication of the 2D character of the H-band ODCR spectrum. It can also be concluded that the ODCR signal originates from microwave heating of the 2D electrons. This experimental result provides unambiguous evidence that the H-band originates from the interface. Furthermore, information about the electron density in each occupied subband together with the subband separations are derived from these ODCR experiments. However, another band at about 8180Å (denoted X in [29]), i.e. close to the GaAs excitons, but also close to the Fermi edge in the notch potential, exhibits a clear 2D character in these ODCR spectra [31]. This result is consistent with observations in PL measurements performed in an applied electric field. The band at

8180Å is interpreted as the Fermi edge singularity [32], as will be further discussed in Sec. 8.

Ossau et. al. [16] have performed magneto-optical studies on p-channel GaAs/AlGaAs heterostructures. In a magnetic field perpendicular to the interface, the corresponding H-band recombination involving, in this case, 2D holes splits into four components. A linear splitting and shift towards higher energies is observed for three of these components, which are interpreted as the first three LLs of the second heavy-hole subband. The experimentally derived splittings and shifts have also been compared with theoretically predicted LL structure and a good agreement is provided [16].

7. PHOTOLUMINESCENCE EXCITATION SPECTROSCOPY

Normal PL measurements provide information about only the electronic states that are populated at the temperature used for the measurement. Information about higher states can only be acquired by performing the PL measurements at elevated temperatures; for example, this has been demonstrated by Kukushkin et. al. [17]. However, an increasing broadening of the lines is always observed with increasing temperatures. A way to avoid such effects is by means of PL excitation (PLE) spectroscopy, in which empty states can also be monitored at the lowest temperatures. Fig. 8 shows typical PLE spectra for a modulation doped heterostructure, measured at 2K, for three different detection energies close to or resonant with the H-band (8240Å) [20]. In these PLE spectra, two well-defined peaks (labelled E^{2lh} and E^{2hh} in Fig. 8) appear. These peaks are clearly intensity correlated with the H-band, as obvious in Fig. 8. Similarly to HB1, the energy positions of the E^{2lh} and E^{2hh} peaks depend on the sample design and the experimental conditions. E.g. both E^{2lh} and E^{2hh} are found to shift towards higher energy with increasing excitation intensity, similarly to HB1, although the shift of the E^{2lh} and E^{2hh} peaks is considerably less [20]. The E^{2lh} and E^{2hh} peaks are interpreted as excitons related to the first excited n = 2 electron state and the n = 1 hh and lh holes, to be compared with the HB1 emission observed in luminescence, which originates from the recombination between holes and 2D electrons in the first (n = 1) subband.

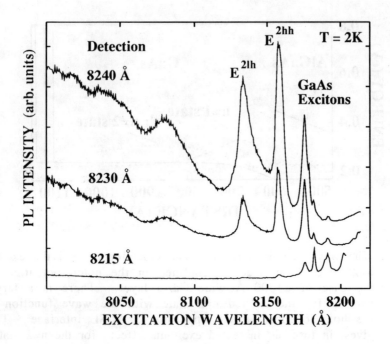

Fig. 8 PLE spectra of a heterostructure with a 500 Å wide GaAs layer for three different detection energies. When the HB1 emission is resonantly detected (the upper spectrum), the peaks denoted E^{2lh} and E^{2hh} reach their maximum intensity. These peaks are interpreted as the excitons originating from the n=2 electron and the n=1 hh (E^{2hh}) and the n=2 electron and the n=1 lh (E^{2lh}). At an off-resonant detection (the lower spectrum), the bulk GaAs related excitons dominate.

The excitonic effects are concluded to be weak for the case of HB1. This is partially due to the large spatial separation between electrons and holes in their ground states. The wave function of the n = 2 electron state, on the other hand, is significantly more delocalized as illustrated in Fig. 9, which in turn increases the overlap with the hole wave fuction. The excitonic effects for HB1 are also weak because of the high electron density, which gives rise to strong correlation effects and many body interactions. The "normal" excitons are quenched at high densities; instead edge

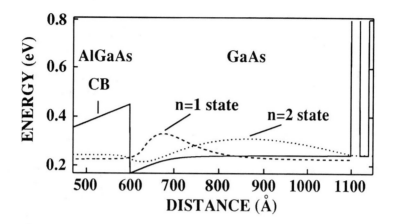

Fig. 9 The wave functions for the n=1 ground state and the excited n=2 state of an electron confined in the notch potential for the case of a 500 Å wide GaAs layer. There is a larger overlap for the excited n=2 state with the wave function of the hole localized at the oppposite GaAs/SL interface. This gives, in turn, an increased excitonic effect for the n=2 state.

singularities [25] are formed, as already mentioned above and will be further discussed in Sec. 8. At higher photon energies in the PLE spectrum (Fig. 8), additional structure is observed, which is interpreted as originating from higher states of the confined electrons and holes.

The sample used in Fig. 8 has an active GaAs layer of width 500Å. Due to band bending effects, the effective width of this layer, as experienced by the holes is further reduced. This means that confinement effects have to be taken into account for the hole. The hole states, which are degenerate in the 3D case, will split into a two-fold degenerate heavy hole (hh) state with $m_j = \pm 3/2$ and a two-fold degenerate light hole (lh) state with $m_j = \pm 1/2$ in the 2D case. Theoretical calculations based on the EMA and a realistic potential in the active GaAs layer result in hh-lh splittings, which are in very good agreement with the observed energy separations between the E^{2lh} and E^{2hh} peaks [20]. Accordingly, for a sample with a thicker GaAs layer, the E^{2lh} and E^{2hh} separation is found to decrease, in agreement with the theoretical predictions.

8. PHOTOLUMINESCENCE WITH AN APPLIED ELECTRIC FIELD

The properties of the H-band luminescence are closely related to the potential in the active layer. For instance, the energy position of the H-band depends strongly on the excitation intensity. A way to simulate different band bendings is by the application of an external electric field perpendicular to the layers. This fact is illustrated in Fig. 10 for the case of HB2, which shifts upwards in energy, when an increasing negative gate voltage is applied, i.e. the potential in the active GaAs layer is flattened out [18]. Both HB1 and HB2 can be shifted in the same energy range with an external field as is the case for varying excitation intensity level. Such an experiment therefore constitutes a critical confirmation of the model proposed to describe the H-band behavior in terms of band bending effects.

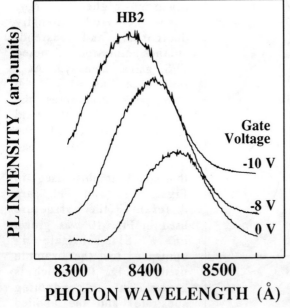

Fig.10 PL spectra showing the HB2 emission measured at 2K with a varying applied gate voltage. An increasing negative gate voltage corresponds to flatter potential in the active GaAs layer. The shift of the HB2 recombination towards higher energy with increasing negative gate voltage is consistent with the proposed band bending model.

However, in addition to the observed H-band shift, when an increasing negative gate voltage is applied, new features appear in the PL spectrum towards higher energies, as illustrated in Fig. 11. At zero field an emission band appears in the PL spectrum at about 1.515 eV, i.e. slightly above HB1 and close to the energy position of the GaAs free exciton. However, in the ODCR experiments described above (Sec. 6), this 1.515 eV emission exhibits a clear 2D character. Also, when the top AlGaAs layer is etched off, the 1.515 eV band disappears in the PL spectrum [18]. The interpretation of this band as the GaAs FE can therefore be excluded. The intensity of the 1.515 eV emission is strongly dependent on the electric field as obvious in Fig. 11. It increases in intensity with increasing field up to about 1.5 V. At this field, the HB1 emission has decreased in

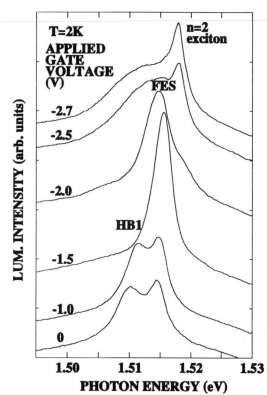

Fig.11 With an increasing negative gate voltage, the HB1 emission shifts towards higher energy. Also the HB1 intensity decreases, and instead, another emission, denoted FES, gains intensity. At a field exceeding ≈1.5V, the FES peak dominates the spectrum. When the field is further increased, the n=2 exciton will instead become dominating. Note that the samples used in Figs. 10 and 11 are different (the structure used in Fig. 10 was grown on a SI substrate in contrast to the sample used in Fig. 11, which is grown on a conducting substrate) and the applied voltages are accordingly not comparable.

intensity and the remainder has merged into the 1.515 eV band. When the electric field is further increased (to about 2.5 V), the intensity of the 1.515 eV emission decreases and instead another line appears in the PL spectrum at about 1.518 eV (Fig. 11).

The 1.515 eV emission is interpreted as the Fermi edge singularity (FES), which up to now only has been experimentally observed and investigated in modulation doped QWs [33 - 37]. The large density of electrons in a n-type modulation doped QW, or heterostructure in our case, can be regarded as a degenerate electron gas. The interaction between the photoexcited electrons and holes, which results in the formation of excitons, is drastically reduced due to screening of the Coulomb interaction. The normal exciton resonances will disappear, when there is a large probability of finding a carrier within the exciton volume. However, Mahan [38] demonstrated that bound states, FES, will exist even at high densities. The existence of such Fermi edge resonances is due to restrictions of the exclusion principle on the electrons' scattering [38, 39]. In fact, an enhancement of the FES has been observed, when the separation between the Fermi edge, E_F, and the next unoccupied subband is less than 5 meV [36, 37]. This enhancement is usually explained in terms of an efficient scattering for electrons at the Fermi edge via the nearly resonant adjacent subband [36, 37, 40], but also other models have been proposed, e.g. that the Fermi edge enhancement occurs due to hole localization caused by interface roughness. The enhancement gets larger for the second subband, because the overlap with the hole wave function is much larger for the second electron subband than for the first subband (cf Fig. 9) [41, 42].

We have in our system the possibility to continuously change the separation between the Fermi edge, E_F, and the next subband within certain limits via the applied field. When the electric field is increased, the Fermi level is raised and the conditions for the FES (at about 1.515 eV) are improved to become optimized at an applied field of 1.5 V in Fig. 11. For even higher fields, the separation between the Fermi edge and next subband gets too small and instead the next n = 2 subband becomes successively more populated. This fact results in an increasing intensity of the band at about 1.518 eV in Fig. 11 interpreted as excitons related to the n = 2 electron and the n = 1 hh states [32]. This observation is consistent

with the PLE measurements described above (Sec. 7), in which this exciton appears at the same energy position.

9. CONCLUSIONS

The first observation of luminescence from a semiconductor heterointerface was observed only a few years ago. We now have presented a large number of experiments for the case of n-type modulation doped heterostructures, which all give consistent information about the 2D electrons in the interface notch. The recombination process between the 2D electrons and free or acceptor bound holes have been observed and investigated by means of luminescence and related techniques.

Information regarding the details of heterointerfaces have been most difficult to acquire, requiring most elaborate transport measurements. Here we have used relatively simple, non-invasive characterization techniques such as PL, PLE, PL decay, PL perturbed by an electric, a magnetic, a microwave-field or combinations of these to gain considerable information about these structures. Such measurements have provided information about the shape and width of the notch, the electronic structure of the 2D electrons, band-offsets, the 2D electron density, and other phenomena that are usually difficult to access. In addition these experiments have provided a great deal of input to our theoretical understanding of recombination processes in this complex system; these processes contain ingredients such as varying spatial separation between the participating carriers, quantum confinement experienced by the carriers, a band structure, which is very sensitive to the exact experimental conditions, important many-body effects, and other complicated phenomena.

10. ACKNOWLEDGEMENTS

The work performed at University of California at Santa Barbara (UCSB) was supported by the NSF Science and Technology Center for Quantized Electronic Structures (QUEST). The authors acknowledge the expert MBE growth of the samples used in this study by M. Sundaram and A.C. Gossard, UCSB. Thanks are also due

Q.X. Zhao, J.P. Bergman, T. Lundström, and W. M. Chen for providing experimental data and for stimulating discussions.

11. REFERENCES

1. R. Dingle, H.L. Störmer, A.C. Gossard, and W. Wiegmann, Appl. Phys. Lett. 33, 665 (1978)
2. H.L. Störmer, R. Dingle, A.C. Gossard, W. Wiegmann, and R.A. Logan, Conf. Ser. - Inst. Phys. 43, 557 (1979)
3. Y. R. Yuan, K. Mohammed, M.A.A. Pudensi, and J.L. Merz, Appl. Phys. Lett. 45, 739 (1984)
4. Y. R. Yuan, M.A.A. Pudensi, G.A. Vawter, and J.L. Merz, J. Appl. Phys. 58, 397 (1985)
5. W. Ossau, E. Bangert, and G. Weimann, Solid State Commun. 64, 711 (1987)
6. I.V. Kukushkin and V.B. Timofeev, Pis'ma Zh. Eksp. Teor. Fiz. 43, 387 (1986) [JETP Lett. 43, 499 (1986)]
7. V.M. Asnin, A.A. Rogachev, A.Yu. Silov, and V.I. Stepanov, Solid State Commun. 74, 405 (1990)
8. See e.g. C. Weisbuch, in Applications of Multiquantum Wells, Selective Doping, and Superlattices, edited by R. Dingle, Chap. 1, in Semiconductors and Semimetals, Vol. 24, edited by R.K. Willardson and A.C. Beer, Academic Press Inc., London, (1987)
9. W. Kohn and L.J. Sham, Phys. Rev. 140, A1133 (1965)
10. G.E.W Bauer and T. Ando, Phys. Rev. B31, 8321 (1985)
11. G.E.W Bauer, Surf. Science 229, 374 (1990)
12. F. Stern, Appl. Phys. Lett. 43, 974 (1983)
13. H.L. Störmer, R. Dingle, A.C. Gossard, W. Wiegmann, and M.D. Sturge, Solid State Commun. 29, 705 (1979)
14. W.M. Chen, Q.X. Zhao, M. Ahlström, and B. Monemar, in "The Physics of Semiconductors", ed. W. Zawadzki (Inst. of Phys., Polish Acad. Sci., Warzaw), 279 (1988)
15. G.D. Gilliland, D.J. Wolford, T.F. Kuech, and J.A. Bradley, Phys. Rev. B43, 14251 (1991)
16. W. Ossau, E. Bengert, and G. Weimann, Solid State Commun. 64, 711 (1987)
17. I.V. Kukushkin, K.v. Klitzing, and K. Ploog, Phys. Rev. B45, 8509 (1988)

18. Q.X. Zhao, J.P. Bergman, P.O. Holtz, B. Monemar, C. Hallin, M. Sundaram, J.L. Merz, and A.C. Gossard, Semicond. Sci. Technol. 5, 884 (1990)
19. J.P. Bergman, Q.X. Zhao, P.O. Holtz, B. Monemar, M. Sundaram, J.L. Merz, and A.C. Gossard, Phys. Rev. B43, 4771 (1991)
20. Q.X. Zhao, Y. Fu, P.O. Holtz, B. Monemar, J.P. Bergman, K.A. Zhao, M. Sundaram, J.L. Merz, and A.C. Gossard, Phys. Rev. B43, 5035 (1991)
21. W. Ossau, T.L. Kuhn, G. Landwehr, and G. Weimann, Proc. of the 19th. Int. Conf. on the Physics of Semiconductors, Warzaw, 1988, ed. W. Zawadzki (Inst. of Phys., Polish Acad. Sci., Warzaw)
22. I. Balslev, Semicond. Sci. Technol. 2, 437 (1987)
23. I.V. Kukushkin, K.v. Klitzing, K. Ploog, and V.B. Timofeev, Phys. Rev. B40, 7788 (1989)
24. B.J. Skromme, S.S. Dose, and G.E. Stillman, J. Electron. Materials, 15, 345 (1986)
25. D. Huang, H.Y. Chu, Y.C. Chang, R. Houdré, and H. Morkoç, Phys. Rev. B38, 1246 (1988)
26. K. Köhler, H.J. Polland, L. Schultheis, and C.W. Tu, Phys. Rev. B38, 5496 (1988)
27. I.V. Kukushkin, K.v. Klitzing, K. Ploog, V.E. Kirpichev, and B.N. Shepel, Phys. Rev. B40, 4179 (1989)
28. A.S. Plaut, I.V. Kukushkin, K.v. Klitzing, and K. Ploog, Phys. Rev. B42, 5744 (1990)
29. R.M. Kusters, J. Singleton, G. Gobsch, G. Paasch, D. Schulze, F.A. Wittekamp, G.A.C. Jones, J.E.F. Frost, D.C. Peacock, and D.A. Ritchie, Superlattices and Microstructures, Vol. 9, No.1, 55 (1991)
30. W.M. Chen, B. Monemar, Q.X. Zhao, P.O. Holtz, M. Sundaram, J.L. Merz, and A.C. Gossard, Proc. of the Eighth International Conference on Electronic Properties of Two-Dimensional Systems, p. 191, Grenoble, 1989
31. W.M. Chen, B. Monemar, E. Sörman, P.O. Holtz, M. Sundaram, J.L. Merz, and A.C. Gossard, Proc. of the 7th International Conference on Hot Carriers in Semiconductors (HCIS7), Nara, Japan, 1991
32. Q.X. Zhao, P.O. Holtz, T. Lundström, J.P. Bergman, B. Monemar, M. Sundaram, J.L. Merz, and A.C. Gossard, Proc. of the International

Meeting on Optics of Excitons in Confined Systems, Sicily, September, 1991
33. M.S. Skolnick, J.M. Rorison, K.J. Nash, D.J. Mowbray, P.R. Tapster, S.J. Bass, and A.T. Pitt, Phys. Rev. Lett. $\underline{58}$, 2130 (1987)
34. J.S. Lee, Y. Iwasa, and N. Miura, Semicond. Sci. Technol. $\underline{2}$, 675 (1987)
35. G. Livescu, D.A.B. Miller, D.S. Chemla, M. Ramaswamy, T.Y. Chang, N. Sauer, A.C. Gossard, and J.H. English, IEEE J. of Quantum Electronics, Vol. 24, 1677 (1988)
36. W. Chen, M. Fritze, A.V. Nurmikko, D. Ackley, C. Colvard, and H. Lee, Phys. Rev. Lett. $\underline{64}$, 2434 (1990)
37. W. Chen, M. Fritze, A.V. Nurmikko, M. Hong, and L.L. Chang, Phys. Rev. B$\underline{43}$, 15738 (1991)
38. G.D. Mahan, Phys. Rev. $\underline{153}$, 882 (1967)
39. S. Schmitt-Rink, D.S. Chemla, and D.A.B. Miller, Phys. Rev. B$\underline{32}$, 6601 (1985)
40. J.F. Mueller, A. Ruckenstein, and S. Schmitt-Rink, Mod. Phys. Lett. $\underline{2}$, 135 (1991)
41. P. Hawrylak, Phys. Rev. B$\underline{44}$, 3821 (1991)
42. P. Hawrylak, Phys. Rev. B$\underline{44}$, 6262 (1991)

Type-I—Type-II Transition in GaAs/AlAs Superlattices

Guo-Hua Li

*National Laboratory for Superlattices and Microstructures
Institute of Semiconductors, Academia Sinica
P.O.Box 912, Beijing 100083, China*

ABSTRACT

The theoretical and experimental studies on the electronic structures and optical properties of GaAs/AlAs superlattices are reviewed. The main attention is paid on the characteristics of the lowest conduction band and the type-I—type-II transition in $(GaAs)_n(AlAs)_n$ short period superlattices. Both theoretical calculations and experimental measurements indicate that there are two types of GaAs/AlAs superlattices. One is type-I superlattice in which the electrons and holes are confined in the same layers. The other is type-II superlattice in which the electrons and holes are confined in adjacent layers. For the $(GaAs)_n(AlAs)_n$ superlattices, the type-I—type-II transition point is in the range of $8 \leq n \leq 14$ though the exact point obtained from different theoretical calculations and various experimental measurements are somewhat divergent. The Γ-X mixing and the X-valley splitting are also reviewed briefly.

1. Introduction

Recent advances in molecular beam epitaxy technology made possible the growth of short period GaAs/AlAs superlattice(SL) with very high quality. Since then the electronic structures and the optical properties of GaAs/AlAs SL have attracted extensive attention in both theoretical and experimental investigations.

Bulk GaAs is a semiconductor with direct band gap. On the other hand bulk AlAs is an indirect semiconductor. Both theoretical calculation and experimental measurement demonstrate that there exist two types of SL when the GaAs and AlAs layer are grown alternately along the [001] direction. One is so called type-I SL in which the electrons and holes are confined in the same layers. The other is type-II SL in which the electrons and holes are confined in adjacent layers. A GaAs/AlAs SL can either be a type-I SL or a type-II SL, depending

on the thickness of GaAs and AlAs layers. In addition some external conditions such as pressure or electric field can also make the conversion of the SL configuration. Therefore the characteristics of GaAs/AlAs SL and the type-I—type-II transition have been the one of the most active topics in recent years.

There have been quite a lot of theoretical investigations on the band structure of GaAs/AlAs SL, especially on the characteristics of the lowest conduction bands in SL. The calculation methods are ranging from the simplest effective mass approximation[1-3] and various empirical methods[4-13] to some first principle calculations[14-18]. The electronic structures and optical properties of GaAs/AlAs SL have also been studied by means of various optical spectroscopies, including photoluminescence and photoluminescence excitation[1,2,19-22,32,33], time-resolved photoluminescence[3,39], absorption[23], photoreflectance[24], photocurrent[25], photoacoustic[26] etc. The hydrostatic pressure[27-31,40], the magnetic field[34], the uniaxial stress[35-37] and the electric field[38] have also been used as an external perturbations to change the band structure of GaAs/AlAs SL in order to obtain more information.

The theoretical and experimental investigations on the type-I—type-II transition in GaAs/AlAs SL are reviewed in this paper. The main attention is paid on the $(GaAs)_n(AlAs)_n$ short period SL where n is the number of monolayer. The crystal structure and related properties of GaAs/AlAs SL are described in section 2. Section 3 and 4 represent the theoretical and experimental results of type-I—type-II transition, respectively. In section 5 the studies about the Γ-X mixing and the X-valley splitting are reviewed briefly.

2. Structure of GaAs/AlAs superlattice

The GaAs/AlAs short period SL discussed in this paper is an one-dimensional periodic structure consisting of alternating GaAs and AlAs thin layers in the [001] direction. As usual we use $(GaAs)_m(AlAs)_n$ or (m,n) to denote the SL with m monolayers of GaAs and n monolayers of AlAs in one period, where one monolayer consists of one Ga (Al) plane and one As plane. The thickness of one monolayer is half of a lattice constant (about 2.83 Å). Since both GaAs and AlAs have the zinc-blend crystal structure with nearly equal lattice constant, all the atoms in SL are located on the sites of a zinc-blend lattice.

The mismatch at the interface can be neglected in the majority of the treatments. In Fig.1 we show the arrangement of some atoms in a GaAs/AlAs SL along [001] direction. Note that the atoms of same element in the alternate planes along the [001] direction do not lie on top of each other. The lattice constant, a, and the thickness of one monolayer, 1ml, are also indicated in this figure.

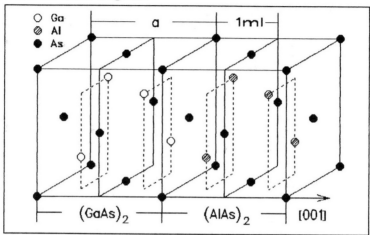

Fig.1 Arrangement of some atoms in the $(GaAs)_2(AlAs)_2$ SL along [001] direction.

The lowest conduction band of bulk GaAs and AlAs consists of Γ, X, and L valleys. The generally accepted viewpoint is that there are three types of one-dimensional square potential well in GaAs/AlAs SL, formed by Γ, X and L band edges of GaAs and AlAs, respectively. Correspondingly, there are also three types of confined carriers with the effective masses of Γ, X and L valleys of GaAs or AlAs and often called as Γ-like, X-like and L-like electrons, respectively. The band edge alignment of Γ and X valleys along [001] direction is schematically drawn in

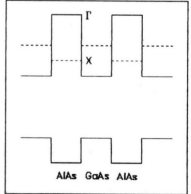

Fig.2 Schematic diagram of the band edge alignment of GaAs/AlAs SL.

Fig.2. The valence band discontinuity at the interface is about 0.55 eV according to the recent PL experiment[1]. Therefore both valence holes and conduction Γ-like electrons are confined in GaAs layers (Γ well). On the contrary, the X-like electrons tend to be confined in AlAs layers (X well). Moreover, two series of X-like state exist in the X well because the effective mass along [001] direction is different for the X valley at (001) and the X valleys at (100) and (010). We label these two series of X-like states by X_z and X_{xy} as do many authors, where the z axis is chosen to parallel to the [001] direction. The energy gap, the effective masses of GaAs and AlAs and the valence band offset[1] are listed in Table I. The parameters used by various authors in their calculations are somewhat divergent, especially for the data of AlAs.

Table I. Band gap, E_g, band offset, ΔE_v, (in eV) and effective mass, m, (unit of m_0). From [1].

		GaAs	AlAs
$E_{g.dir}(\Gamma_{8v}-\Gamma_{6c})$		1.519	3.1
$E_{g.ind}(\Gamma_{8v}-X_{6c})$		1.98	2.23
$E_{g.ind}(\Gamma_{8v}-L_{6c})$		1.81	2.5
$m_n(\Gamma)$		0.067	0.13
$m_n(X)$	m_t	0.23	0.19
	m_l	1.3	1.1
m_{hh}		0.45	0.5
ΔE_v		0.53	

The band alignment picture based on the band structure of bulk GaAs and AlAs becomes doubtful in the SL with ultrathin GaAs and AlAs layers. Another method for investigating GaAs/AlAs short period SL is to treat the SL as a new material with the unit supercell consisting of one period of SL. Consequently, the energy band can be calculated using the usual band structure analysis method. The symmetry properties of the unit supercell have been also investigated in detail[5]. There are two types of Bravais lattice for GaAs/AlAs SL consisting of m monolayers of GaAs and n monolayers of AlAs. When m+n is even

the Bravais lattice is simple tetragonal with the basis vectors as $(1,1,0)a/2$, $(-1,1,0)a/2$ and $(0,0,m+n)a/2$. When m+n is odd the Bravais lattice is body-centered tetragonal with the basis vectors as $(1,1,0)a/2$, $(-1,1,0)a/2$ and $(0,1,m+n)a/2$. Because of the inequivalence of the Ga and Al planes, the T_d symmetry associated with the bulk GaAs and AlAs is lowered. The point group of GaAs/AlAs SL is D_{2d}, and the space group is D_{2d}^5 for the case of even m+n and D_{2d}^9 for the case of odd m+n, respectively. The Brillouin zone for $(GaAs)_m(AlAs)_n$ SL (SBZ) with m+n=3 (odd) and m+n=2 (even) are shown in Fig.3 together with the Brillouin zone of bulk GaAs (BBZ). Some points of high symmetry which are of particular interests are also indicated in figure. We put a bar over the symbol for the symmetry points in the SBZ to distinguish it from that in the BBZ. The SBZ can be obtained by folding the BBZ along the z direction. Thus the Γ-X_z line of BBZ is folded to the $\bar{\Gamma}$-$\bar{Z}$ line of SBZ and the X_z point is folded onto the $\bar{\Gamma}$ point for cases with m+n=even and onto the $\bar{Z}$ point for cases with m+n=odd. On the other hand, the X_x and X_y points of BBZ are related to the $\bar{M}$ point of SBZ for the cases of m+n=even and $\bar{X}$, $\bar{Y}$ points of SBZ for the cases of m+n=odd.

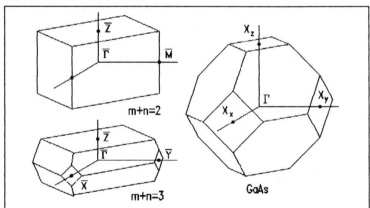

Fig.3 Brillouin zone of $(GaAs)_m(AlAs)_n$ SL with m+n=2 (even) and m+n=3 (odd) together with the Brillouin zone of bulk GaAs.

Since the Γ-like and X-like electrons are confined in different layers and the X_z and X_{xy} valleys are folded to different point of SBZ, the optical transition in GaAs/AlAs SL have three different situations:

When Γ conduction subband is the lowest band, both electrons and holes are confined in the same GaAs layers and at the same $\overline{\Gamma}$ point in SBZ. Therefore the transition is direct in both real and k space. The system is a type-I SL.

When X_z is the lowest valley the electrons and holes are confined in different layers but they are still at the same $\overline{\Gamma}$ point in SBZ, so that the transition is direct in k space though it is indirect in real space. The system is type-II SL and the transition is often called as pseudodirect transition.

When X_{xy} is the lowest valley the conduction minimum is at the $\overline{M}$ or $\overline{X}(\overline{Y})$ point in SBZ but the holes are at the $\overline{\Gamma}$ point of SBZ. Therefore the system is type-II SL and the transition is indirect in both real and k space.

3. Theoretical calculation

The energy band structures of GaAs/AlAs short period SL have been calculated by many authors using various methods. These calculation methods can be classified into two approaches. In the first approach, the SL is considered as a layered structure consisting of alternating GaAs and AlAs layers, which still maintain the band structure of bulk material. The electron states are obtained by jointing two types of wave function across the interface. In the second approach, the SL is treated as a new material with the unit supercell including both GaAs and AlAs layers. The band structures are then calculated using conventional energy band theory. Some of these calculations are empirical methods with the empirical parameters based on the bulk materials, the others are the first principle calculations.

The structures of valence bands calculated using various method are somewhat similar. However, the conduction band structures calculated from different method show different characteristics, especially for the lowest conduction subband. Therefore the main attention in this paper are lied on the conduction band structure of SL.

3.1. Effective mass approximation

The effective mass approximation is the simplest method for the calculation of the band structure of SL. It is sometimes called as the envelope function approximation or Kronig-Penney model. In this approximation, the band structure of GaAs and AlAs layers in SL are the same as that of bulk materials. The

band edges have a discontinuity across the interface, forming an one-dimensional square potential well. The carriers in each layer can be described by the effective mass approximation and their effective masses are the same as those of bulk materials. The energy level of subband is then obtained by jointing the wave function in the wells and barriers across the interface, using proper boundary conditions.

The energy level of Γ, X_z and X_{xy} states in $(GaAs)_n(AlAs)_n$ SL's calculated by Nagle et al[2] using the Kronig-Penney model are shown in Fig.4. The main result is that the Γ-like state is the lowest level when n>11 and the X_z-like state is the lowest state at n<11. The X_{xy} state is always higher than the X_z state due to the smaller effective mass of X_{xy} state along the [001] direction.

The Fig.5 shows the boundary between type-I and type-II SL for $(GaAs)_m(AlAs)_n$ ($1 \leq m,n \leq 20$) SL's, calculated using the room

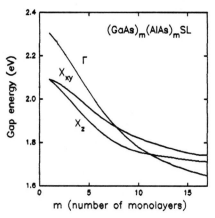

Fig.4 Calculated energies of the confined Γ, X_z and X_{xy} states by using the Kronig-Penney model. From [2].

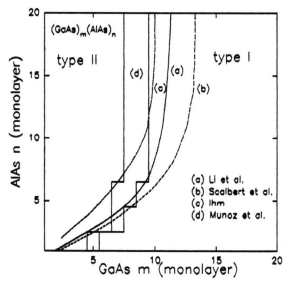

Fig.5 Boundaries between type-I and type-II SL. (a) and (b) are calculated using K-P model[29,3]. (c) and (d) are tight-binding calculation[4,7].

temperature parameters. The low temperature results calculated by Scalbert et al.[3] are also shown in the figure. It can be seen that the SL is type I when m<11 (300K) or m<13 (4K) for SL's with thick AlAs layers. For the SL's with thin GaAs and AlAs layers the type-II SL is obtained when n/(m+n) is larger than 0.36.

3.2. Tight-binding method

In the tight-binding method four(sp^3) or five(sp^3s^*) local orbitals for each atom are used as the basis functions and the unit supercell is composed of m layers of GaAs and n layers of AlAs. The tight-binding interaction parameters are chosen to reproduce the bulk band and the band edge offset. The parameters associated to interface are taken to be averages of the bulk values.

Ihm[4] has calculated the lowest conduction state of $(GaAs)_m$ $(AlAs)_n$ SL's using a nearest-neighbor tight-binding method and obtained the boundary of type-I and type-II SL as shown in Fig.5 by curve (c). He indicated that the electronic state deviates from the Kronig-Penney model significantly for the SL's with sufficiently thin GaAs and AlAs layer. According to his calculation the X_{xy} levels are always lower than the X_z level. However this order is not valid in practice since the next-nearest neighbor interaction is completely neglected in his calculation.

Lu and Sham[5,6] have calculated the subband structure of GaAs/AlAs SL's along the [001] direction through the center of the SBZ and through the zone boundary point at (100) using a second-neighbor tight-binding method. The symmetry property of

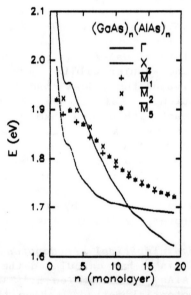

Fig.6 Energies of lowest two level at $\bar{M}$ point. The lowest two level at $\bar{\Gamma}$ point (Γ and X_z) are also plotted for comparison. From [6].

SL and the parity of states are also discussed in detail in his paper. The calculated lowest Γ level, lowest X_z level and X_{xy} level (denoted as $\overline{M}$ in the figure) of $(GaAs)_n(AlAs)_n$ SL's are shown in Fig.6. According to their calculation the lowest Γ level in $(GaAs)_n(AlAs)_n$ SL's has a even parity and the parity of lowest X_z level is the same as that of n. Therefore the Γ-X anticross occurs at n=12 due to the same parity of two levels. In addition the X_z level lies below the X_{xy} level, except the case of n=1. The double degenerate X_{xy} state is split to two states ($\overline{M}_1$ and $\overline{M}_2$ in Fig.6) in the case of even n and retain degenerate ($\overline{M}_5$ in Fig.6) in the case of odd n.

Muñoz et al.[7] have also calculated the electronic structure of GaAs/AlAs SL's using a sp^3s^* empirical tight-binding model and the surface Green-function matching method. The calculated boundary of Type-I and type-II SL is also drawn in Fig.5 by curve (d).

3.3. Wannier orbital model

Ting and Chang[8] have calculated the conduction band structure of $GaAs/Al_xGa_{1-x}As$ and $Al_xGa_{1-x}As/AlAs$ SL's using a one-band Wannier orbital model. The parameters are obtained by fitting the lowest conduction band to the results of pseudo-potential calculation over the entire Brillouin zone. In contrast with Lu and Sham's results, their calculated pMwity of the lowest X_z level is opposite to that of n, and the X_{xy} state is split in the case of even n. Afterward, Xia and Chang[9] employed a second-neighbor tight-binding method to calculate the conduction band of GaAs/AlAs short period SL. They argued that the same parity feature as that of the Wannier model can also be obtained by using the tight-binding method through the properly selecting the interaction parameters.

3.4. Empirical pseudopotential method

Xia[10] has calculated the electronic structures of $(GaAs)_m$ $(AlAs)_n$ short period SL's with use of the empirical pseudo-potential method, taking the Hamiltonian of the alloy with x=n/(m+n) as the zero-order Hamiltonian and the wave function of the alloy as basis functions. The form factors of pseudo-potentials are determined from the fit to the experimental energy band of bulk GaAs and AlAs, the valence band offset and the PL results of $(GaAs)_4(AlAs)_4$ SL[2]. The difference of the pseudopotential form factors of Ga and Al atoms is then taken

as the perturbation potential. The calculated direct and indirect energy gap of $(GaAs)_n$ $(AlAs)_n$ SL's are shown in Fig. 7, where $C\Gamma1$, $CX1$ and $HH1$ represent the lowest energy level with Γ and X symmetry and the first heavy hole state, respectively. According to his calculation only when n>10 can the $C\Gamma1$ level become lower than the $CX1$ level. In addition his calculated square optical transition matrix elements of $CX1-HH1$ are much smaller than those of $C\Gamma1-HH1$, denoting that although $C\Gamma$ and CX states are all at the center of the SBZ they still largely keep their properties of alloy states.

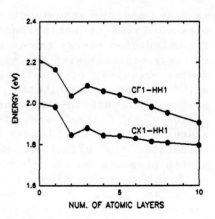

Fig.7 Variation of direct and indirect energy gaps of $(GaAs)_n$ $(AlAs)_n$ SL with n, calculated using the empirical pseudopotential method. From [10].

In Gell's pseudopotential calculation[11,12] a perturbation potential describing the changes of atomic potentials in SL is added to the Hamiltonian of a perfect infinite crystals such

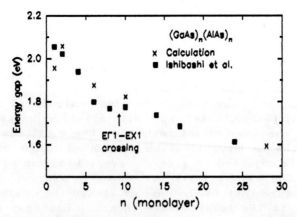

Fig.8 Calculated variation of energy gap of $(GaAs)_n(AlAs)_n$SL's together with the experimental results of Ishibashi et al.[19] From [12].

as GaAs, and the wave function of GaAs is taken as basis function. The calculated energy gap is shown in Fig.8 together with the experimental results of Ishibashi et al.[19]. They attributed the plateau near n=8 to the crossing region of Γ-like and X-like states. Gell et al.[13] have also calculated the effect of hydrostatic pressure on the electronic and optical properties of GaAs/$Al_xGa_{1-x}As$ SL's. The calculated variance of the conduction band with the hydrostatic pressure for a GaAs(51Å)/AlAs(6Å) SL is shown in Fig.9. The state with the lowest energy at zero pressure is the Γ-related state EΓ1 whilst state EX1 directly above it in energy is an X-related state. However, when pressure beyond some value the EX1 become the lowest state therefore the Γ-X cross can be achieved by application of pressure. The another feature of Fig.9 is that apart from the anticrossing region, the energy of all states responds more or less linearly with pressure.

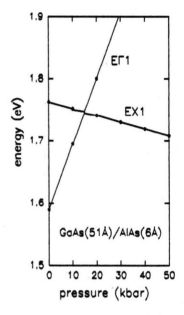

Fig.9 Variation with hydrostatic pressure of the energies of the lowest conduction states in the GaAs(51Å)/AlAs(6Å) SL. From [13].

3.5. First-principle calculation

Nakayama and Kamimura[14,15] have calculated the band structure of $(GaAs)_n(AlAs)_n$ SL's with n=1-10 by using the self-consistent pseudopotential method. In their calculation the α value in the Xα approximation for an exchange correlation potential is adjusted to give a correct band gap of bulk GaAs and AlAs. The main results are shown in Fig.10. It is seen from the figure that in the case of 2≤n≤10 the conduction band state at Γ is the lowest state thus the band gap is direct. However, in the case of n=1 the conduction band state at L is the lowest so that the band gap becomes indirect.

Gopalan et al.[16] have investigated the character of

electron and hole states in $(GaAs)_m(AlAs)_n$ SL's with $m,n \leq 3$ by means of the first principle linear-muffin-tin-orbital method. They obtained that the lowest conduction state are at the $\bar{R}(L)$ point of SBZ in the case of n=1. The other first principle calculations[17,18] have also gave the similar results, which is different from that of other model and have to be checked by further experiments.

The calculated type-I—type-II tra-nsition point for $(GaAs)_n$ $(AlAs)_n$ SL's by several authors are colle-cted in Table II. The majority of the results are in the range of $8 \leq n \leq 14$ although the accurate value obtained from various methods are somewhat different.

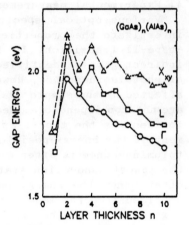

Fig.10 Gap energies of various minima of (n,n) SL as a function of n. From [15].

Table II. Theoretical results of type-I—type-II transition point in $(GaAs)_n(AlAs)_n$ SL's.

author	method	point	Ref.
Nagle et al.	Effective mass	n=11	[2]
Li et al.	Effective mass	n=11	[29]
Scalbert et al.	Effective mass	n=13	[3]
Ihm	Tight binding	n=10	[4]
Lu and Sham	Tight binding	n=12	[5]
Muñoz et al.	Tight binding	n=8-9	[7]
Xia and Chang	Tight binding	n=12	[9]
Gell et al.	Pseudopotential	n=8	[11]
Xia	Pseudopotential	n>10	[10]
Nakayama and Kamimura	Self-consistent	n=2	[14]
Gapalan et al.	First principle	n>7	[16]

4. Experimental measurements

Various optical spectroscopic methods have been used to investigate the properties of GaAs/AlAs SL's and the type-I —type-II transition. In the earlier work[19] the direct and indirect gap is identified mainly according to the intensity of photoluminescence. However the emission intensity is also affected by the sample quality, therefore the obtained results are not too dependable. Recently, the main experimental method to identify the type-I and type-II SL is the direct measurement of the energy position of Γ-like and X-like states. Photoluminescence is often used to measure the energy position of the lowest conduction state (either Γ-like or X-like). On the other hand, the energy position of the Γ-like state can be obtained from the measurement of the absorption coefficients, regardless of whether it is the lowest state or not. The decay time can also be used to characterize the direct and indirect gap since the decay time is much larger for the indirect transition. In addition some external perturbation methods have also been used to change the band structure in order to obtaine more information.

4.1. Photoluminescence and photoluminescence excitation

Photoluminescence(PL) and photoluminescence excitation(PLE) are widely used to investigate the type-I—type-II transition in GaAs/AlAs SL's[1,2,19-22,32,33]. In practice, the PLE spectra represent the change of the absorption coefficient so it can be used to measure the direct transition from valence Γ point to conduction Γ point. If the threshold energy of PLE is higher than the PL peak energy, the SL is the type-II with Γ-like state located higher than the X-like state. On the other hand, the SL is the type-I when the PLE threshold energy and the PL peak energy are nearly equal.

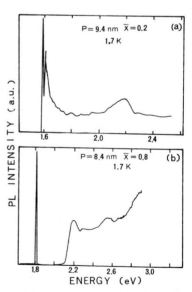

Fig.11 PLE spectra of (a)direct, (b)indirect samples at 1.7K. From [1].

The typical PLE spectra show the different character of type-I and type-II GaAs/ AlAs SL's[1] are presented in Fig.11, where the sharp peaks at lower energy are due to the PL. In Fig.11(a) the PL peak and the PLE peak are at nearly same position, showing the typical situation of type-I SL. In Fig.11(b), the PLE threshold is located at the energy position much higher than that of PL peak, so that it is a type-II SL.

Jiang et al.[20] have reported a systematic investigation of $(GaAs)_l(AlAs)_m$ SL's with l and m ranging from 1 to 73 based on the PL and PLE measurements. The PLE spectra of some representative samples are shown in Fig. 12,

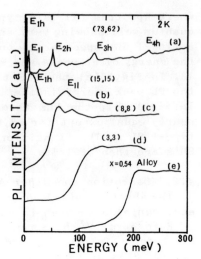

Fig.12 PLE spectra of $(GaAs)_l$ $(AlAs)_m$ SL's at 2K. The energy range is measured from the detection energy. From [20].

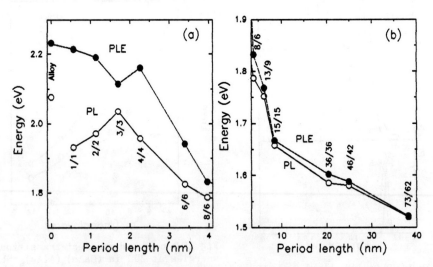

Fig.13 PLE threshold energy and PL peak energy of $(GaAs)_l(AlAs)_m$ SL's as a function of the period length. From [20].

where spectrum (a) is a typical spectrum of GaAs/AlAs multi-quantum well showing very sharp peaks, spectrum (b) represents a type-I SL, spectra (c) and (d) represent the type-II SL's. the energy of PL peaks and PLE threshold are collected in Fig. 13. In Fig.13(a) the PLE threshold energies are higher than the PL peak energy indicating the character of type-II SL's. In Fig.13(b), the PLE threshold and the PL peak energy are nearly equal except for (8,6) SL. Therefore those samples are type-I SL's. The transition point between type-I and type-II SL's are then at l=13.

4.2. Absorption and other spectroscopies

The absorption(AB), photoreflectance(PR), photocurrent(PC) etc. can also be used to identify the type of SL.

Kato et al.[23] have removed the GaAs substrate from samples by preferential etching and measured the AB spectra of $(GaAs)_m$ $(AlAs)_m$ SL's with m=4-18. The measured spectra at 20K for m= 10, 12 and 14 are shown in Fig.14, in which the solid and dotted lines indicate the PL and

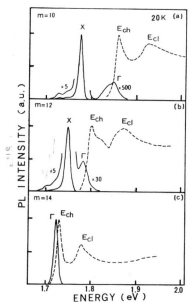

Fig.14 PL(solid line) and AB spectra(dotted line) of $(GaAs)_m$ $(AlAs)_m$ SL's at 20K. From [23].

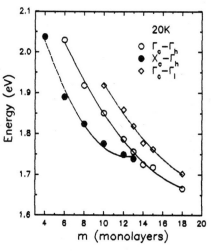

Fig.15 Direct and indirect transition energies at 20K in $(GaAs)_m(AlAs)_m$ SL obtained from PL(open circles and full dots) and AB(squares) measurements. From [23].

AB spectra, respectively. The AB peak is in close with the PL peak in the case of m=14, giving a evidence that the $(GaAs)_{14}$ $(AlAs)_{14}$ is a type-I SL. On the other hand, the AB peaks are located at the higher energy than the PL peaks in the case of m=10 and m=12 SL's, indicating that these SL's are type-II. The measured direct (Γ_c-Γ_h, Γ_c-Γ_l) and indirect (X_c-Γ_h) transition energies for the SL's with m from 4 to 18 are shown in Fig. 15. It can be seen that the Γ_c-X_c crossover point occurs at the vicinity of m=14.

Fujimoto et al.[24] have measured the PR and PL spectra of $(GaAs)_n$ $(AlAs)_n$ SL's with n ranging from 1 to 15 at room temperature. The modulated PR spectra are analyzed by a third-derivative formula to obtain the critical point energies. The spectra of samples with n=8, 10 and 12 are shown in Fig.16, where the dot-dashed curves are the PL spectra, the solid curves are the PR spectra. The vertical arrow indicates the lowest interband transition energy obtained from the fitting procedure. In the spectra of SL's with n=10 and 12, the lowest critical point energies obtained from PR are very close to the photon energies corresponding to the PL peaks. Therefore they concluded that the SL's with n>10 have direct energy gaps. On the other hand, the lowest transition energy is higher than the PL peak energy for n=8 SL. Similar results are also obtained for the SL's with n=5 and 3. These results suggest that the $(GaAs)_n(AlAs)_n$ SL's with n<10 are indirect.

Barrau et al.[25] have used a PC

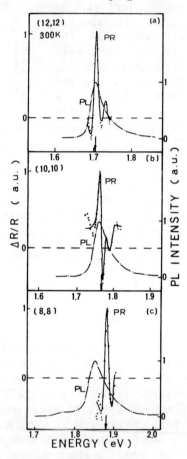

Fig.16 Room-temperature PR (circles) and PL (dot-dashed curve) spectra. The solid curve is determined by best-fitting procedure. From [24].

spectroscopy technique for the study of the direct-indirect band gap transition in GaAs/AlAs short period SL's. A built-in electric field perpendicular to the interface is set up between the Schottky barrier on the GaAs top layer of the SL and the In ohmic contact on the n⁺-type GaAs substrate. The PC spectra are then related to the variation of the absorption coefficient with the wavelength of incident optical beam. One of the PC spectra are shown in the Fig.17. The main structures are due to the heavy hole, light hole and hole of spin-orbit splitting band to electron transitions and are labeled by $hh_1-e_1(\Gamma)$, $lh_1-e_1(\Gamma)$ and $s.o.h_1-e_1(\Gamma)$, respectively. A much less intensive structure labeled $hh_1-e_1(X)$ is related to the transition from heavy hole subband to the X_z minimum of the conduction band.

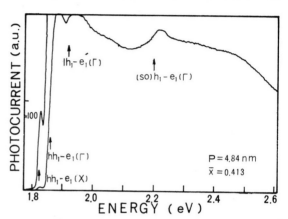

Fig.17 The photocurrent spectrum of the GaAs/AlAs SL with period length P=4.84 nm and mean composition $\bar{X}$=0.413. From [25].

Cingolani et al.[26] have observed the transition from type-I to type -II in $(GaAs)_n$ $(AlAs)_n$ SL's by means of photoacoustic saturation spectroscopy(PAS) and high excitation intensity PL (HEL). The PAS is based on the heat transfer induced by the absorption of photon in a crystal and then represents the dependence of the reflectance factor on the incident light frequency. Fig.18 shows the normalized PAS spectra of several $(GaAs)_m(AlAs)_m$ SL's with m ranging between 2 and 15, measured at 300K. Two broad maxima labeled Γ and X become nearly coincident in the n=12 SL. The HEI spectra measured at 10K are plotted in Fig.19. The exciting power density employed in the measurements is about 300KW/cm² and the pulse duration is about 5 ns. The estimated band filling in the energy space is of the order of 50 meV under this condition. An emission related to the transition from X point appears as a shoulder

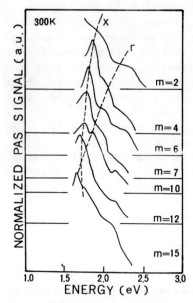

Fig.18 Normalized photoacoustic saturation spectra of several $(GaAs)_m(AlAs)_m$ SL's. From [26].

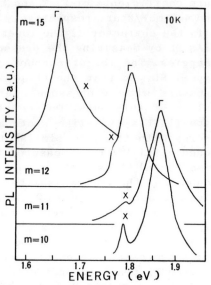

Fig.19 Photoluminescence spectra under high-excitation intensity (300KW/cm^2) for $(GaAs)_m(AlAs)_m$ SL's with 10<m<15 at 10K. From [26].

on the high energy side in the spectrum of n=15 due to the band filling effect. Then both direct and indirect emission in type-I SL can be seen in the HEI spectra. The crossover point identified from the HEI spectra is also around n=12.

4.3. Photoluminescence under hydrostatic pressure

The normal PL and PLE measurements are difficult to distinguish the type-I SL and the SL just at the crossover point between type-I and type-II since the X-like state in the type-I SL's are hard to be detected. Although in the HEI spectra both the Γ- and X- related emission can be observed, the significant broadening of the peak still makes the accurate determination of the energy position of the X-like state much more difficult.

The measurements of PL under hydrostatic pressure provide a powerful tool for the investigation of the type-I—type-II transition in the GaAs/AlAs short period SL's. Under hydrostatic pressure the Γ, X and L valleys of bulk GaAs and other

III-V compounds move at different rate. For example the pressure coefficients of Γ, X and L valleys of GaAs are 10.7, -1.3 and 4.5 meV/kbar, respectively. Therefore, it is easy to identify the character of the lowest conduction subband of a GaAs/AlAs SL by measuring the pressure coefficients of the PL peak energies. The lowest conduction subband is Γ-like state for type-I SL, so that the PL peak moves to higher energy under pressure with a pressure coefficient similar to that of Γ valley of bulk GaAs. On the other hand, the lowest state in type-II SL's is X-like state and the PL peak moves to lower energy with the rate similar to that of bulk GaAs X valley. Jones et al.[27] have measured the low temperature PL spectra of a $(GaAs)_9(AlAs)_9$ SL in the pressure range of 0-4 kbar and found that the PL peak shifted to the low energy with increasing pressure at a rate of -2meV/kbar. From this finding, they concluded that the lowest conduction band state should be X-like and the $(GaAs)_9(AlAs)_9$ SL is type-II. Danan et al.[1] and Takarabe et al.[28] have also performed optical measurements under hydrostatic pressure and assigned the Γ-related and X-related PL peaks according to their pressure coefficients.

Moreover, a type-I SL can be converted to the type-II SL by application of the hydrostatic pressure and than the X-related emission in the type-I SL can be observed when pressure beyond some threshold value. By extrapolating the pressure dependence of X-like state to the normal pressure, the energy position of X-like state and the separation between Γ-like and X-like states in type-I SL's can be obtained.

Li et al.[29,30] have systematically investigated the PL spectra of $(GaAs)_n(AlAs)_n$ ($4 \leq n \leq 17$) short period SL's under hydrostatic pressure in the range of 0-50 kbar at room and liquid nitrogen temperature. The room temperature PL spectra of three representative SL's are shown in Fig.20(a). The pressure dependance of the peak energies are shown in Fig.20(b). For the n=6 SL, two peaks are observed at atmospheric pressure. The peak on the high energy side (E^Γ) has a rapid blue shift with increasing pressure. The pressure coefficient is nearly the same as that of the Γ valley of bulk GaAs. On the contrary, the peak on the lower energy side (E^X) shifts only slowly to lower energy. Its pressure coefficient is similar to that of the X valley of bulk GaAs. Thus these two peaks can be identified to the transition from conduction Γ-like and X-like electrons to heavy hole subband, respectively. The $(GaAs)_6$

$(AlAs)_6$ SL is type-II since the E^Γ is higher than E^X at atmospheric pressure. For the n=11 and n=17 samples, only one peak is observed at atmospheric pressure, which is attributed to E^Γ in the light of their pressure coefficients. When pressure is beyond a certain value, the peak E^X begins appear on the lower energy side. The energy position of E^X at atmospheric pressure is deduced by a linear extrapolation of its pressure dependence. It can be seen from Fig.20(b) that the n=11 and n=17 SL's

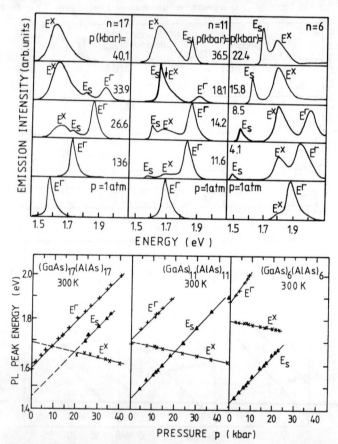

Fig.20 (a) Room temperature PL spectra under different pressure and (b) pressure dependence of the peak energies of three representative $(GaAs)_n(AlAs)_n$ SL's. From [30].

represent two characteristic cases. For n=17 SL, the E^Γ is lower than E^X at atmospheric pressure, thus the $(GaAs)_{17}(AlAs)_{17}$ is a type-I SL. For the case of n=11, the E^Γ and E^X are nearly at the same energy at atmospheric pressure, thus the $(GaAs)_{11}(AlAs)_{11}$ is just at the transition point between type-I and type-II SL's.

The 77K PL spectra of these samples are shown in Fig.21. The situation is similar to that of 300K except the E^X peak is the main peak in the PL spectra as expected from the Boltzmann

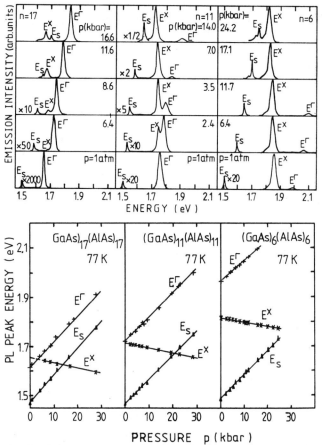

Fig.21 (a) 77K PL spectra and (b) pressure dependence of peak energies of three representative samples. From [30].

distribution. The measured energy separation of Γ-like and X-like states at atmospheric pressure for SL's with different layer thicknesses are shown in Fig.22 together with some PL and PLE results. The solid curve is a calculation results based on the Kronig-Penney model and room temperature band parameters

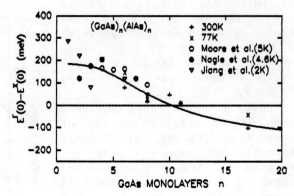

Fig.22 Dependence of the energy separation $E^\Gamma(0)-E^X(0)$ on the layer thickness n. The solid line is calculated results using the K-P model. From [30].

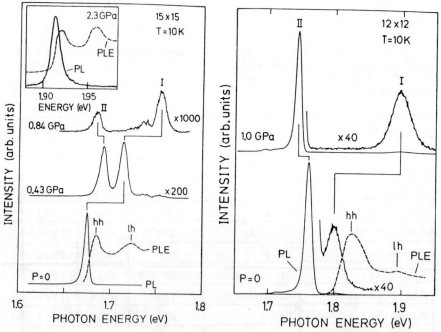

Fig.23 PL spectra of (15,15) and (12,12) at various pressure, The inset shows the PL and PLE spectra at 2.3 GPa. From [31].

of bulk GaAs and AlAs. It is clear that the PL and PLE date can be obtained only in the type-II region but the PL under pressure can provide the energy separation in both type-I and type-II regions.

Holtz et al.[31] have measured the 10K PL and PLE spectra of some $(GaAs)_n(AlAs)_n$ SL's under hydrostatic pressure. The PL spectra at various pressure are shown in Fig.23, where the n=12 sample is type-II SL with peak I (direct) higher than peak II (indirect) and n=15 sample is type-I SL. The pressure dependence of PL peak energy is shown in Fig.24, from which they concluded that the type-I—type-II transition will occur at n slightly less than 14 at 10K.

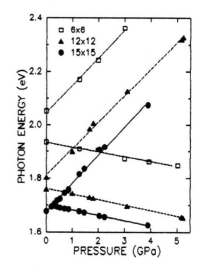

Fig.24 Pressure dependence of the direct (I) and indirect (II) transition energies. From [31].

Table III. Experimental results of the type-I—type-II transitions in $(GaAs)_n(AlAs)_n$ SL's

author	method	tem.	point	Ref.
Ishibashi et al.	Intensity	4K	n=2	[19]
Jiang et al.	PL & PLE	2K	n=13	[20]
Finkman et al.	PL & PLE	1.7K	n>8	[32]
Moore et al.	PL & PLE	5K	n>8	[33]
Kato et al.	PL & AB	20K	n=14	[23]
Fujimo et al.	PL & PR	300K	n=10	[24]
Li et al.	PL under HP	300K 77K	n=11	[30]
Holtz et al.	PL under HP	10K	n=14	[31]
Cingolani et al	PAS	300K	n=12	[26]
Cingolani et al.	HEI	10K	n=12	[26]

The experimentally obtained transition point between type-I and type-II SL's for symmetric $(GaAs)_n(AlAs)_n$ SL's are listed in Table III. It can be seen that the transition point is also in the range of $8 \leq n \leq 14$ except the earlier work of Ishibashi et al.[19]. This range is in agreement to the theoretical results. The divergence of the experimental results may be not only due to the different measurement methods but also due to the fluctuation of layer thickness in the practical samples.

5. Γ-X mixing and X-valley splitting
5.1. Order of X_z and X_{xy} valleys

There are two series of X-like states in GaAs/AlAs short period SL's. One is folded from the X_z valley and at the $\bar{\Gamma}$ point of SBZ, the other is folded from the $X_x(X_y)$ valleys and at the $\bar{M}$ or $\bar{X}(\bar{Y})$ point of SBZ. The transition between holes and X_z electrons can be detected in the PLE spectra since the transition is direct in the k space. Finkman et al.[32] have measured the PL and PLE spectra of a GaAs(19Å)/AlAs(19Å) SL (about 8 monolayer). The results are shown in Fig.25, where a

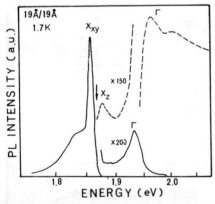
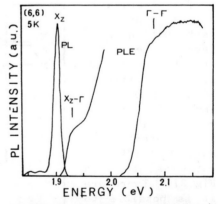

Fig.25 PL(full lines) and PLE(dashed lines) spectra for GaAs (19Å)/AlAs(19Å) SL. PLE spectra were monitored at the wavelength indicated by the arrows. From [32].

Fig.26 PL and PLE spectra of $(GaAs)_6 (AlAs)_6$ SL. Note that there is an increase in the gain of the detection system in the X_z-Γ region. From [33].

week structure observed in PLE spectra is attributed to the X_z-related absorption, which locates at 20 meV higher than PL peak. They attributed the PL peak to the X_{xy}-related emission mainly according to the nonexponential decay behavior and then

included that the X_{xy} is the lowest state. Moore et al.[33] have also observed similar spectra in a $(GaAs)_6(AlAs)_6$ sample, which are drawn in Fig.26. However they identified the PL peak as X_z-related transition according to the K-P model calculation and the increase of the PL peak with the increase of temperature. They then indicated that the X_z state is the lowest state. Thus the order of X_z and X_{xy} states in type-II GaAs/AlAs SL's has attracted much theoretical and experimental attention. Kesteren et al.[34] have investigated the order of the X conduction band valleys in type-II GaAs/AlAs SL's by the optically detected magnetic resonance (ODMR) measurements. They obtained that the X_z conduction band valley is the lowest valley for the SL's with AlAs layer thinner than ~55Å while X_{xy} is the lowest valley for the SL with thicker AlAs layers. They refereed this result to the consequence of the lattice mismatch splitting of the AlAs X conduction band, which dominates the confinement splitting in the SL's with AlAs layers thicker than ~55Å. The similar results are also obtained by Scalbert et al.[3] in their time resolved PL measurements.

The uniaxial stress is an ideal perturbation to investigate the symmetry of X-like state in type-II SL's. Under [001] and [110] oriented stress, the X_z and X_{xy} states will move in different rates as shown in Fig.27. Therefore it is possible to identify the character of the PL peak based on the behavior under uniaxial stress. Lefebvre et al.[35,36] have investigated the piezospectroscopy of some GaAs/AlAs SL's and obtained that in the GaAs(17Å)/AlAs(26Å) SL the X_z state is the lowest conduction valley, which is in agreement with the above results. In addition they found that for the GaAs(4Å)/AlAs(6Å) sample the X_{xy}

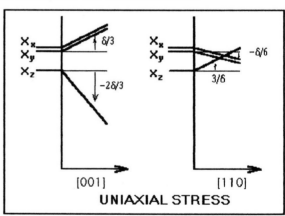

Fig.27 Schematic behavior of the three zone-edge-related X conduction bands under both direction of the stress. From [35].

valley is the lowest state. Ge et al.[37] have also used the PL under uniaxial stress to determine the lowest conduction subband in the $(GaAs)_n(AlAs)_n$ (1≤n≤4) SL's. They obtained that the lowest conduction subband is X_{xy} for n≤3 and is X_z for n=4. These results indicate that for the SL with ultrathin layers, the X_{xy} may be the lowest conduction state.

5.2. Γ-X mixing

The theoretical calculation indicated that the Γ-X crossover will be an anticross when the Γ-like and X-like states have same parity. The anticrossing was observed by Meynadier et al. in their PL investigation under application of an electric field[38]. The sample used for the measurements is a GaAs(35Å)/AlAs(80Å) SL embedded in a p-i-n structure. The SL is designed to be indirect under zero voltage, and can be switched to direct by application of an axial electric fields. The measured PL peak energy under various applied field is shown in Fig.28, where an anticross with about 2.5 meV splitting occurs at the electric field about 4.5×10^4 V/cm². The potential responsible for the Γ-X mixing is therefore of the order of 1 meV.

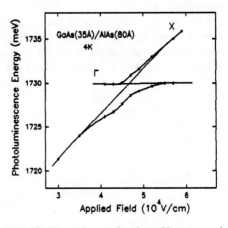

Fig.28 Energies of the direct and indirect transitions at high fields showing the anticrossing between the X and Γ transitions. From [38].

The existence of Γ-X mixing can also be probed by the measurement of the time decay of the indirect luminescence. The perturbation theory told us that the radiative recombination rate of type-II transition ought to be inversely proportional to the square of energy separation of Γ-like and X-like states. Dawson et al.[39] have measured the PL decay time of a series of type-II GaAs/AlAs SL's with 25Å GaAs layers and 19-41 Å AlAs layers. The obtained decay time of the samples with 19, 28 and 41 Å AlAs thickness are 0.135, 1.5 and 6.6 μs, respectively, confirming that the decay time are increasing with the increase of the energy separation of Γ- and X-like

states (47, 117 and 169 meV for three samples). The Γ-X mixing potential deduced from the decay time is about 3 meV. Skolnick et al[40] have investigated the variation of the PL decay time with the Γ-X splitting by application of the hydrostatic pressure. Fig.29 shows the measured decay time as a function of pressure at 2K. A strong increase in decay time with pressure is observed, from ~400 nsec at 3 kbar to ~12 μsec at 36 kbar. The increase is nearly quadratic when the pressure lower than about 15 kbar.

Li et al.[30] have investigated the pressure dependence of the relative intensity of type II PL peaks. The measured results are presented in Fig.30. The decrease of the intensity is approximately proportional to $(E^\Gamma-E^X)^{-2}$ as expected from the perturbation theory. The mixing potential deduced from 77K data are 14, 7 and 1.4 meV for $(GaAs)_n(AlAs)_n$ SL's with n=8, 11 and 17, respectively. It indicated that the mixing potential increases slightly with the decrease of the period.

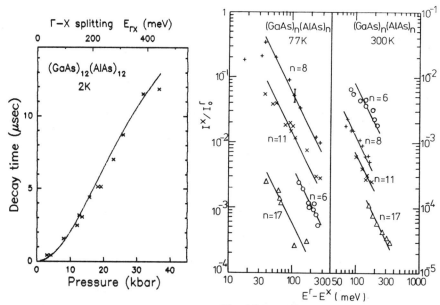

Fig.29 PL lifetime as a function of pressure at 2K. From [40].

Fig.30 Dependence of the relative PL intensity I^X/I_0^Γ on the energy separation $E^\Gamma-E^X$. From [30]

6. Conclusion

All the theoretical investigations, including the effective mass approximation, the various empirical methods and some first principle calculations have shown that two types of SL exist in GaAs/AlAs short period SL. When GaAs and AlAs layer is thin enough, the $(GaAs)_n(AlAs)_n$ SL is type-II SL. On the other hand, the $(GaAs)_n(AlAs)_n$ SL with thick layer become type-I SL. The transition point between type-I and type-II SL's are in the range of $8 \leq n \leq 14$, though the accurate point obtained from different method are somewhat divergent.

The existence of type-I and type-II SL's have been confirmed by many optical measurements. Various optical spectroscopies have been used to investigate the type-I—type-II transition. The majority of the resulting transition points fall in the same range as indicated by theoretical calculations. Again, the results obtained using different methods are somewhat scattered.

Though many theoretical and experimental work have done, there seems still exist some questions left to be clarified.

Acknowledgements

The author would like to express sincere thanks to professor De-Sheng Jiang and Professor Jian-Bai Xia for useful discussions. Financial support from the National Science Foundation of China is acknowledged.

Reference

[1] G.Danan, B.Etienne, F.Mollot, R.Planel, A.M.Jean-Louis, F.Alexandre, B.Jusserand, G.Le Roux, J.Y.Marzin, H.Savary and B.Sermage, *Phys. Rev.* **B35** (1987) 6207.
[2] J.Nagle, M.Garriga, W.Stolz, T.Isu and K.Ploog, *J.Physique* **C5** (1987) 495.
[3] D.Scalbert, J.Cernogora, C.B.Guillaume, M.Maaref, F.F.Charfi and R.Planel, *Solid State Commun.* **70** (1989) 945.
[4] J.Ihm, *Appl. Phys. Lett.* **50** (1987) 1068.
[5] Y.T.Lu and L.J.Sham, *Phys. Rev.* **B40** (1989) 5567.
[6] L.J.Sham and Y.T.Lu, *J. Lum.* **44** (1989) 207.
[7] M.C.Muñoz, V.R.Velasco and F.G.Moliner, *Phys. Rev.* **B39** (1989) 1786.
[8] D.Z.Y.Ting and Y.C.Chang, *Phys. Rev.* **B36** (1987) 4359.
[9] J.B.Xia and Y.C.Chang, *Phys. Rev.* **B42** (1990) 1781.
[10] J.B.Xia, *Phys. Rev.* **B38** (1988) 8358.
[11] M.A.Gell, D.Ninno, M.Jaros and D.C.Herbert, *Phys. Rev.* **B34** (1986) 2416.
[12] M.A.Gell, M.Jaros and D.C.Herbert, *Superlattices and Microstructures* **3** (1987) 121.

[13] M.A.Gell, D.Ninno, M.Jaros, D.J.Wolford, T.F.Keuch and A.Bradley, *Phys. Rev.* **B35** (1987) 1196.
[14] T.Nakayama and H.Kamimura, *J.Phys. Soc. of Jpn.* **54** (1985) 4726.
[15] H.Kamimura and T.Nakayama, *Comments Cond. Mat. Phys.* **13** (1987) 143.
[16] S.Gopalan, N.E.Christensen and M.Cardona, *Phys. Rev.* **B39** (1989) 5165.
[17] S.H.Wei and A.Zunger, *J. Appl. Phys.* **63** (1988) 5794.
[18] S.B.Zhang. M.S.Hybertsen, M.Cohon, S.G.Louie and D.Tomanek, *Phys. Rev. Lett.* **63**, (1989) 1495.
[19] A.Ishibashi, Y.Mori, M.Itabashi and N.Watanabe, *J. Appl. Phys.* **58** (1985) 2691.
[20] D.S.Jiang, K.Kelting, T.Isu, H.J.Queisser and K.Ploog, *J.Appl. Phys.* **63** (1988) 845.
[21] K.J.Moore, G.Duggan, P.Dawson and C.T.Foxon, *Phys. Rev.* **B38** (1988) 5535.
[22] M.Nakayama, I.Tanaka, I.Kimura and H.Nishimura, *Jpn. J. Appl. Phys.* **29** (1990) 41.
[23] H.Kato, Y.Okada, M.Nakayama and Y.Watonabe, *Solid State Commun.* **70** (1989) 535.
[24] H.Fujimoto, C.Hamaguchi. T.Nakazawa, K.Taniguchi, K.Imanishi, H.Kato and Y.Watanabe, *Phys. Rev.* **B41** (1990) 7593.
[25] J.Barrau, K.Khirouni, Th. Amand, J.C.Brabant, B.Brousseau, M.Brousseau, P.H.Binh, F.Mollot and R.Planel, *J.Appl. Phys.* **65** (1989) 350.
[26] R.Cingolani, L.Baldassarre, M.Ferrara M.Lugara and K.Ploog, *Phys. Rev.* **B40** (1989) 6101.
[27] E.D.Jones, T.J.Drummond, H.P.Hjalmarson and J.E.Schirber, *Superlattices and Microstructures* **4** (1988) 233.
[28] K.Takarabe, S.Minomura, M.Nakayama and H.Koto, *J. Appl. Phys. of Jpn.* **58** (1989) 2242.
[29] G.Li, D.Jiang, H.Han, Z.Wang and K.Ploog, *Phys. Rev.* **B40** (1989) 10430.
[30] G.Li, D.Jiang H.Han, Z.Wang and K.Ploog. *J.Lum.* **46** (1990) 261.
[31] M.Holtz, R.Cingolani, K.Reimann, R.Muralidharan, K.Syassen and K.Ploog, *Phys. Rev.* **B41** (1990) 3641.
[32] E.Finkman, M.D.Sturge and M.C.Tamargo, *Appl. Phys. Lett.* **49** (1986) 1299.
[33] K.J.Moore, P.Dawson and C.T.Foxon, *Phys. Rev.* **B38** (1988) 3368.
[34] H.W.van Kesteren, E.C.Cosman, P.Dawson, K.J.Moore and C.T.Foxon, *Phys. Rev.* **B39** (1989) 13426.
[35] P.Lefebvre, B.Gil, H.Mathieu and R.Planel, *Phys. Rev.* **B39** (1989) 5550.
[36] P.Lefebvre, B.Gil, H.Mathieu, R.Planel, *Phys. Rev.* **B40** (1989) 7802.
[37] W.Ge, M.D.Sturge, W.D.Schmidt, L.N.Pfeiffer and K.W.West, *Appl. Phys. Lett.* **57** (1990) 55.
[38] M.H.Meynadier, R.E.Nahory, J.M.Worlock, M.C.Tamarga, J.L.de Miguel and M.D.Sturge, *Phys. Rev. Lett.* **60** (1988) 1338.
[39] P.Dawson, K.J.Moore, C.T.Foxon, G.W.'t Hooft and R.P.M.van Hal, *J. Appl. Phys.* **65** (1989) 3606.
[40] M.S.Skolnick, G.W.Smith, I.L.Spain, C.R.Whitehouse, D.C.Herbert, D.M.Whittaker and L.J.Reed, *Phys. Rev.* **B39** 1989 11191.

PHOTOLUMINESCENCE STUDIES OF INTERFACE ROUGHNESS IN GaAs/AlAs QUANTUM WELL STRUCTURES

D. GAMMON, B.V. SHANABROOK, and D.S. KATZER
Naval Research Laboratory
Washington DC, 20375 USA

Abstract

We review the use of photoluminescence to characterize the structural disorder of the interfaces in GaAs/AlAs quantum well structures. In the highest quality samples structural disorder exists as monolayer-high islands. We show the types of possible luminescence spectra which can occur in quantum wells in which the bottom and top interfaces have differently-sized islands. It is shown how luminescence spectra are simultaneously sensitive to both large and small island structures — whether they occur on the same interface or occur separately on the top and bottom interfaces.

I Introduction

The exciting possibility of artificial semiconductor structures with atomic perfection remains a goal of today's semiconductor science and technology community. In the pursuit of this goal, considerable attention has been focussed on the (001) GaAs/AlAs hetero-interface, and in particular, on the roughness at the interface (IF). This is in large part because the GaAs/AlAs system is at the most advanced state of growth, and also because it has served as the medium for an enormous amount of fundamental and applied studies of semiconductor heterostructures. Furthermore, there is very little strain between GaAs and AlAs, resulting in the absence of associated complications. As a result, this system now represents to a considerable degree the prototype for the growth, the physics, and the technology of semiconductor heterostructures. However, in spite of the relatively advanced state of growth and the large amount of work involving this heterostructure, our knowledge of the IF is still modest. Even our *structural* picture of the IF is incomplete. In fact, there has been considerable surprise and controversy over experimental results concerning the qualitative structure of the disorder at the IF in the highest quality samples which can be grown today[1, 2, 3, 4, 5].

What is known about the IF roughness is due in large part to luminescence studies in which the exciton confined to a quantum well formed by two GaAs/AlAs IFs serves as a probe. The use of, and the results obtained with, the luminescence techniques (photoluminescence and cathodoluminescence) have been reviewed up to the year 1989[6]. In the present paper we review the current state of understanding as obtained through photoluminesence studies. To this end we focus on our own photoluminescence work which has been done mostly in the period since then. We clarify and illustrate through examples of experimental results some of the ideas of what we will call the standard island model. This model assumes that the highest quality IFs consist of approximately one monolayer of disorder. It is thought that under suitable growth conditions the atoms on the growing surface migrate long distances to the edges of islands. In this way the crystal grows layer by layer. An IF is formed when the Ga and Al fluxes are interchanged. Thus, the IF is characterized by a combination of Ga and Al islands, and in this high quality regime, disorder is defined in terms of the sizes of these islands. This model of layer by layer growth is supported by reflection high-energy electron diffraction (RHEED) studies in which oscillations of the RHEED spots are observed with the same period as a monolayer deposition time[7, 8]. In addition, very good evidence for the existence of islands is provided by luminescence studies[9]. The luminescence evidence is especially valuable in that it shows that the island sizes can become large — much larger than the exciton Bohr diameter (200Å). This evidence, which will be discussed in this paper, is the observed splitting of the exciton luminescence peaks from GaAs/AlAs quantum wells (QWs) into multiplets with energy splittings which correspond to approximately one monolayer difference in QW width. There also have been cathodoluminescence studies which have mapped out the islands, indicating that the islands can be microns in diameter[6].

The standard island model is successful in explaining many experimental observations. But it is somewhat incomplete as it was described above. This was first suggested by results from a group using a type of high resolution transmission electron microscopy (which was called chemical lattice imaging [CLI])[1]. In their study, a sample, in which large island structures were known to exist from earlier luminescence experiments, was shown to have also a significant amount of small scale structure at the IFs. The results of this experiment make it clear that the islands have a broader distribution in sizes than has been generally appreciated. In fact, it has been suggested [2] that the distribution is bimodal, with a peak in the distribution at very large island sizes and also one at very small sizes with a minimum in the vicinity of the exciton. Whether or not the distribution takes this form or not, it seems clear that it is necessary to describe the IF in terms of an island *distribution*. This is also clear from photoluminescence and Raman scattering measurements [5] which were recently used to establish the simultaneous existence of both large and small scale structure at the IF. The luminescence results will be described in detail in this paper.

Other possible complications which are often ignored is the deviation from perfect (001) orientation of the sample (*i.e.*, the extent to which the surface is vicinal), and also the anisotropy of the surface which is present even in samples which have

negligible offset from (001). Vicinal surfaces have a series of terraces which can become quite narrow even for small offsets from (001). For example, a 0.5 degree offset corresponds to an average terrace width of 300Å. This is comparable to the exciton diameter, and would force the IF to appear *rough* to the exciton[10, 11]. Presumably, samples in which the luminescence spectra show exciton multiplets (implying that the island sizes are much larger than the exciton diameter) require that the offset be much smaller than 0.5 degrees. However, the observation with cathodoluminescence imaging of extremely large island diameters ($8\mu m$) implies the incredibly small offset of 0.002 degrees[12]. Furthermore, one cathodoluminescence group reports evidence of terrace step bunching[12]. We will not consider these complications, and assume that in our samples the offsets from the (001) orientation are much smaller than 0.5 degrees. Even in perfectly oriented surfaces, however, the (110) and the ($1\bar{1}0$) directions have different bond configurations [13]. This leads to an anisotropic surface migration of the atoms, which in turn could cause the islands to be rectangular. Again, we will not consider this possibility and assume that both axes of the rectangle can become larger than the exciton diameter. We believe that these assumptions are reasonable in view of the consistent production of samples which show exciton multiplets in the luminescence spectra. Nevertheless, these are important characteristics which will certainly receive further study.

In the remainder of the paper we will discuss our own results. We will begin in Sec. II by describing the growth technique. In Sec. III we show the types of exciton spectra that are obtained with different types of IFs. In our own work we have separately controlled both interfaces through growth interrupts, and thus we are able to illustrate the spectra which can occur with a variety of two different IFs[14, 15]. This discussion is presented within the standard island model. However, an IF can have simultaneously fine scale structure and large island structures, a possibility which has been confirmed with luminescence spectroscopy[5]. We review this work in Sec. IV.

II Growth details

The GaAs/AlAs quantum well structures discussed in this paper were grown in a Vacuum Generators V80H molecular beam epitaxy (MBE) system. Our MBE system, it must be emphasized, does not have an exceptionally high-quality vacuum. In fact, during most of this study the residual gas analyzer (RGA) gave indications of an oxygen peak at mass 32 and a high background water level (the 18 peak was as much as twice as high as the 28 peak) even when the shrouds have been cooled for a substantial period of time. A 32 peak in an RGA spectrum is a "smoking pistol" indication of an air leak in the system[16], and high background levels of water and O_2 in the vacuum are reported to degrade AlGaAs and AlAs films grown by MBE[17]. However, no leak was ever found even with very careful helium leak checking and testing of the RGA by the manufacturer. In addition, the base pressure of the system was routinely below 5×10^{-11} mbar after a 3 day bakeout at 200 C. Since the growth

of the samples discussed here, we have upgraded our system with the addition of a cryopump and several new Knudsen cells. Since the upgrades, the water signature has practically vanished but the 32 peak remains. We do not yet know if there was a water leak in one of the replaced water cooled Knudsen cells, or if the cryopump is simply pumping the residual water better than the shrouds. We also do not know if the 32 peak is an artifact of our particular RGA. In short, although we have been able to grow state-of-the-art GaAs/AlAs quantum wells, we do not believe it is a consequence of exceptional vacuum quality.

The MBE system was pumped by 400 liter/s ion pumps, titanium sublimation pumps, and liquid nitrogen cooled cryoshrouds during the time the samples discussed here were grown. The base vacuum in the growth chamber, measured by an ion gauge outside the cryoshrouds near the inlet of the growth chamber ion pump with all relevant sources, except arsenic, at the growth temperature was below 2×10^{-9} mbar. The system had the original (non-conical) Knudsen cells and magnetically-coupled shutters. To conserve liquid nitrogen, and to lessen the danger of the aluminum cell cracking if the power to the cell was lost[18], we let the shrouds warm to room temperature at the end of the day. Dry nitrogen was blown through the shrouds to minimize water condensation inside the shrouds.

The samples discussed in this study were (001) oriented undoped semi-insulating GaAs from M/A-COM and Sumitomo. Small indium-bonded pieces of the substrates, or quarters of 3" diameter wafers were used in 2" indium-free mounts. Sapphire backing wafers were used in the indium-free mounts. The substrates are prepared in a fairly standard way: a 15 minute degrease of detergent in 18 MΩ DI water with ultrasonic agitation; a 2 to 5 minute running DI rinse; a 2 minute soak in acetone with ultrasonic agitation; a 2 minute soak in methanol with ultrasonic agitation; and a 5 minute running DI rinse. The wafers were then dried with concentrated sulfuric acid for 1 minute. The acid was then decanted and the wafer was etched with room temperature 7:1:1 (concentrated sulfuric acid:30% hydrogen peroxide:DI water) for 3.5 minutes. The etch was quenched in running DI water to minimize uneven etching when decanting. The samples were rinsed thoroughly and spun dry, then mounted in the wafer mounts.

The source materials were the purest we could find. 99.999999% (8-nines) pure gallium from Alcon, 5-nines aluminum from IHT, and a 7-nines slug of arsenic from Furukawa were used for the source material. To be conservative, we generally loaded the gallium source material in a new, unetched pyrolytic boron nitride crucible when we loaded new gallium. We always used a new crucible when recharging the aluminum cell. The crucibles for the gallium, aluminum and arsenic sources were outgassed at 1200 C for 1 hour before loading the source material.

The substrate temperature was estimated by noting the oxide desorption temperature (580 C) [19] in RHEED. We consider the appearance of the 2x4 reconstruction in RHEED within 1 - 2 minutes at a fixed temperature as being an indication of the substrate being at 580 C. We also took the 2x to 4x transition temperature when the arsenic shutter was closed as being indicative of 640 C. That is, if the 2x reconstruction changes to the 4x reconstruction in 3 seconds after the arsenic shutter is

closed, then we would estimate that the true temperature is near 640 C. (Note that the 2x to 4x transition temperature is a function of the arsenic flux.) In addition, optical pyrometry using an Ircon Series V pyrometer gave similar results with an emissivity setting of 0.700. Although temperature measurements precise to 1 C are possible with this instrument, we were limited by the large measured spot size (2.3 cm in diameter) of our particular instrument. Thus, the pyrometer was susceptible to stray light from the substrate heater on the indium-free mounted samples, and sampled the molybdenum block on the indium mounted samples. From the reasonable agreement between the techniques discussed above (+/-10 C), we estimate that our temperatures are accurate to +/-10 C.

All of the samples grown for this study were grown at temperatures much lower than traditionally used for GaAs/AlAs or GaAs/AlGaAs quantum wells. Although we have not done an exhaustive study of the substrate temperature's effect on the quantum well quality, we have found that it is not necessary to use temperatures greater than about 580 C to achieve excellent quality GaAs/AlAs single quantum wells. Recently we have presented data that indicate that we are able to grow nearly optically-equivalent quality GaAs/AlAs quantum wells with substrate temperatures between about 500 C and 610 C [20]. In addition, we use growth rates which are much lower than those traditionally used for MBE growth of heterostructures. We generally used GaAs growth rates of 0.3-0.4 monolayers/s (mL/s) and AlAs growth rates of 0.2 - 0.3 mL/s. The use of low growth temperatures and low growth rates have several advantages for the growth of high-quality structures: 1) excessive outgassing of the regions surrounding the Knudsen cells and the substrate manipulator is minimized, leading to improved vacuum quality; 2) outgassing of the contaminants of the gallium and aluminum source materials is minimized; 3) diffusion of impurities and defects from the substrate is minimized; and 4) control of shutter sequencing is easier using low growth rates since the shutter timing does not then need to be controlled with great accuracy. The use of low growth rates does, however, require that the vacuum be pure enough so that impurities incorporated in the growing film from the vacuum environment do not degrade the material.

We originally believed that the low arsenic-to-gallium and arsenic-to-aluminum flux ratios were very important for achieving high-quality material [15]. Recently, however, we have presented data which indicate that this ratio can vary substantially across the substrate and yet give quantum wells with nearly equivalent quality over most of the sample as measured by photoluminescence spectra [5, 20].

Although much of the early literature on the growth of GaAs/AlGaAs quantum wells indicated that smoother interfaces can be achieved using growth interruptions, it was also stated that growth interruptions caused degradation of the optical quality of the quantum wells due to increased impurity incorporation[6]. Although this may have been the case in the past, it has not seemed to be a limiting factor in our quantum wells. To illustrate this fact we grew a GaAs/AlAs quantum well structure and interrupted the growth for 3 minutes in the center of the GaAs quantum well region. The low temperature photoluminescence of this QW was as good or better than a similar structure grown without the interrupt in the center of the well [15].

While this is not conclusive proof that the highly reactive AlAs surface is not degraded during long growth interrupts, we have not seen any evidence such as optically active impurity centers visible in the PL of these samples (which used growth interrupts up to 2 minutes in duration), unlike the QWs discussed in Ref. [6].

During the course of this study, we grew a series of GaAs/AlAs quantum well samples and varied the interrupt time at the bottom (GaAs on AlAs) interface. Using RHEED observations during the growth of these samples, we were able to correlate the intensity and duration of the RHEED intensity oscillations in each step of the growth on a particular sample. Since the diffraction conditions did not change during the growth, we could say with authority that the changes in the RHEED observations were a direct consequence of the growth conditions. This *in situ* work, discussed in more detail in Ref. [15], coupled with *ex situ* photoluminescence measurements permitted us to make an unambiguous determination of the bottom interface smoothness as a function of interrupt time. This work clearly demonstrated that, contrary to previous assertions, it is possible to achieve AlAs surfaces with large islands simply using growth interruptions at 580 C. We thus showed that migration-enhanced epitaxy [21], substrate temperature switching [22], or other more complicated techniques are not required to achieve bottom interfaces with large islands.

III Excitons as probes of the IF

Perhaps the most powerful probe of the structural disorder or roughness of the interface (IF) in GaAs/AlAs heterostructures is the exciton. The exciton is easily measured with photoluminescence (PL) spectroscopy [6]. With PL one excites a quantum well with a laser energy which is above the band gap energy of the quantum well. The resulting luminescence is dominated by the recombination of free excitons confined to the quantum well. Note that we are considering only *type I* excitons (*i.e.*, an exciton consisting of a Γ electron and hole both confined to the GaAs QW). The exciton is shifted up in energy from the bulk GaAs value by the confinement energy of both the electron and hole. In addition, the increased wave function overlap of the electron and hole increases the binding energy of the exciton, and also the recombination rate. It is found that, unlike the case of bulk GaAs, the PL intensity from the free excitons dominates that involving impurities in nominally undoped QWs. As mentioned in the previous section this remains true even in QW samples with long growth interrupt times at the IFs [23].

Because the exciton confinement energy depends on the width and shape of the quantum well, the exciton's spectral energy and lineshape will contain information on the interface roughness. Moreover, when the quality of the IF becomes such that the width of the interface in the growth direction becomes limited to approximately one monolayer, the exciton becomes uniquely important as a probe of the scale of the roughness in the lateral dimensions within the monolayer. This is possible because the exciton averages over structure within the plane of the QW which is much less than the Bohr diameter (200 Å) [24, 25]. However, interface structure which is comparable

in size creates a nonhomogeneous broadening. Thus, the exciton provides a lateral length scale (Fig. 1), making it possible to characterize the lateral size of the disorder.

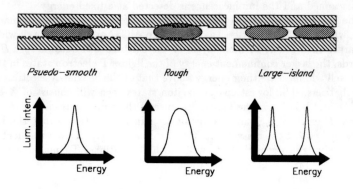

Figure 1: A schematic illustration of the standard island model. The PL classification scheme includes the a) *psuedo-smooth*, b) *rough*, and c) *large-island* regimes.

This is very important since modern growth techniques are at the state where further improvements in the quality of the IF involve increasing the size of the lateral structure (island sizes). As illustrated in Fig. 1, the natural length scale of the exciton has given rise to a useful classification. When the island sizes on both interfaces in the QW are much smaller than the exciton diameter, the exciton averages over the roughness, and the exciton spectrum has a single, narrow peak. This type of IF is called *psuedo-smooth*. However, when the island sizes become much larger than the exciton diameter the exciton spectrum splits into two or three narrow peaks, each corresponding to a region of the quantum well in which the well width has been assumed to be an integral number of monolayers. We will call this the *large-island* regime. When the islands become comparable in size to the exciton the spectrum shows a single, broad peak because of the wide distribution in energy states available to the exciton. This regime is called *rough*. This classification is useful, but the nomenclature is somewhat simplistic. In particular, as discussed below, even when the IFs are in the *large-island* regime there may still be considerable small scale disorder within the large islands, and the IF is not abrupt. Furthermore, in this discussion we have assumed that the two interfaces have the same character. In fact the two IF in the QW may be any combination of the types of interfaces, and even within this simple model the possible luminescence spectra will be more complicated [14]. These possibilities will be discussed in more detail later. When there is a roughness-induced

distribution in exciton energies, thermal occupation effects must be considered. It should be kept in mind that the excitons sensed in PL are thermalized to the lower energy states of the energy distribution. As a result the peak measured in PL may be shifted down in energy from the peak in an absorption measurement such as that provided by photoluminescence excitation spectroscopy (PLE) [6]. In PLE the laser energy is scanned and the luminescence is detected at a fixed energy.

For the highest sensitivity to the interface roughness the well width is made small. However, with pure AlAs barriers (as opposed to AlGaAs) the GaAs well width must remain larger than about 14 monolayers or the excitons will become *type II* [26]. In other words, the larger confinement energy of the lighter Γ electron states in the GaAs quantum well will increase their energy above that of the heavier X electron states in the AlAs barriers. The lowest energy exciton states then will consist of X electrons in the AlAs and Γ holes in the GaAs. Because of the complications in the character of the *type II* exciton, it is more difficult to gain information about the interface. Furthermore, because the mass of the X electrons are larger, the confinement energies are smaller, giving less sensitivity to well width fluctuations. To our knowledge there have not yet been investigations into the nature of the interface using lineshapes and energies of *type II* excitons comparable to investigations involving *type I* excitons.

As mentioned above, the lineshapes of the exciton peaks in the PL spectra have been very valuable in determining that large islands can exist at the interface. It is also generally assumed that the height of the islands is one monolayer, and that the extent of the disorder in the growth direction is confined to one or two monolayers. Otherwise it is difficult to conceive of islands forming at all otherwise. More information is obtained from PL with a technique which was recently introduced [14]. In this technique the sample is grown such that the average GaAs layer thickness varies considerably across the sample due to cation nonuniformity. With such a sample the laser spot can be scanned across the sample in small steps, producing a set of spectra in which the well width changes. Such a set of spectra is shown in Fig. 2 for a nonhomogeneous sample in which a 200μ m laser spot was moved in 300μ m steps across the sample.

This sample was grown with long growth interrupts at each interface which produced large islands such that both IFs are in the *large-island* regime. Thus, each spectrum shows a doublet in which the splitting corresponds to a difference in well width of one monolayer. PLE spectra performed on this class of samples do not reveal any noticeable Stokes shifts relative to the peak energies measured in PL. As the average well width is changed by moving the laser spot across the face of the sample, the energy spectrum changes. As we will show, the way that the spectrum changes can allow one to tell if there are differences between the two interfaces in the QW. In this sample the spectrum changes in a quantized way. In other words, as the well width is decreased the higher energy peak of the doublet increases in intensity at the expense of the lower energy peak until the lower energy peak disappears and a new peak starts to grow about one monolayer higher in energy. This happens with very little change in the energy of the peaks. Thus, there are a discrete set of exciton energies associated with this QW. This is evidence that both interfaces are in the

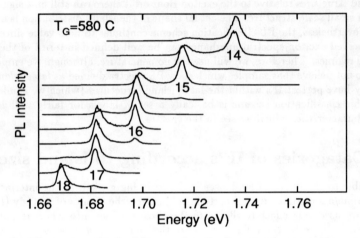

Figure 2: PL spectra obtained by moving the laser spot across the sample. The spectra from this QW illustrate the quantized exciton energies which are indicative of two *large-island* IFs (obtained through long growth interrupts).

large-island regime. We will discuss this in detail below, but first we make another point. The sample of Fig. 2 contains pure AlAs barriers. If the IF islands are much larger than the exciton diameter (which they are) and if there is no other disorder, the absolute exciton energies should be determined only by the fundamental material parameters of GaAs and AlAs. Thus the set of exciton energies from a given QW should be the same as that from any other QW on the same sample, or any other sample. Note that if the barriers consist of the alloy, AlGaAs, this is not true because of uncertainties in the Al concentration. We find that the absolute exciton energies often do match in different QW's grown under the same conditions. However, by changing the growth conditions the QW's can be grown such that there are significant shifts in absolute energy without changes in linewidths [5]. We show below that this is very good evidence for the existence of microroughness on top of the islands as first reported by Ourmazd, et.al. with CLI measurements [1]. We emphasize that by measuring the absolute exciton energies the PL technique gains sensitivity to the short scale structure of the IF in addition to the large scale island structure.

We conclude this section with another caution on nomenclature. The lineshapes in luminescence spectra are sensitive to how the sizes of the largest island structures at the IF compare to that of the Bohr orbit of the exciton. Therefore, it is useful when characterizing the IF with PL to use the *psuedo-smooth, rough* and *large-island* classification. However, in general, the lateral structure must have a distribution in sizes. The PL classification scheme tends to imply that this distribution is a single-peaked function with a fairly narrow width. This is not necessarily true. In fact, PL and Raman results [5] and the earlier TEM results [1] show that even in IFs with

large island structures relative to the exciton Bohr orbit there can still be a significant amount of small scale structure. The actual shape of the island distribution is not yet known. Nevertheless, the PL classification scheme continues to hold value since most of the observed exciton spectral lineshapes can be well defined in terms of the large island size regimes. Therefore, we will use the nomenclature, although, we emphasize that we do not believe that samples which show PL spectra defined as *large-island*-like necessarily have perfect IFs within the large island structures (which we think must exist). This classification scheme is basically a set of labels for fairly well-defined spectral characteristics which we see in the spectra.

IV Catagories of IF's according to island sizes

It is possible to change the island sizes by interrupting growth at the interfaces for different amounts of time. In this section we describe the results of a study [14, 15] in which we used this effect to illustrate the three regimes into which island sizes are naturally divided by the Bohr diameter of the exciton: *psuedo-smooth*, *rough* and *large-island*. Furthermore, we illustrate the types of exciton spectra which occur for the different combinations of two IF's. In Fig. 3 are shown the spectra obtained from four samples in which the interrupt times at the two IF's of a single quantum well structure were varied. The four samples have interrupt times at the (bottom, top) IF of (0,0), (0,90), (5,90), and (30,90) seconds, respectively.

In Fig. 3a are shown spectra from a QW in which there was no interrupt at either IF. The different spectra were obtained by moving the 100 μm laser spot across the sample in increments of 300 μm. In each spectrum there is a single, broad peak. As the average well width is decreased by moving the laser spot across the sample this peak moves continuously to higher energies. This behavior can be compared to that of the (0,90) QW in which the growth was interrupted at only the top IF for 90 s as shown in Fig. 3b. Each spectrum now shows a doublet in which the splitting corresponds approximately to a difference in well width of one monolayer. However, as the well width is decreased the peaks in the spectra from this sample show a continuous shift to lower energy superimposed on an upward quantized movement (previously described for the spectra in Fig. 2 above). We interpret the spectra from Figs. 3a and 3b as follows. The broad single peak in the spectra from the (0,0) sample in Fig. 3a implies that at least one of the IFs is *rough*. When the top IF is interrupted, but not the bottom, as for the (0,90) sample in Fig. 3b, a doublet is obtained with narrow linewidths. From this behavior we deduce that the top interface has become *large-island*-like due to the long growth interrupt. However, the peaks in this doublet are narrow and show a continuous shift. We conclude that the bottom IF, without any interrupt, is *psuedo-smooth*. This implies that the QW grown without any interrupts has a *psuedo-smooth* bottom IF and a *rough* top IF. If the laser spot is moved along a different direction on the sample, the continuous shift is found to go in the same way as the discrete shift — to higher energy. The sign of the continuous shift relative to the quantized shift apparently depends on the slope of the *psuedo-smooth* bottom IF

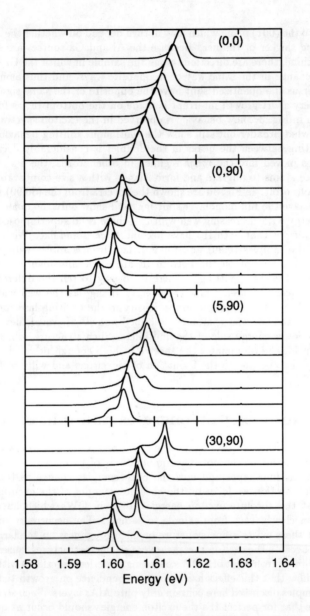

Figure 3: PL spectra from four samples in which the growth interrupts on the (bottom, top) interfaces are as follows: (a) (0,0) s; as discussed in the text the IFs are (*psuedo-smooth, rough*) (b) (0,90) s; (*psuedo-smooth, large-island*), (c) (5,90) s; (*rough, large-island*), and (d) (30,90) s; (*large-island, large-island*).

with respect to the (001) plane which may in turn depend on whether the AlAs barrier layer is getting thicker or thinner. Because the Al and Ga sources are separated in the MBE machine, there are directions along the sample in which the GaAs and AlAs layer widths change in the same way or in opposite ways, and thus along which the continuous versus the quantized shift have the opposite or the same sign.

When a very short growth interrupt is made on the bottom IF, as for the (5,90) sample shown in Fig. 3c, new behavior is observed in the exciton spectra. The peaks become somewhat broader but still show the continuous shifting behavior. However, now the splitting between the peaks in the doublet is considerably decreased. The bottom IF has moved into the *rough* regime. In other words, the 5 s interrupt has allowed surface atoms to migrate and form islands with a size comparable to that of the exciton Bohr orbit. In Fig. 3d are shown the spectra from the (30,90) sample. The doublets now have the full monolayer splitting, and show only the quantized upward shifting in energy with decreasing well width — there is no superimposed continuous shift. This indicates that both IFs have become *large-island*-like. We note that the spectra from the samples (0,90) and (30,90), i.e. from samples with no interrupt and with a long interrupt on the bottom IF are essentially identical with respect to linewidths and energy splittings. There is no way to tell the difference between the two cases except by seeing how the spectra change as the well width changes. We also note that with increasing power density in the exciting laser beam there is no difference in the spectral character described above. However, new features do appear on the lower energy side of the multiplets. Such structure, which barely can be observed in Fig. 3d, dominates the spectra at high enough power densities. This effect is believed to be due to the formation of biexcitons and will not be discussed further.

V Absolute exciton energies and microroughness

In the previous section we catalogued the types of exciton spectra which occur in narrow QWs in terms of the possible combinations of two IFs. This classification involved both the lineshapes and how the peak energies change as the well width changes. In this section we focus on the absolute energies of the exciton peaks. We will show that the absolute exciton energies, even in QWs which have two *large-island* IFs, can clearly vary from sample to sample. The occurrence of such shifts indicates that there exists a fine structure or microroughness on the large islands at the IF which, because the exciton averages over this microstructure, causes the exciton energy to be shifted from that of a QW consisting of perfect interfaces within the large islands. We show that this effect has a strong dependence on growth temperature.

All the samples discussed here contain only pure AlAs layers. Such structures have the advantage that for perfect IFs the exciton energies should occur at quantized energies which depend only on the discrete number of monolayers of GaAs within the QW, and on the fundamental properties of GaAs and AlAs. We observe that QWs which are grown with long growth interrupts at both IFs often show the spectral

characteristics seen in Figs. 2 and 3d. In particular, the spectra show doublets with very narrow components. Furthermore, when the laser spot is scanned across a non-homogeneous sample so that the average well width changes, the spectra change in a quantized fashion. From each QW there are a discrete set of exciton energies. It is very compelling to take this as evidence that the excitons exist in regions of the QW which have perfect IFs, and that these sets of exciton energies are the fundamental energies of the perfect GaAs/AlAs QW. We show below that this is an incorrect conclusion. However, if a structure is grown in which there are three independent QWs (*i.e.*, separated by thick GaAs regions), we find that the three sets of exciton energies from the three QWs are nearly identical. The spectra from such a sample is shown in Fig. 4. In this sample, the three QWs have QW widths which are in the ratio of 2:1.5:1. Thus the doublets from the three QWs are well separated in energy in any given spectrum. By sweeping the laser spot across the sample, however, a set of quantized exciton energies can be generated for each QW. There is a region in energy where these sets overlap (shown by the arrows in Fig. 4). In fact, we find that the energies of the excitons from the three QWs line up to within the linewidths.

Before continuing, there are two additional points which should be made about these spectra. First of all, this set of data was obtained by moving the laser spot across the entire length of a 3.5 cm sample. We often find that on such a large sample there are regions where the spectra indicate that one IF has become *rough*. Such spectra are seen in the top left hand corner of Fig. 4. We suggest a possible explanation for this as follows. The island sizes are a function of the cation surface mobility which in turn is a function of cation to anion ratio (among other parameters). In the MBE machine both the cation and the anion fluxes are nonhomogeneous, and the ratio varies across the sample. Thus, a possible explanation for the appearance of *rough* spectra in an otherwise *truly-smooth* sample involves a reduction in the largest island sizes from much larger than the Bohr orbit of the exciton down to approximately the same size because of an increasing anion to cation ratio. Apparently, there is one region of this particular sample where one of the IFs has become *rough*. Note that we are able to interpret this behavior in terms of the standard island model. Secondly, we point out that the excitons go through the *type I* to *type II* transition at the highest energies shown in this figure. This is accompanied by the appearance of new features (*type II* excitons) and abrupt decreases in intensities. As in all the figures in this paper, each doublet in each spectra has been independently normalized so that the changes in intensity are not shown. These features will not be discussed further.

The spectra in Fig. 4 were taken from a sample which was grown at $T_G = 580$ C (including all three QWs). A second sample which is identical except that the growth temperature is higher at $T_G = 610$ C shows very similar behavior: the spectra show the *large-island* behavior described in the previous section, and the exciton energies from the three independent QWs on this sample are the same to within the linewidth. However, as shown in Fig. 5 there is a significant shift of the absolute exciton energies between the two samples grown at two different temperatures. This shift occurs with no significant change in lineshapes, and gives convincing evidence that there is indeed a significant amount of fine structure within the large island structure.

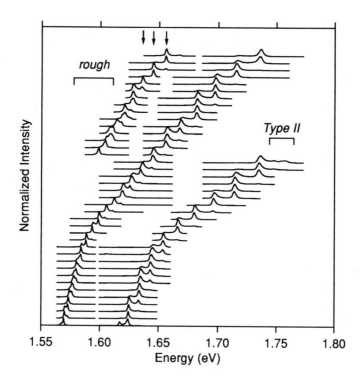

Figure 4: PL spectra from a sample with three independent QWs grown at a constant growth temperature, $T_G = 580$ C. The quantized exciton energies from the three QWs line up in energy.

The shift, which is about 40% of the monolayer splitting, is constant across the entire well width range which has been measured. Note that because the shift tracks with the monolayer splitting it is good evidence that the origin of the shift is due to a difference in IF disorder and not to some morphological effect such as a difference in strain between the GaAs QWs. An effect such as strain would show up as a constant shift in absolute energy independent of QW width.

Further evidence that the shifts are due to a difference in microroughness density due to a change in growth temperature is given by the following result. Additional samples were grown which were identical to the two samples described above except that the growth temperature was changed during growth. The first two QWs on each sample were grown at approximately $T_G = 580$ C. The growth temperature was than changed by a small amount to a new value during the growth of the thick GaAs buffer layer before the third QW. We find in this type of sample that the exciton

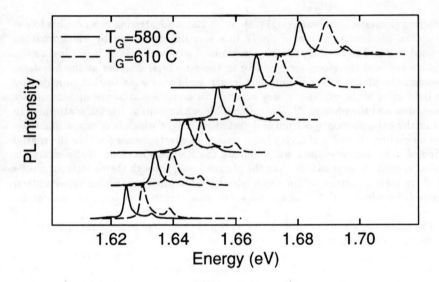

Figure 5: Two sets of PL spectra obtained from QWs from two samples grown at different temperatures, T_G.

energies from the first two QWs consistently line up with each other, but are shifted from the third QW (which was grown at a different temperature). From these results we conclude that within the large island structures (which must exist at both IFs because of the splitting and quantization of the exciton spectral features) there also exists a significant amount of structure which is much smaller than the exciton's Bohr orbit. Thus, the existence of large island structure as required by monolayer splittings and quantized exciton energies, or by other experimental evidence (i.e. cathodoluminescence imaging) is not sufficient to prove that the IFs are perfectly abrupt. In fact, there can exist large *psuedo-smooth* islands.

VI Conclusion

In this paper we have illustrated how photoluminescence can be used to characterize the interface roughness. We have not attempted to be complete, and have restricted our discussion considerably. There has been an enormous amount of research involving luminescence over the last decade which has attempted to provide a description of the interface, and coupled with growth studies, of how to modify it. For example, there have been cathodoluminescence experiments which have provided images of the large scale island structure [6]. There has been also considerable effort with time-resolved PL [23]. This technique provides additional information on how the excitons relax and move between and within the islands [27]. Another optical technique is Raman

scattering (inelastic light scattering) [28]. With Raman scattering it is possible to use the optical phonons as probes of the IF in a way similar to that by which excitons are used in PL. The optical phonons are confined within the GaAs layers just as the excitons are, and therefore, are sensitive to the structural disorder at the interface. However, the phonons lack an intrinsic length scale analogous to that provided by the Bohr orbit of the exciton. A way around this limitation is to use simultaneously the exciton and the phonon [5]. This is possible with resonant Raman scattering. In this method the phonon spectrum is detected with light which is resonant in energy with an exciton. Because of a very large resonance effect, phonons within the spatial region of the resonant exciton will dominate the Raman spectrum. If the exciton is chosen to exist on large islands than the phonons will, through their spectrum, probe the short scale structure within those islands. This provides an alternative optical approach by which both the large and short scale structure of the interface can be studied.

The existence of the microroughness on top of even large island structures creates a difficulty in determining the QW width. One might hope from the quantization of the exciton energies in some samples (Figs. 4 and 2) that it may be possible to use the exciton energies to precisely determine the well width. Of course, the shifting of the exciton energies from sample to sample creates an uncertainty in the energy, and obviously, because of the roughness the well width is not well defined to begin with. One approach that might be taken is to use the number of oscillations measured with RHEED during growth to define the width of the layer in monolayers. The exciton energy can then be calibrated. The result of such an approach is shown in Fig. 6. The exciton energies were all measured from a single QW. Another sample was grown to be as homogeneous as possible and the number of RHEED oscillations was recorded. The PL spectrum then provided the exciton energy to associate with this QW width, and the horizontal axis of the plot was shifted to line up the other set of data with this point. We emphasize that this QW width definition is arbitrary and of only limited use.

The question remains as to the character of the IF (*i.e.*, the form of the island distribution) in the highest quality IFs. Furthermore, what is it about the growth process that makes it that way. It seems to us that neither one of these questions can be answered very well yet. The splitting of the exciton spectral features by energies corresponding to monolayer differences in QW width, and also cathodoluminescence images, provide definitive evidence that islands can become very large. It also seems very clear that there can exist a significant density of very small islands. However, the form of the overall distribution is still unknown. The difficulty is that the fields of view of available techniques are very limited. PL is limited by the size of the exciton. It can only determine whether that structure is much larger or much smaller. Cathodoluminescence can image only the very large islands. On the other hand chemical lattice imaging can only image small areas ($\leq 200\text{Å}$).

The growth mechanisms which dominate are also not well known. It is clear that the atoms can migrate large distances on the growing surface under appropriate growth conditions. This surface diffusion length is determined in part by the growth

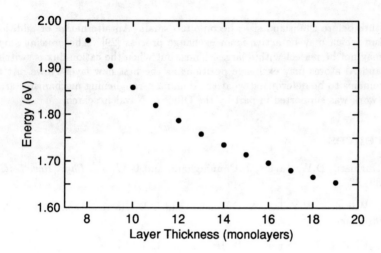

Figure 6: The measured exciton energies from a single, nonhomogeneous QW as a function of QW width. RHEED oscillations were used for the absolute calibration of the QW width.

temperature and the anion-to-cation ratio. Large diffusion lengths obviously allow the growth of large islands. But how is it possible for there to be also a high density of very small structures? The PL results may give a clue. It was shown above that the absolute exciton energies are very sensitive to the growth temperature, implying that the density of small scale structure is very temperature dependent. On the other hand, this density is apparently not strongly sensitive to the anion-to-cation ratio. This follows from the fact that the exciton energies consistently line up if the growth temperatures are the same, and yet, the anion-to-cation ratio changes a great deal across the surface of our samples. One way to interpret this is that the small scale structure is not especially sensitive to the migration length (which presumably is changing with the anion-to-cation ratio). On the other hand, it is possible that the growth temperature plays a direct role in the generation of small scale structure through a second mechanism: the thermal activation of cations from the last layer of the lattice into the adsorbed layer on top. There would then be two, somewhat independent mechanisms which determine the structure of the growing surface: the migration of the adsorbed atoms and the thermally-activated vacancies. To our knowledge, there have been no theoretical studies which could assist us in considering such possibilities. However, if we assume for the moment that this is the correct way to view the growing surface, we can consider the approach that one might take to create a perfect IF. Of course it would be necessary to lower the growth temperature to reduce the vacancy density on the growing surface. The problem with this is that the migration lengths would also be reduced. The question, then, is whether the thermally activated vacancies would disappear with decreasing growth

temperature before the island sizes become too small. An alternative or additional mechanism which may be active is an exchange process [29]. The *growing surface* may or may not be perfect within large islands, but when the cation flux is switched, the Ga and Al atoms may exchange positions as the first new layer grows. At this time it remains to be determined what the dominant roughening mechanisms are.

This work was supported in part by the Office of Naval Research.

References

[1] A. Ourmazd, D.W. Taylor, J. Cunningham, and C.W. Tu, *Phys. Rev. Lett.* **62** (1989) 933.

[2] C.A. Warwick, W.Y Jan, A. Ourmazd, and T.D. Harris, *Appl. Phys. Lett* **56** (1990) 2666.

[3] R.K. Kopf, E.F. Schubert, T.D. Harris, and R.S. Becker, *Appl. Phys. Lett.* **58** (1991) 631.

[4] B. Deveaud, B. Guenais, A. Poudoulec, A. Regreny, C. d'Anterroches, *Phys. Rev. Lett.* **65** (1990) 2317 and A. Ourmazd and J. Cunningham, *Phys. Rev. Lett.* ibid. p.2318.

[5] D. Gammon, B.V. Shanabrook, D.S. Katzer, *Phys. Rev. Lett.* **67** (1991) 1547.

[6] M.A. Herman, D. Bimberg, and J. Christen, *J. Appl. Phys.* **70** (1991) R1.

[7] J.H. Neave, B.A. Joyce, P.J. Dobson and N. Norton, *Appl. Phys. A* **31** (1983) 1.

[8] J.M. Van Hove, C.S. Lent, P.R. Pukite and P.I. Cohen, *J. Vac. Sci. Technol.* **1** (1983) 741.

[9] B. Deveaud, J.Y. Emery, A. Chomette, B. Lambert and M. Baudet, *Appl. Phys. Lett.* **45** (1984) 1078.

[10] J. Massies, C. Deparis, C. Neri, G. Neu, Y. Chen, B. Gil, P. Auvray and A. Regreny, *Appl. Phys. Lett.* **55** (1989) 2605.

[11] K. Kanamoto, K. Fujiwara, Y. Tokuda, N. Tsukada, M. Ishii and T. Nakayama, em Appl. Surf. Sci. **41/42** (1989) 526.

[12] K. Wada, A. Kozen, Y. Hazumi and J. Temmyo, *Appl. Phys. Lett.* **54** (1989) 436.

[13] M.D. Pashley, *J. Crystal Growth* **99** (1990) 473.

[14] D. Gammon, B.V. Shanabrook, D.S. Katzer, *Appl. Phys. Lett.* **57** (1990) 2710.

[15] D.S. Katzer, D. Gammon, B.V. Shanabrook, and B. Tadayon, *Superlattices and Microstructures* **8** (1990) 19.

[16] P. Danielson, *Semiconductor International* Feb. 1989, p. 94.

[17] M. Ilegems, in *Properties of III-V Layers," in The Technology and Physics of Molecular Beam Epitaxy* ed. by E.H.C. Parker (Plenum Press, NY, 1985) p.110-113.

[18] N. Chand, *J. Crystal Growth* **97** (1989) 418.

[19] A.J. SpringThorpe, S.J. Ingrey, B. Emmerstorfer, and P. Mandeville, *Appl. Phys. Lett.* **50** (1987) 77.

[20] D.S. Katzer, D. Gammon, and B.V. Shanabrook, presented at the 11th MBE Workshop, Austin, TX, 16 Sept. 1991. To appear in J. Vac. Sci. Technol.

[21] Y. Horikoshi, M. Kawashima, and H. Yamaguchi, *Japan J. Appl. Phys.* **27** (1988) 169.

[22] M. Tanaka and H. Sakaki, *Superlattices and Microstructures* **4** (1988) 237.

[23] M. Gurioli, A. Vinattieri, M. Colocci, A. Bosacchi and S. Franchi, *Appl. Phys. Lett.* **59** (1991) 2150, and references therein.

[24] C. Weisbuch, R.C. Miller, R. Dingle, A.C. Gossard, and W. Wiegmann, *Solid State Commun.* **37** (1981) 219.

[25] J. Singh, K.K. Bajaj and S. Chaudhuri, *Appl. Phys. Lett.* **44** (1984) 805.

[26] M.K. Jackson, D.Z.-Y. Ting, D.H. Chow, D.A. Collins, J.R. Soderstrom and T.C. McGill, *Phys. Rev. B* **43** (1991) 4856, and references therein.

[27] For example, see M. Zachau, J. A. Kash, and W.T. Masselink, *Phys. Rev. B* **44** (1991) 8403.

[28] For a review, see B. Jusserand and M. Cardona, in *Light Scattering in Solids V*, edited by M. Cardona and G. Guntherodt (Springer-Verlag, Heidelberg, 1989) p. 49.

[29] J.M. Moison, C. Guille, F. Houzay, F. Barthe and M.Van Rompay, *Phys. Rev. B* **40** (1991) 6149.

Optical and Magneto-Optical Properties of Narrow $In_xGa_{1-x}As$-GaAs Quantum Wells

D. C. Reynolds
Wright State University
University Research Center
Dayton, Ohio 45435

K. R. Evans
Wright Laboratory
Solid State Electronics Directorate (WL/ELRA)
Wright-Patterson Air Force Base, Ohio 45433-6543

Abstract

Heterostructures in the lattice-mismatched $In_xGa_{1-x}As$-GaAs quantum well system can be grown without misfit dislocations, i.e., pseudomorphically, with in-plane lattice-constant matching, provided the film thickness is below some mismatch-dependent critical value. High quality pseudomorphic quantum wells and superlattices have been grown on GaAs substrates by molecular beam epitaxy (MBE). In such structures the InGaAs is under biaxial in-plane compression while the InGaAs lattice constant is increased in the growth direction. When thick GaAs overlayers are used, the strain is confined to the InGaAs layer for thicknesses less than critical. Details of the MBE growth of high quality, pseudomorphic InGaAs-GaAs strained layer quantum wells will be presented.

Direct coupling between the GaAs free-exciton in the barrier with the free-excitons in the well for narrow ($6 Å \leq L_z \leq 50 Å$) $In_{0.1}Ga_{0.9}As$-GaAs quantum wells will be reported. Enhanced emission due to heavy-hole free-exciton collapse in narrow, single $In_{0.1}Ga_{0.9}As$-GaAs quantum wells was observed when the excitation energy was resonant with the GaAs barrier free-exciton energy. Magnetic field measurements lend support to the direct coupling interpretation. Additionally, donor-bound exciton transitions and bound-to-bound transitions will be reported for the same system. Magnetic field splittings and diamagnetic shifts will be reported for magnetic fields applied normal to the growth direction. Confinement energies of the conduction and valence subbands will be discussed as well as the type (I or II) of confinement of the light-hole subband.

Contents
1. Introduction
2. Pseudomorphic $In_xGa_{1-x}As$-GaAs Heterostructures
   A. Molecular Beam Epitaxy Growth
   B. Band Alignment
   C. Interface Characteristics
3. Photoluminescence Properties of Narrow QuantumWells
   A. Excitonic Transitions
   B. Bound-to-Bound Transitions
4. Direct Coupling of GaAs Barrier Excitons with InGaAs Well Excitons
5. Magneto-Optical Effects
6. Coupled Quantum Wells

1. Introduction

Pseudomorphically strained $In_xGa_{1-x}As$-GaAs quantum well (QW) structures have received considerable attention both from the basic science point of view [1-6] and from the point of view of potential solid state device applications [7-13]. This system has the advantage of a transparent GaAs substrate which makes it attractive for optoelectronic device applications. Tailoring of the band gap for optimal fiber optic applications is one example. Spatial light modulators and optical bistability are other examples. The strained layers result from the lattice mismatch of the GaAs and $In_xGa_{1-x}As$ components. The strain can either be shared between the two layers or be confined totally to the $In_xGa_{1-x}As$ layer. The shared strain is accomplished through the use of an $In_xGa_{1-x}As$ buffer layer matched to the unsupported in-plane lattice constant of the strained layer system. Strain confined to the $In_xGa_{1-x}As$ QW layer is achieved through the use of a GaAs buffer with an in-plane lattice constant matching bulk GaAs. This results in an in-plane compression with an out of plane extension in the well layer. The strain removes the degeneracy of the light- and heavy-hole valence bands, producing a substantial valence band splitting in addition to the splitting due to well quantization. The bandgap energy also increases with increasing strain. The lifting of the degeneracy between the light- and heavy-hole bands allows the heavy-hole to retain much of its bulk character in the QWs. The heavy-hole effective mass will be heavy along the growth direction but light parallel to the growth planes. This property led to speculation of enhanced hole transport [14-17], suggesting high potential for device applications. Since this review focuses on the optical properties of this system, transport properties will not be covered.

The investigation of narrow $In_xGa_{1-x}As$-GaAs QWs is attractive since the photoluminescence (PL) emission lines can have extremely narrow linewidths [6,18,19]. As the QW becomes narrower, more of the QW wave function extends into the GaAs barrier region. Since the barrier is a high purity binary material, the PL linewidths due to QW transitions approach the linewidths observed in bulk GaAs. As the wells become wider, the emission lines are broadened by alloy scattering, characteristic of a ternary system. The very narrow

emission lines in narrow QWs permit the observation of phenomena that would otherwise be masked by broadened lines characteristic of wider wells. An example is the observation of spin splitting of the heavy-hole free-exciton (HHFE) in an applied magnetic field. The splitting is small and in order to resolve the splitting, very narrow lines are required. Another example is the observation of donor-bound excitons in this system. The binding energy of the exciton to the donor in this system is small (≈ 1 meV). As the wells get wider and the lines broaden, resolution of the free-exciton from the donor-bound exciton is lost. Narrow emission lines are even more critical for observing the bound-to-bound transition which is energetically closer to the HHFE than the donor-bound exciton. Narrow wells are also helpful in observing direct coupling of the exciton wave function in the QW with the exciton wave function in the GaAs barrier. As the wells become wider, the exciton wave function becomes more confined in the well, and the coupling with the exciton wave function in the barrier is greatly diminished.

In this chapter, the molecular beam epitaxial (MBE) growth and structural properties of the strained layer $In_{0.1}Ga_{0.9}As$-GaAs system will be discussed. The optical properties of pseudomorphically grown narrow QWs will be emphasized. The intrinsic and extrinsic properties of narrow wells will be discussed as well as the direct coupling of the QW exciton wave functions with barrier exciton wave functions. A discussion of the magneto-optical effects in narrow wells will also be included. Finally, the properties of coupled $In_{0.1}Ga_{0.9}As$-GaAs QWs will be reviewed.

2. Pseudomorphic $In_xGa_{1-x}As$-GaAs Heterostructures

Heterostructures in the lattice-mismatched $In_xGa_{1-x}As$-GaAs quantum well system can be grown without misfit dislocations, i.e., pseudomorphically, with in-plane lattice-constant matching, provided the film thickness is below some mismatch-dependent critical value. High quality pseudomorphic quantum wells and superlattices have been grown on GaAs substrates by molecular beam epitaxy (MBE). In such structures, the InGaAs is under biaxial in-plane compression while the InGaAs lattice constant is increased in the growth direction. When thick GaAs overlayers are used, the strain is confined to the InGaAs layer for thicknesses less than critical. Although other growth techniques such as metalorganic chemical vapor deposition are used to produce $In_xGa_{1-x}As$-GaAs structures, most high quality structures have been produced by molecular beam epitaxy, and thus we focus on that technique herein. In this section we present details of the molecular beam epitaxial growth used to produce high quality, pseudomorphic $In_xGa_{1-x}As$-GaAs heterostructures. We then discuss characteristics of the resulting $In_xGa_{1-x}As$-GaAs interfaces. We conclude with a brief discussion of the present knowledge regarding conduction and valence band lineup at the heterointerfaces.

A. Molecular Beam Epitaxy Growth

A complete review of the molecular beam epitaxy (MBE) growth process is beyond the scope of this chapter; for an excellent treatise of the present state of MBE technology, see

Ref. 20. In this section we present details of the MBE approach which are specific to the $In_xGa_{1-x}As$-GaAs system. Most of the structures discussed in this chapter were produced according to this approach.

Conventional solid source MBE growth of $In_xGa_{1-x}As$-GaAs structures utilizes elemental indium, gallium, and arsenic sources. The source beams are therefore comprised of atomic indium and gallium, and either tetrameric or dimeric arsenic, depending on whether or not an arsenic "cracker" is used. The use of dimeric arsenic is preferred over the use of tetrameric because of increased arsenic source utilization efficiency and reduced intrinsic defect concentration; however, high quality structures can be grown with the tetrameric beam. The structures reported on herein were all grown on (001) oriented GaAs substrates, or GaAs substrates which were misoriented by 2° from (001) towards (110). The growth rates used were 0.90 and 1.00 monolayers/sec for GaAs and $In_{0.1}Ga_{0.9}As$, respectively.

Growth of $In_xGa_{1-x}As$-GaAs structures by MBE is complicated by the fact that GaAs and $In_xGa_{1-x}As$ have differing optimal substrate temperatures. High quality GaAs layers with smooth surfaces are usually grown between 580 C and 600 C, while high quality $In_xGa_{1-x}As$ layers with x < 0.3 are usually grown near 540 C. Lower growth temperatures than 540 C can produce rough growth fronts, while higher growth temperatures result in a non-unity sticking coefficient for the incident atomic In beam. When the In sticking coefficient deviates from unity, control of layer composition and thickness becomes difficult. In fact, when the In sticking coefficient deviates from unity, strong evidence exists that the resulting In composition (x-value) varies with position from the $In_xGa_{1-x}As$-GaAs heterointerface [21]. Since $In_xGa_{1-x}As$ and GaAs have differing optimal growth temperatures, use of a constant growth temperature represents a compromise in resulting material quality. Our approach, which is similar to that previously reported by Kirby et al. [6], is to grow most of the GaAs at 580 C and all of the InGaAs at 540 C and to ramp the substrate temperature from 580 C to 540 C during the last 350 Å of GaAs preceding an InGaAs layer. Upon completion of growth of the InGaAs layer, the substrate temperature is ramped as quickly as possible back up to 580 C so that all but approximately the first 50 Å of the next GaAs layer is grown under optimal conditions. Growth interruption techniques are not usually employed because of the fear of unwanted impurity incorporation. Additionally, growth interruption upon completion of the growth of an $In_xGa_{1-x}As$ layer is expected to result in segregation of In at the surface.

B. Band Alignment

Knowledge of the band offsets in any QW system is essential to predict the device potential of that system. A large band offset is desirable for increased carrier confinement. In view of this, the valence band offset of the strained $In_xGa_{1-x}As$-GaAs system has received considerable attention [22-31]. In spite of this effort there is still a lack of consensus concerning the confinement of the light-hole in this system. A number of theoretical [32-34] and experimental [35-40] methods have been used to determine the offset; however, the only

usable method to date is to deduce the offset semi-empirically. One approach is to treat the band offset as an adjustable parameter and use it to fit the transition energies obtained from a band structure calculation to photoluminescence excitation (PLE) data. Several reports [22, 23, 28, 31] have compared calculations with the results of PLE measurements to conclude that light-holes are confined in the GaAs layers, while electrons and heavy-holes are confined in the $In_xGa_{1-x}As$ layers; i.e., electrons and heavy-holes exhibit type I behavior, and light-holes exhibit type II behavior. In another report [24], inter-subband light-scattering and PLE were used to conclude that both heavy- and light-holes exhibit type I behavior. In a recent calculation [41] it was shown that the transition energies are relatively insensitive to the offset. In that model a four-band k•P formalism incorporating strain was used to calculate the superlattice band structure. It was shown that oversimplification of the k•P formalism can lead to an erroneous sensitivity of the transition energy to band offset. Recently electron Raman scattering (ERS) data has been published [29]. This technique probes the electronic transitions between the electronic subbands. In contrast to PLE the ERS results are sensitive to the magnitude of the conduction band discontinuity and thus to the offset ratio. It has subsequently been shown [41] that the offset can be deduced from ERS data with the aid of an accurate band structure theory. Using this theory the band structure of two superlattices of Ref [24] was computed and compared with ERS data for the two samples. The results could be interpreted in terms of band offsets that are indium concentration (x-value) dependent. This is contrary to what is observed in the $Al_xGa_{1-x}As$-GaAs QW system, where the band offsets are independent of x-value. It was shown experimentally that the light-hole has type II behavior for an x-value of 0.15 [42].

In Ref [43] experimental measurements were made of the HHFE and light-hole free-exciton (LHFE) transition energies in a series of 2, 4, 6, 10, and 17 monolayer (ML) wide, single $In_xGa_{1-x}As$-GaAs QWs with nominal x-values of 0.1. This study was designed to determine experimentally whether the light-hole subband is confined in the QW or whether it is in the barrier. The experimental results were compared to calculations based on a generalized four-band k•P theory modified to include effects of lattice-mismatch strain [41]. Calculations of the transition energies from the heavy-hole and light-hole subbands to the first conduction subband were made for all of the above well sizes. Calculations were also made of the confinement energies of the conduction subband and the heavy- and light-hole subbands in the same series of wells. Additionally, x-ray analysis

Lz (ML)	HHFE eV	LHFE eV
17	1.4646	1.4910
10	1.4897	1.5053
6	1.5047	1.5116
4	1.5093	1.5149
2	1.5132	1.5146

Table I. *The observed LHFE and HHFE energies for five different well sizes.*

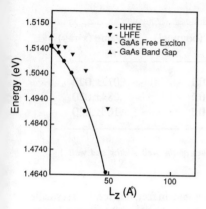

Figure 1. Observed energies of the LHFE and HHFE transitions as a function of well size. Also shown is the GaAs free exciton transition energy and the GaAs bandgap.

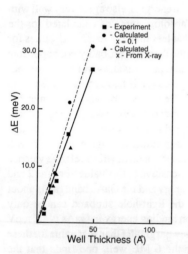

Figure 2. The difference in observed LHFE and HHFE transition energies for the five QW structures (solid curve). Also shown is calculated difference in light-hole and heavy-hole subband energies 1) assuming x = 0.10 (dashed line) and 2) using the actual x-values from x-ray measurements (open circles).

was used to measure the exact well thicknesses and compositions. The HHFE and LHFE transition energies were measured by PL and PLE. These transition energies for all five QWs are given in Table I and are plotted as a function of well size in Fig. 1. The solid curve merely connects the experimental HHFE transition energies. The GaAs free-exciton energy as well as the GaAs band gap energy is also shown. As the wells become narrower, it is clear that the LHFE and HHFE energies tend towards degeneracy, approaching the GaAs free-exciton energy, as expected. Since the radius of the HHFE is on the order of 100 Å, HHFEs in the narrow ($L_z \leq 50$ Å) wells investigated reside in an environment which becomes increasingly GaAs-like as the well width decreases.

The energy separation between the LHFE energy and the HHFE energy is plotted as a function of well size in Fig. 2. The solid curve gives the results for the experimentally measured energies. The light- and heavy-hole subband energies for the same well sizes were calculated, assuming that the x-values were all equal to 0.10, and the differences in these calculated energies are shown as the dashed curve in the same figure. It is seen that for the wider wells the calculated differences are significantly greater than the measured differences. X-ray diffraction measurements were performed on the 4, 6, and 10 monolayer samples taken from adjacent regions of the as-grown wafers to verify the x-values; the results are shown in Table II. Here it is seen that the observed x-values fall short of the nominal value of 0.10. Although surprising when first observed, this result is now understood on the basis of a thickness-dependent indium incorporation rate as discerned from in-situ desorption studies performed during MBE growth of pseudomorphic InGaAs on GaAs [21]. Using the x-ray determined x-values, the energy differences were recalculated and are plotted as o's for the 4, 6, and 10 monolayer wells. The calculated energy difference for the 10

Lz (nominal)	Lz (x-ray)	x-value (nominal)	x-value (x-ray)
10	9.7±0.2	0.1	.072±.020
6	5.8±0.1	0.1	.084±.020
4	4.0±0.04	0.1	.100±.020

Table II. The nominal and x-ray determined values of the well widths and well compositions for three of the wells investigated.

monolayer well is now much closer to the observed value, and in fact, it is somewhat smaller than the measured value indicating that the true x-value may be somewhat larger than the 0.072 value used in the calculation. The light- and heavy-hole subband energies were calculated as well as the confinement energies for both the conduction and valence subbands. The calculated absolute energies agree very well with the experimental energies for well widths greater than approximately ten monolayers. The difference energies, i.e, the difference between the heavy-hole and light-hole subband energies, also agree very well with the measured energies for the narrower wells. The absolute energies calculated for the conduction to valence subband transitions are somewhat less than the measured values for the narrow wells. This is believed due to the fact that the theory is probably not applicable to the very narrow wells since the envelope wave function represents an average that does not contain variations on the scale of a monolayer. Consequently, it is less accurate when there are few lattice sites contained in the well. The energy differences, however, will somewhat cancel these effects and thereby give a meaningful value for comparison with the experimentally observed energy differences. The calculated energy differences between the light- and heavy-hole subbands show the same trend as the measured values and agree very well with the measured values. The best fit to the experimental data is obtained with a relative valence band offset of 0.4, although the results are somewhat insensitive to this value. For the 2 and 4 ML wells, the energy difference between the LHFE energy and the GaAs bandgap is about 4.8 meV. It is quite clear for these size wells that the light-hole subband can be only marginally confined, if confined at all, since the exciton binding energy in GaAs is 4.2 meV and will likely not be very different and most likely somewhat greater than that value for these very narrow wells. Using the same calculation for the 6 ML well, one finds that the confinement energy of the conduction subband plus the binding energy of the LHFE is 7.9 meV. The calculated confinement energy of the conduction subband is 7.6 meV. Thus, the binding energy of the LHFE would have to be 0.3 meV or less for the light-hole to be confined. Since we expect that binding energy is significantly greater than 0.3 meV, then the light-hole must not be confined. For the 10 ML well, the calculated transition energy using the x-value determined from x-ray measurements, from the conduction subband to the heavy-hole valence subband is 1.4952 eV. The experimental HHFE energy was 1.4897 eV. This gives

an exciton binding energy of 5.5 meV, which is less than the calculated HHFE binding energy by ~2.5 meV [44]. The calculated energy for the transition from the conduction subband to the light-hole subband is 1.5084 eV. The experimental LHFE energy was 1.5053 eV. This would give a LHFE binding energy of 3.1 meV, which is less than the HHFE binding energy. This would result from the greater extension of the LHFE wave function into the barrier causing the binding energy of the LHFE to approach the binding energy of the exciton in GaAs for wider wells than would be the case for the HHFE. The fact that it is less than the exciton binding energy in GaAs by 1 meV is likely due to small experimental and theoretical uncertainties. The confinement energies for the conduction subband and the valence subbands were also calculated. The confinement energy for the conduction subband was 10.7 meV, for the heavy-hole subband it was 13.1 meV, and for the light-hole subband it was 0.06 meV. The difference in energy between the GaAs bandgap and the LHFE is 14.2 meV. If one subtracts from this the conduction subband confinement energy of 10.7 meV a difference of 3.5 meV is obtained. This is close to the LHFE binding energy given above. If the LHFE binding energy is the same as the GaAs exciton binding energy then the light-hole subband in the 10 ML well is not confined. An x-value of 0.08 was assumed for the 17 ML well, since x-ray measurements were not made for that sample. Calculation of the light- and heavy-hole subband energies for the 17 ML wide well resulted in an energy difference of 23.3 meV, which is closer to the experimental value shown in Fig. 6, but is about 3 meV less than the experimental value indicating that the true x-value is likely somewhat greater than 0.08. From the calculated electron and hole subband energies, the transition energies from the heavy-hole subband and the light-hole subband to the conduction subband were found to be 1.4751 eV and 1.4984 eV, respectively. The experimental HHFE and LHFE energies for this sample were 1.4646 eV and 1.4910 eV, respectively. This gives a HHFE binding energy of 10.5 meV and LHFE binding energy of 7.4 meV. These binding energies are substantially higher than those calculated for the 10 monolayer well and are higher than the calculated value of the HHFE binding energy by ~1.5 meV [44]. It is noted, however, that the LHFE binding energy is less than the HHFE binding energy as was observed for the 10 ML well and for the same reason. A larger x-value would result in a larger difference between the light-hole and heavy-hole subband energies, which is suggested by the experimental energy difference between the LHFE and the HHFE. A larger x-value would also result in smaller energies for the C1-H1 and C1-L1 transitions. This would result in smaller binding energies for both the LHFE and the HHFE which would be more consistent with what would be expected for this system. It is suggested therefore that the x-value for the 17 ML well is between 0.08 and 0.10.

For the 2 ML and 4 ML quantum wells, one can conclude from the differences in energy between the LHFE and the GaAs bandgap that the light hole subband, if confined at all, can only be marginally confined. This conclusion is based strictly on experimental results. For the 6 ML well, the difference in energy between the LHFE and the GaAs band gap is 7.9 meV. The calculated confinement energy of the conduction subband is 7.6 meV; thus the light-hole must not be confined in the GaAs or else the binding energy of the LHFE

would have to be less than 0.3 meV. The conclusion here is based on two experimentally observed energies plus the calculated value of the conduction subband confinement energy. Similar results are obtained for the 10 ML well where the observed difference in energy between the GaAs bandgap and the LHFE is 14.2 meV. The calculated confinement energy for the conduction subband for this well is 10.7 meV. This gives a value of 3.5 meV for LHFE binding energy. Again, one reaches the conclusion that the light-hole subband is not confined. This is also based on two experimental energies and one calculated energy, the confinement energy of the conduction subband. No conclusions were drawn from the 17 ML well, since the correct x-value was not established. It is safe to conclude for the x-values and well sizes that we have investigated, that the light-hole subband can be only marginally confined in the InGaAs layer, if confined at all. This would be expected for a valence band offset of ~ 0.4 deduced from the theoretical calculations of Ref [45].

C. Interface Characteristics

The quality of the interface between two semiconducting layers in several heterostructure-based electronic devices, such as quantum well lasers and modulation doped field effect transistors (MODFETs), plays an important role in determining their performance characteristics. The materials most commonly used to produce these devices are GaAs-GaAlAs. To gain a better understanding of these effects, electronic [46-48] and optical [49, 50] properties associated with these interfaces have been studied. Mobility enhancement has been observed for GaAs-AlGaAs interfaces for cases in which the GaAs layer is grown before the AlGaAs layer [46, 47]. Interfaces grown in the reverse order show a lesser mobility enhancement [48]. Weisbuch et al. [49] have studied the optical emission from GaAs GaAlAs multiple-quantum well (MQW) structures grown by molecular beam epitaxy. They made their measurements at 1.6 K and were able to explain their data by assuming that there were no well-to-well thickness fluctuations along the direction of growth. However, thickness fluctuations (islands) were present in each well at the heterointerfaces. Their analysis supported a model of an island-like interface with a height of a monolayer and a lateral island size of approximately 300 Å. An essentially similar model was also proposed by Goldstein et al. [50]; they performed PL measurements at 77 K and at room temperature on MQW structures consisting of alternate layers of GaAs and AlAs. Later studies [51, 52] showed fine structure features on both the free exciton transitions and the donor-bound exciton transitions. The fine structure was observed both in PL and PLE. The observed fine structure was explained in terms of changes in the average well size. In this case well size variation of less than 1 ML was observed. The emission linewidths were explained in terms of the interface island heights and the lateral extent of the islands. The linewidths were very narrow, some as narrow as 0.1 meV, and were explained in terms of island heights of one ML with lateral extent from 20 Å to 80 Å. In the AlGaAs-GaAs system it was determined that the dominant line broadening mechanism was interface scattering, therefore the linewidth was a reflection of the interface quality.

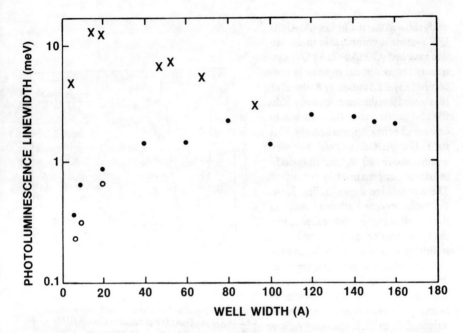

Fig. 3. Comparison of quantum-well photoluminescence linewidth (FWHM) from $Al_{0.38}Ga_{0.62}As/GaAs/Al_{0.38}Ga_{0.62}As$ wells (x) and $GaAs$-$In_{0.11}Ga_{0.89}As$-$GaAs$ SQW's (o, • represent the FWHM of the free-exciton and donor-bound-exciton emission, respectively). Data for $Al_{0.38}Ga_{0.62}As$-$GaAs$-$Al_{0.38}Ga_{0.62}As$ wells from Bertolet et al. (Ref. 20) (Kirby et al. [6]).

Monolayer or sub-monolayer fluctuations have not been observed in the $In_xGa_{1-x}As$-GaAs QW system. Fluctuations in coupled $In_xGa_{1-x}As$-GaAs QWs have been observed and will be discussed in a later section., The emission lines in narrow $In_xGa_{1-x}As$-GaAs QWs are very sharp, with full-width at half-maximum (FWHM) values of 0.1 - 0.2 meV, therefore ML fluctuations should be readily observed if they are present. It has been suggested [6] that the presence of strain in the pseudomorphic system does not allow terracing at the heterointerfaces. To achieve terracing, the $In_xGa_{1-x}As$ unit cell would need to be constrained along two axes. Since the unit cell has an in-plane compression and an extension in the growth direction, Poisson's extension [6] allows only one axis to be constrained. One concludes that in the $In_xGa_{1-x}As$-GaAs system the dominant line-broadening mechanism is alloy scattering. Therefore, the emission linewidth is not a measure of interface quality.

3. Photoluminescence Properties of Narrow Quantum Wells

A strained layer QW system such as $In_xGa_{1-x}As$-GaAs might be expected to show a degraded optical quality, reflected in line broadening due to the inherent strain. This would be particularly true if strain gradients were present, giving a spatial extent to the induced strain shifted line. Studies of this system by both PL and PLE show that this is not the case. One

study showed that the $In_xGa_{1-x}As$-GaAs QW system is comparable to the lattice matched AlGaAs-GaAs QW system as far as optical quality is concerned [53]. PL studies by Kirby et al. [6] showed that the emission linewidths FWHM decreased as the well widths decreased below approximately 40 Å for $In_xGa_{1-x}As$-GaAs QWs. For well widths above 40 Å the linewidths remained approximately constant. These trends are shown in Fig. 3. As the wells become narrower, more of the HHFE wave function extends into the GaAs barrier region. The barrier in this system is a binary, so the linewidths approach those that are characteristic of the bulk binary GaAs. The higher energy subband transitions at low temperatures are primarily determined from PLE studies. At low temperatures the higher lying states are not occupied, preventing observation by PL.

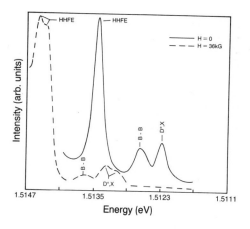

Fig. 4. PL spectrum for the undoped, two monolayer wide $In_{.1}Ga_{.9}As$-GaAs QW structure showing HHFE and B-B transitions (Reynolds et al. [58]).

A. Excitonic Transitions

Although there have been a few reports [54-57] on the binding energy of excitons to neutral donors in GaAs-AlGaAs quantum well (QW) structures, only one such study for the InGaAs-GaAs system has been reported [58]. Likewise, there are no calculations which describe the binding energy of donors (E_D) or the binding energy of excitons to donors (E_{BE}) as well as their dependence on well size or indium content in the InGaAs-GaAs system. Kirby et al. [6] observed transitions in PL measurements on undoped $In_{.11}Ga_{.89}As$-GaAs QWs which they tentatively identified with the collapse of neutral donor-bound excitons (D°,X). Although E_{BE} was not the focus of that work, many of our observations bear similarity to theirs.

Since the spatial extent of the ground state exciton wave function is on the order of 200 Å in diameter, very narrow $In_{.1}Ga_{.9}A$-/GaAs QWs should result in values of E_{BE} which are near that obtained for bulk GaAs, while very wide quantum wells should result in values of E_{BE} which approach that of bulk $In_{.1}Ga_{.9}As$. The value of E_{BE} previously obtained [59] for bulk GaAs is 1.2 meV, while the value for bulk $In_{.1}Ga_{.9}As$ has not been measured but is

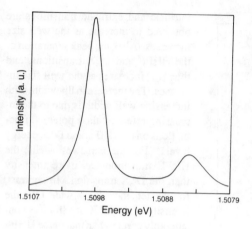

Fig. 5. PL spectrum for donor-doped, four monolayer wide $In_{.1}Ga_{.9}As$-GaAs QW structure showing the HHFE and D°, X transitions (Reynolds, et al. [58]).

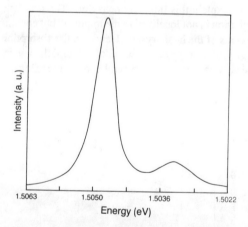

Fig. 6. PL spectrum for donor-doped, six monolayer wide $In_{.1}Ga_{.9}As/GaAs$ QW structure showing the HHFE and D°, X transitions (Reynolds, et al. [58]).

expected to be somewhat smaller than 1.2 meV because of its lower bandgap and lower effective carrier masses. Intermediate values of quantum well width are expected to give rise to an increase in E_{BE} because: 1) E_D, the donor binding energy for donors located in QWs, is known to increase due to confinement effects, and 2) E_{BE} is known to vary approximately linearly with E_D [60]. Thus, it is expected that E_{BE} will reach a maximum at some intermediate value of quantum well width. Indeed, this effect has previously been observed for $Al_{.25}Ga_{.75}As$-GaAs QWs, for which a maximum in E_{BE} was found to occur for a well width of approximately 100 Å [61].

The PL spectrum from a two monolayer wide, donor-doped $In_{.1}Ga_{.9}As$-GaAs QW is shown in Fig. 4 and shows three features: the HHFE, a transition interpreted as a bound-to-bound (B-B) transition, and the donor-bound exciton (D°,X) transition. From the energy separation of the HHFE and D°,X transitions the binding energy E_{BE} of the exciton to the donor is determined. Thus, $E_{BE} = 1.2$ meV for this size well. This value is identical to the value of E_{BE} previously observed [59] for bulk GaAs. The similarity in value of E_{BE} for the narrow $In_{.1}Ga_{.9}As$ well and bulk GaAs is not unexpected since the QW HHFE wave function extends well into the barrier region for this size well. Thus, the HHFE experiences an environment which is predominately of GaAs-like character, and this is reflected in the observed value of E_{BE}.

The PL spectra obtained for four, six, and ten monolayer wide donor-doped $In_{.1}Ga_{.9}As$-GaAs QW structures are shown in Figs. 5, 6, and 7, respectively. The linewidths of both the

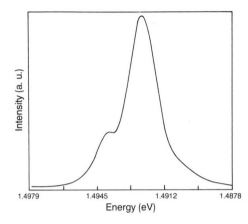

Fig. 7. PL spectrum for the donor-doped, ten monolayer wide In_xGa_yAs-GaAs QW structure showing the HHFE and $D°, X$ transitions (Reynolds et al. [58]).

intrinsic and extrinsic transitions are observed to increase as the well size increases. Also, the peak separation of the HHFE and $D°,X$ transitions, and thus E_{BE}, increases as the well size increases. The increase in linewidth with increasing well width is due to the increasing effect of alloy potential fluctuations on the HHFE and $D°,X$ energy levels. For the narrowest wells, the HHFE transition has more intensity than the $D°,X$ transition. In contrast, for the ten monolayer wide QW the intensity of the $D°,X$ transition is considerably greater than that of the HHFE transition. If the concept of the giant oscillator strength for impurities and defects in semiconductors as proposed by Rashba and Gurgenishvili [62] is applicable to QWs, then that concept may explain this intensity behavior. The principle behind this concept is that the optical excitation is not localized in the impurity, but encompasses a number of neighboring lattice points of the host crystal. Hence, in the absorption process, light is absorbed by the entire region of the crystal consisting of the impurity and its surroundings. In the present study for the very narrow wells, the number of surrounding lattice sites associated with a specific donor will be relatively few. As the wells get larger, the number of surrounding lattice sites will increase leading to an increased oscillator strength for the impurity-related transition.

In Fig. 8 the values of E_{BE} are plotted for the two, four, six, and ten monolayer wide QWs as a function of the HHFE transition energy. This plot shows a linear increase in the binding energy as the well size increases up to the ten monolayer QW. It is evident from Figs. 4 through 7 that a precise determination of E_{BE} becomes increasingly difficult for larger wells. It is expected that the increase in E_{BE} with

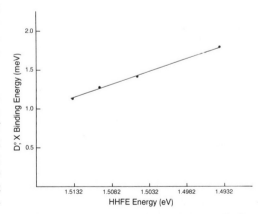

Fig. 8. Variation of E_{BE} with HHFE transition energy for the two, four, six, and ten monolayer wide In_xGa_yAs-GaAs QW structures (Reynolds et al. [58]).

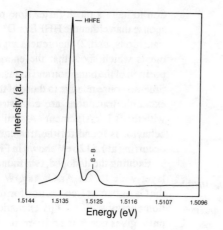

Fig. 9. PL spectrum for the donor-doped, two monolayer wide $In_.1Ga_.9Al/GaAs$ QW structure showing the HHFE, B-B, and $D°, X$ transitions Reynolds et al. [58]).

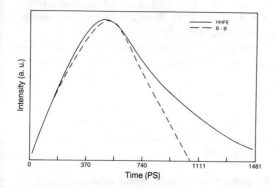

Fig. 10. Time-resolved PL decay curves for the HHFE (solid) and B-B (dashed) transitions for the undoped, two monolayer wide $In_.1Ga_.9As$-GaAs QW structure (Reynolds et al. [58]).

well size as shown in Fig. 8, will continue until a maximum is reached at some critical well size. As mentioned above, such a critical well size was previously found for the AlGaAs-GaAs system to be about 100 Å [61]. For still larger well sizes, the value of E_{BE} is expected to decrease, tending toward that of bulk $In_.1Ga_.9As$. However, it must be noted that the notion of very large $In_.1Ga_.9As$ well widths is confounded by the fact that eventually the critical thickness for strain relief will be reached and the $In_.1Ga_.9As$ may no longer grow pseudomorphically. It is expected that the presence of strain in the $In_.1Ga_.9As$ does not have an appreciable effect on the value of E_{BE} since strain increases the effective bandgap and decreases the effective heavy-hole mass; these two factors have opposite effects on E_{BE} and therefore are expected to cancel each other somewhat.

B. Bound-to-Bound Transitions

The PL spectra from an undoped, two monolayer $In_.1Ga_.9As$-GaAs QW structure is shown in Fig. 9. Two features are observed: The HHFE transition occurs at 1.5133 eV, and another transition is observed to occur 0.7 meV lower in energy at 1.5126 eV. This energy difference of 0.7 meV is considered to be too small to be associated with the binding energy of an exciton to a neutral donor. Also, it occurs at too high an energy to be a free-to-bound transition. A similar transition has been observed [63] in AlGaAs-GaAs QWs and is now associated with a bound-to-bound (B-B) transition. The identification of the B-B transition in AlGaAs-GaAs QWs was based on: 1) time-resolved spectra which showed the B-B transi-

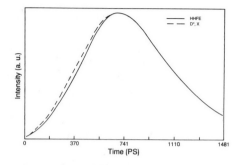

Fig. 11. *Time-resolved PL decay curves for the HHFE (solid) and D°, X (dashed) transitions for the donor-doped, four monolayer wide $In_{.1}Ga_{.9}As$-GaAs QW structure (Reynolds et al. [58]).*

tion to have a much faster time response than either the HHFE or D°, X transitions, and 2) theoretical arguments which show that the energy position of the transition and its faster time-response relative to those of the excitonic transitions are consistent with the B-B assignment. A similar behavior is found for the transition occurring at 1.5126 eV shown in Fig. 9. Exciting the undoped, two monolayer wide $In_{.1}Ga_{.9}As$-GaAs QW at the GaAs barrier free-exciton formation energy with a 70 ps excitation pulse gives rise to the time-resolved PL decay curves shown in Fig. 10. The solid and dashed curves are the time-responses of the HHFE and B-B transitions, respectively. Here it is seen that the B-B transition has a significantly faster time-response than that of the HHFE transition, which is qualitatively similar to our results obtained on AlGaAs-GaAs QWs. Fig. 11 shows time-responses of the HHFE and D°, X transitions in the donor-doped, four-monolayer wide $In_{.1}Ga_{.9}As$-GaAs QW. Both transitions have similar time-responses. Thus, both the time-response and the energy position of the B-B peak shown in Fig. 9 preclude it from being excitonic. The simplest bound-to-bound transition would involve the recombination of a shallow bound-electron and a shallow bound-hole, although the exact nature of the centers involved in the observed B-B transition is not known. The faster time-response of the B-B transition in comparison to the HHFE transition suggests that the centers involved in the B-B transition give rise to a complex in which the electron and hole are more closely spaced than they are in the HHFE. This would result in a greater overlap of the electron and hole wave functions which greatly enhances the recombination rate, thereby resulting in a much faster time-response [63].

4. Direct Coupling of GaAs Barrier Excitons with InGaAs Well Excitons.

The low temperature photo-luminescence (PL) spectrum of a high quality, undoped AlGaAs-GaAs quantum well structure is usually dominated by emission due to collapse of heavy-hole free-excitons (HHFEs) located in the well region. Emission from the barrier region is largely suppressed, even if the excitation energy lies at or above the barrier bandgap, since carrier transfer from the barrier into the well is very efficient. When the excitation energy is resonant with a quantum well transition which lies above the HHFE in energy, efficient transfer between the laser excited state and the HHFE may result, and enhanced HHFE emission may be observed. Kusano et. al. [64] showed that laser excitation of the light-hole free-exciton (LHFE) leads to enhanced emission due to HHFE collapse. Using time-

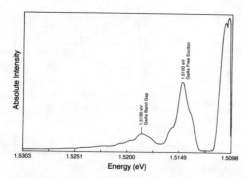

Fig. 12. *Selective excitation for the six monolayer In_xGa_yAs-GaAs single quantum well structure. The detector hold position is the HHFE energy of 1.5093 eV (Reynolds et al. [96]).*

resolved luminescence techniques with picosecond resolution, they showed that the kinetics governing the formation of HHFEs from LHFEs differ significantly from those governing the formation of HHFEs from free electrons and holes. Since the wave function overlap between the initial state (LHFE) and final state (HHFE) is great, the LHFE to HHFE transition is a favorable process. Energy conservation is accommodated by phonon emission.

In quantum well structures for which the barrier is a binary compound, for example the $In_xGa_{1-x}As$-GaAs system, an additional resonant excitation pathway arises: direct excitation of excitons in the barrier may lead to enhanced emission from the quantum well. The PL spectra for each QW sample showed two main features: the normal GaAs bandedge emission peaks and the InGaAs HHFE emission peak. The penetration of the quantum well exciton wave function into the barrier was previously demonstrated in the InGaAs-GaAs system by Kirby et al. [6]. They found that the HHFE linewidth decreased as the well width decreased and concluded that the linewidth was primarily influenced by alloy broadening in the InGaAs well. As the well size decreases, more of the wave function extends into the barrier, which in this case is a binary, thus reducing the alloy scattering and thereby resulting in narrower lines. The same trends were observed here: the HHFE PL linewidths FWHM for 6, 4, and 2 monolayer $In_{0.1}Ga_{0.9}As$-GaAs quantum wells were 0.54, 0.22, and 0.11 meV, respectively. The different structures studied differed only in the width of the quantum well, and the results of our observations were very similar for each of these structures. For this reason, below results for only the 6 monolayer quantum well sample will be discussed.

Fig. 12 shows the selective excitation spectra obtained for the 6 monolayer single quantum well (SQW) structure with the detector

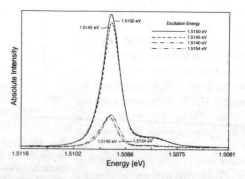

Fig. 13. *Resonant excitation spectra of the six monolayer In_xGa_yAs-GaAs single quantum well structure obtained using direct excitation of the GaAs barrier free-exciton at 1.5150 eV (Reynolds et al. [96]).*

hold position at the HHFE energy of 1.5093 eV. The peak at 1.5190 eV is the GaAs bandedge, and the peak at 1.5150 eV is the GaAs free-exciton creation energy. The large increase in intensity in the lowest energy region is due to the increase in scattered laser light which occurs when the excitation energy approaches that of the detector hold position. Both the GaAs bandedge and the GaAs exciton peaks are seen to resonantly excite the In-GaAs well HHFE transition; however, excitation at the GaAs exciton energy is significantly more efficient in producing subsequent HHFE emission.

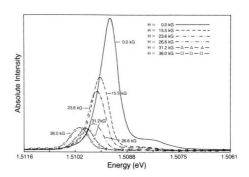

Fig. 14. Resonant excitation of the six monolayer In_xGa_yAs-GaAs single quantum well structure showing the intensity variation of the HHFE transition as a function of applied magnetic field strength (Reynolds et al. [96]).

In Fig. 13 the resonant excitation spectra of the 6 monolayer SQW structure is shown for various excitation energies. The data show strong emission due to HHFE collapse when the excitation energy is resonant with the GaAs exciton formation energy, while excitation energies just a few tenths of an meV both above and below the GaAs free-exciton energy results in considerably less HHFE emission, demonstrating the resonant nature of the excitation. Since free carriers are not produced at the GaAs free-exciton formation energy, the resonant behavior observed is due to a transition involving the GaAs barrier free-exciton and the quantum well HHFE. If the HHFE formation process from the GaAs free-exciton involved recombination of the GaAs free-exciton followed by absorption of the emitted photon to create free carriers in the well, followed by subsequent HHFE formation, then the overall (GaAs free-exciton —> HHFE) process would be expected to be very inefficient, in contrast to our observations. It is most likely that the GaAs free-exciton is directly coupled to the quantum well HHFE level.

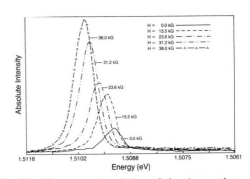

Fig. 15. Non-resonant excitation of the six monolayer In_xGa_yAs-GaAs single quantum well structure showing the intensity variation of the HHFE transition as a function of applied magnetic field strength (Reynolds et al. [96]).

Resonant excitation spectra for the 6 monolayer SQW structure are shown in Fig. 14 in the HHFE emission region for different applied magnetic field strengths. The excitation energy is resonant with the GaAs free-exciton formation energy, 1.5150 eV. Strong emission is observed at zero applied magnetic field, while increased applied magnetic fields lead to a dramatic decrease in HHFE emission, in contrast to usual observations regarding excitonic transition strengths in magnetic fields. The observed behavior is understood on the basis of the proposed, direct (GaAs barrier free-exciton —> HHFE) process as follows: The magnetic field compresses the wave functions of both the GaAs barrier free-exciton and the InGaAs well HHFE, leading to a reduced coupling between the two exciton levels due to the reduction in wavefunction overlap. Thus, the observed magnetic field dependent resonant excitation behavior of the HHFE transition is consistent with the proposed mechanism of direct HHFE formation from the GaAs barrier exciton.

In marked contrast, under non-resonant excitation conditions, the HHFE emission strength is found to increase with increasing applied magnetic field strength. Fig. 15 shows the variation of HHFE emission strength as a function of applied magnetic field strength using a non-resonant excitation energy of 1.5253 eV, which is 6.3 meV above the GaAs barrier bandgap. In this case, free electrons and holes are created which migrate to the well and generate the HHFE in the well. The presence of an applied magnetic field probably does not markedly affect the rate at which the free carriers migrate into the well region and form HHFEs. However, once the HHFE is formed, its recombination rate is enhanced because its wave function is compressed by the applied magnetic field. The observed differences in magnetic field dependent behavior between the cases of resonant and non-resonant excitation is consistent with the model of direct coupling of the GaAs barrier free-exciton with the quantum well HHFE (when resonant excitation is used).

We have also found that when several InGaAs quantum wells of different sizes ($L_z \leq 50$ Å) are grown in the same sample, resonant excitation of the GaAs free-exciton leads to simultaneous resonant excitation of all of the InGaAs QWs. The situation is different for the AlGaAs/GaAs system for which each quantum well size requires a different resonant excitation energy.

The GaAs barrier free-exciton is coupled to the quantum well HHFE via the wave function penetration of the HHFE into the barrier region. Such a coupling also must satisfy energy and momentum conservation rules. Since the total angular momentum of holes associated with both the GaAs barrier free-exciton and the quantum well HHFE states is J=3/2, total angular momentum conservation is ensured.

The situation is somewhat similar for coupling between the LHFE and HHFE levels in AlGaAs/GaAs quantum wells. The wave function overlap is obviously great because the LHFE and HHFE occupy the same region in the quantum well and do not differ markedly in size. However, the light- and heavy- hole levels differ in m_j value ($m_j = \pm 1/2$ and $m_j = \pm 3/$

2, for light- and heavy-holes, respectively). Since the hole associated with the GaAs barrier free-exciton has mostly heavy-hole character [65], a change in m_j value is not required for the (GaAs free-exciton —> HHFE) process studied here for the InGaAs/GaAs system.

The observed resonant excitation behavior, along with the marked difference in the effect of an applied magnetic field on the HHFE emission strength for the cases of non-resonant and resonant excitation, is consistent with a model of direct coupling between the GaAs barrier free-exciton level and the InGaAs quantum well heavy-hole free-exciton level.

5. Magneto-Optical Effects

The effects of applied magnetic fields on the properties of excitons confined in QWs have received considerable attention in the last few years. Most reports on this subject use magnetic fields applied parallel to the growth axis [66-70], while less effort has been devoted to magnetic fields applied normal to the growth direction [70-73]. All of these reports have focused on the GaAs-AlGaAs system. When the magnetic field is applied parallel to the growth axis of a QW structure, it is also parallel to the confining electric field. As a result, the Hamiltonian can be separated into an electric part which results in subbands and a magnetic part which splits the subband levels into Landau levels. The planes of the resulting carrier cyclotron orbits are parallel to the QW layers. When the magnetic field is applied normal to the growth axis, the Hamiltonian cannot be separated in the above manner, and the problem becomes more complicated. The planes of the resulting carrier cyclotron orbits are perpendicular to the layers, and carrier orbits tend to intersect both wells and barriers. The translational invariance is broken by the quantum well potential, thus removing the degeneracy of the discrete Landau levels and forming a band. Therefore, new excitonic properties are expected.

Recently, Platero and Altarelli have theoretically investigated the case where the magnetic field is applied parallel to the layers [74]. They have shown that the magnetic field oriented in this direction lifts the spin degeneracy and produces in addition to the diamagnetic shift, a spin splitting of the subbands. In a previous study [75], both the diamagnetic shift and the magnetic field splitting of both light- and heavy-hole free-excitons in AlGaAs-GaAs multi quantum well (MQW) structures were measured in a magnetic field applied parallel to the growth layers. These results are compared with the predictions of Platero and Altarelli [74].

There are no published results of these same measurements being made on InGaAs--GaAs QWs with the same magnetic field orientation. Previously, five different well sizes for the InGaAs-GaAs system (two monolayers to seventeen monolayers) were investigated, showing that the diamagnetic shift of the HHFE increases as the well size decreases, tending toward the value obtained in GaAs for the narrowest well size [76]. This is opposite to what is observed in the AlGaAs-GaAs system (diamagnetic shift decreases as the well size

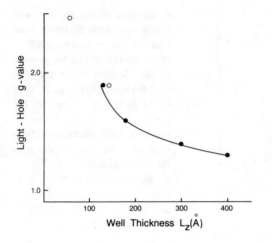

Fig. 16. *The light-hole g-value as function of well thickness. The solid dots are experimental points taken from Ref. 6. The O's are determined from theoretical curves of Ref. 74 (Reynolds et al. [18]).*

decreases) and is explained by the greater extension of the heavy-hole free-exciton wave function into the barrier as the well size decreases thus reflecting more of the GaAs character. Only one of the QWs, the narrowest well, showed a magnetic field splitting of the HHFE. The splitting observed was 0.13 meV at a field of 36 kG. The HHFE line width, full-width at half-maximum (FWHM) for this well was 0.11 meV. The neutral donor-bound exciton (D°,X) was also observed to split for a donor-doped two monolayer wide well. The magnitude of the observed splitting of the D°,X is found to be different than that of the HHFE. The next larger well, four monolayer well, had a FWHM for the HHFE of 0.22 meV. It becomes clear why magnetic field splitting is not observed in the wider wells, the line width increases and even if the magnetic field splitting does not decrease, the splitting will not be resolved. We have examined the HHFE line width for all five well sizes and it increases from 0.11 meV for the two monolayer well to 3.0 meV for the seventeen monolayer well. It may be possible to observe magnetic field splitting of the HHFE in somewhat larger well sizes at higher magnetic fields; however, when the linewidths approach 1.0 meV, it is unlikely that magnetic field splitting of the HHFE in this system will be observed.

In the earlier magneto-optical studies of AlGaAs-GaAs MQW structures with the magnetic field applied parallel to the growth layers [75], both the diamagnetic shift and magnetic field splitting of the LHFE and the HHFE were observed. This experiment showed that the diamagnetic shift for both the LHFE and the HHFE decreased with decreasing well size and that the LHFE had a smaller diamagnetic shift than the HHFE (see Fig. 5 of Ref. 75). It is clear that the trends observed in Ref. 75 are predicted by the theoretical results of Ref. 74. Theory [74] and experiment [75] both show the same trends for the diamagnetic shift of HHFE and LHFE in the GaAs-AlGaAs system: the LHFE has a smaller diamagnetic shift than the HHFE for a given well size, and the diamagnetic shift increases with well size. From the magnetic field splittings of Ref. 75, a light- and heavy-hole g-value was deduced. This was done by assuming that the magnetic field splittings of the free excitons in MQW structures are analogous to those in bulk semiconductors with nondegenerate valence bands (i.e., II-VI semiconductors with wurtzite structure), where the light-hole free-exciton splits

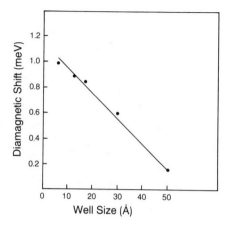

Fig. 17. *The diamagnetic shift of In_xGa_yAs-GaAs single quantum wells as a function of well thickness for well sizes varying from two to sixteen monolayers. Magnetic field strength is 36 kG (Reynolds et al. [18]).*

as the sum of the electron and hole g-values and the heavy-hole free-exciton splits as their difference. Assuming that the g-value of the electron is the same in GaAs/AlGaAs QWs as in bulk GaAs (0.5), the g-values of the holes are determined from the magnetic field splittings. We can now compare these results with the theory of Ref. 74. In Fig. 16 we show as solid dots the light-hole g-value as a function of well size determined from the magnetic field splitting of the LHFE. Also shown as x's are the light-hole g-values determined from the theoretical curves in Figs. 3 and 4 of Ref. 74 for the magnetic field splittings of the LHFE for the 57 Å and 142 Å QWs. The solid line drawn between the solid dots is merely to aid the eye and has no theoretical significance. It should be pointed out that the experiment monitors exciton transitions, while the theory is concerned with subband to subband transitions and does not take into account the exciton binding energy or its dependence on magnetic field. It is believed that the dependence of exciton binding energy on magnetic field is small; as a result the agreement between theory (x's) and experiment (solid dots) is quite respectable. It is difficult to compare theory and experiment for the magnetic field splitting of the HHFE since the magnetic field splittings are much smaller than for the LHFE, which is due to the fact that the HHFE splitting depends on the difference of the electron and hole g-value.

In the case of InGaAs-GaAs QWs, the diamagnetic shift, as observed in small well sizes, is opposite to that observed in the AlGaAs-GaAs system, namely, the diamagnetic shift increases as the well size decreases. This feature is demonstrated in Fig. 17, where the diamagnetic shift of the QW HHFE transition for an applied magnetic field of 36 kG is shown as a function of well size for well sizes between 6 and 50 Å (between 2 and 17 monolayers). The diamagnetic shift is seen to increase roughly linearly with decreasing well size. The solid line connecting the data points extrapolates at zero well width to 1.2 meV which is the observed diamagnetic shift for bulk GaAs for the same applied magnetic field [75]. This is understood on the basis that as the well size decreases, the HHFE wave function penetrates further into the GaAs barrier and thus the HHFE takes on more and more GaAs character. As the well size increases beyond 50 Å it is expected that a minimum in the diamagnetic shift

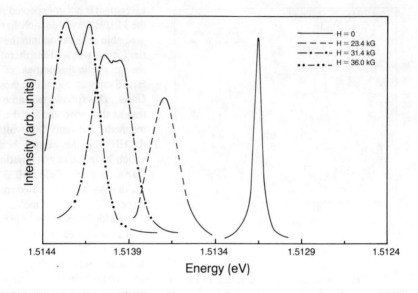

Fig. 18. The Pl spectra of two monolayer $In_.1Ga_.9As$-GaAs quantum well applied magnetic fields of 0, 23.4, 31.4, and 36 kG (Reynolds et al. [18]).

will be reached and with further increase in well size, the diamagnetic shift will tend towards the value observed in bulk InGaAs.

The HHFE in the two monolayer $In_.1Ga_.9As$-GaAs QW shows a magnetic field splitting. The PL spectra for this QW are shown in Fig. 18 for different applied magnetic fields. The HHFE splitting as a function of magnetic field is shown in Fig. 19. The magnetic field splitting at an applied field of 36 kG is 0.13 meV. If one now makes the same assumptions as were made for the AlGaAs-GaAs system, and the same electron g-value (0.5) is used, a heavy-hole g-value of 1.1 is obtained. This agrees very well with the effective hole g-value of 1.1 ± 0.1 which was previously observed for bulk GaAs [77, 78]. Such close

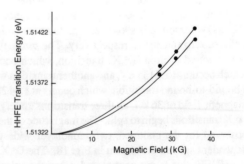

Fig. 19. The magnetic field splitting as a function of magnetic field for the two monolayer $In_.1Ga_.9As$-GaAs quantum well giving rise to the PL spectra shown in Fig. 18 (Reynolds et al. [18]).

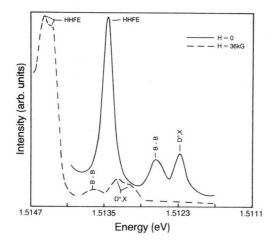

Fig. 20. *Magnetic field splitting of the HHFE and $D°$, X transitions in a donor-doped, two monolayer $In_{.1}Ga_{.9}As$-GaAs single quantum well (Reynolds et al. [18]).*

agreement is not unexpected since the HHFE wave function has appreciable penetration into the barrier for this size well and therefore, the magnetic properties of the HHFE are very much like those in GaAs. This follows the same pattern as that observed for the diamagnetic shift and line width of the HHFE in this sample, both of which have the characteristics of GaAs. The zero field HHFE line width FWHM for this two monolayer well was 0.11 meV. The line-width for the next larger size well investigated (four monolayers) was 0.22 meV, for which a magnetic field splitting was not observed. The increased line-width prevented resolution for the magnetically split components at fields up to 36 kG, the maximum field for our system. At higher fields, the splitting would probably be observed. As the well size increases, the line-widths increase, and thereby prevent the observation of spin splitting in this system for well sizes much greater than six monolayers, assuming the g-values do not change appreciably with well size.

We have not observed the LHFE in these wells - in the narrow wells LHFE is most likely not confined. The PL spectra for the donor-doped, two monolayer wide $In_{.1}Ga_{.9}As$-GaAs QW structure is shown in Fig. 20 as the solid and dashed lines for applied magnetic fields of zero and 36 kG, respectively. The zero field spectra show three QW transitions: the donor-bound exciton $(D°,X)$ transition, which occurs at 1.5123 eV, the HHFE transition, which occurs at 1.5135 eV, and another transition which has previously been identified with a bound-to-bound transition, which occurs at 1.5127 eV [63]. It is seen that with the applied magnetic field of 36 kG, all three transitions undergo a blue shift and both the HHFE and the $D°,X$ transitions begin to split. The magnitude of the splitting of the HHFE is identical to that observed for the undoped, two monolayer wide well structure for which magnetic field dependent spectra are shown in Fig. 18. The $D°,X$ shows a somewhat larger splitting than the HHFE; this is not unexpected, since the splitting mechanism is different for the free exciton compared to that for $D°,X$. Splitting is not resolved for the B-B transition although the linewidth increases with magnetic field, indicating that some splitting may be occurring.

6. Coupled Quantum Wells

It has been shown that both monolayer [79-82] and submonolayer [88, 89] variations in well size are reflected in the PL and PLE spectroscopy of $Al_xGa_{1-x}As$-GaAs quantum well systems. These variations appear as sharp structure on the HHFE and LHFE transitions, with the energy separation of the line structure corresponding to well width fluctuations of an integral number of monolayers, or in some cases, energy

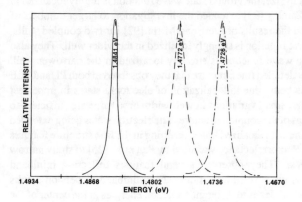

Fig. 21. HHFE energy for a 10 ML coupled In_xGa_xAs-GaAs quantum well for barrier widths of 120 Å (solid curve), 30 Å (dashed curve), and 20 Å (dot-dashed curve).

separations of less than a monolayer. These fine structure features were explained in terms of inhomogeneities within or between isolated QWs in single or multi-quantum well (MQW) systems. Similar fine structure features were not observed from PL or PLE measurements of single $In_xGa_{1-x}As$-GaAs QW structures. More recently, analogous fine structure features have been observed in double QW and superlattice $In_xGa_{1-x}As$-GaAs systems [90, 91]. In the coupled system, the degenerate states of each well mix and split into discrete states, each being made up of a linear combination of the envelope functions of the isolated wells. The lowest energy transition corresponds to the recombination of electron and heavy-hole states associated with the symmetric isolated well eigen functions. The higher energy heavy-hole excitonic transition is associated with the recombination of electron and heavy-hole states associated with the antisymmetric isolated well eigen functions.

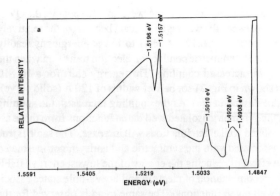

Fig. 22a. LHFE energy 1.5010 eV for the quantum well in Fig. 21 with 120 Å barrier.

Investigating a model consisting of two coupled rectangular wells, Lang and Mishi [92] showed that by varying the width of one of the wells by a small amount, while

keeping the other well constant in size, the ground state wave function is highly localized in the wider well. Littleton and Camley [93] extended the investigation to larger numbers of coupled wells. They confirmed the results of Lang and Mishi [92] for two coupled wells, showing that the symmetric wave function is strongly localized in the wider well. They also showed that the antisymmetric wave function is strongly localized in the narrower well. Moore et al. [90] interpreted the detailed fine structure that they observed in both PL and PLE of $In_xGa_{1-x}As$-GaAs samples as being due to localization of electronic states in imperfect coupled structures. In particular, small variations in well width or x-value are sufficient to localize holes in different regions of the coupled structure. The electron states being extended over the whole system recombine with localized holes, resulting in the fine structure features observed in both PL and PLE. Photoreflectance was used by Pan et al. [25] to study narrow $In_xGa_{1-x}As$-GaAs coupled QWs. They observed several features reflecting miniband dispersion of both confined and unconfined transitions. The miniband dispersion results from coupling between wells which leads to different transition energies at the center of the mini-Brillouin zone, and at the edge of the zone in the growth direction. Their calculated transition energies agreed well with experiment for appropriate growth parameters, and for a valence band offset of 0.3 plus/minus 0.05. PL and PLE were used by Moore et al. [94] to follow the evolution of electronic states in the same coupled QW system, into superlattice minibands as the number of wells were increased from two to twenty. In going from two to five wells, their spectra showed strong $\Delta n = 0$ exciton transitions as well as transitions from confined states to continuum states. When the number of wells was increased to twenty, superlattice properties were observed. The separation between electron states was approximately 1 meV, small enough to be considered a miniband, rather than twenty individual eigenstates. With the formation of a superlattice the wave vector in the growth direction becomes a good quantum number. Momentum conserving optical transitions were observed at the center of the mini Brillouin zone, and at the edge of the zone.

In the above two studies, rather wide barriers of 100 Å were used to separate the QWs. Even with the wide barriers coupling between the wells is significant because of the relatively small confinement energies for an x-value of 0.12. A shift to lower energies is readily observed in the symmetric heavy-hole to symmetric conduction electron transition when the barrier width is reduced, resulting in increased coupling. This energy shift for a 10 ML coupled $In_{0.1}Ga_{0.9}As$-GaAs QW is shown in Fig. 21 for barrier widths of 120 Å (solid curve), 30 Å (dashed curve), and 20 Å (dot-dashed curve). As the coupling increases, the splitting of electron states associated with the symmetric isolated well eigen functions from those associated with the antisymmetric isolated well eigen functions will increase. The same is true of the splitting of the hole states associated with the symmetric and antisymmetric isolated hole wave functions. This has the effect of reducing the energy of the lowest energy HHFE transitions, which results from the recombination of electron and heavy-hole states associated with the symmetric isolated well eigen functions. The same trend is observed for the LHFE, as shown in Fig. 22. The LHFE transition results from the recombination of electron and light-hole states associated with the symmetric isolated light-hole and electron eigen

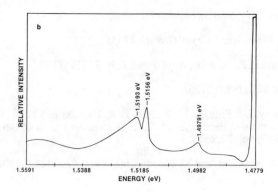

Fig. 22b. *LHFE energy (1.4979 eV) for the same quantum well with 30 Å barrier.*

functions. Figs. 22a, b, and c show the shift of the LHFE transition to lower energies as the barrier width is reduced from 120 Å to 30 Å to 20 Å, respectively. The transition at 1.4918 eV in Fig. 22a, results from the electron and heavy-hole states associated with the antisymmetric isolated well eigen functions. It is noted that this transition is missing in Fig. 22b and 22c. As the barrier width decreases and the coupling increases, the splitting of the electron states associated with the symmetric and antisymmetric isolated well eigen functions increases dramatically. For a barrier width of 120 Å, this splitting for the QW described above is approximately 10 meV [96]. As the barrier is reduced to 20 Å, the same splitting is approximately 40 meV [95]. It is likely, for these narrow barriers, that the electron state associated with the antisymmetric isolated well eigen functions is no longer confined in the QW. This would explain the presence of the peak at 1.4918 eV in Fig. 22a, but its absence in Figs. 22b and 22c.

Acknowledgements

The authors would like to thank C. E. Stutz, J. E. Ehret, and E. N. Taylor for crystal growth support, Lt. Col. K. Soda, R. E. Walline, B. Jogai, and G. L. McCoy for technical support, P. S. Woosley for manuscript layout and D. K. Oblinger for editing support. This work was partially supported by AFOSR. Author DCR was supported, during this work, under USAF Contract No. SCEEE-AFAL/1405ELR.

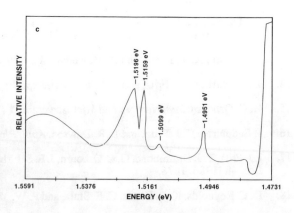

Fig. 22c. *LHFE energy (1.4951 eV) for the same quantum well with 20 Å barrier.*

References

1. J. W. Matthews and A. E. Blakeslee, *J. Cryst. Growth* **27** (1974) 118.

2. P. L. Gourley, I. J. Fritz, and L. R. Dawson, *Appl. Phys. Lett.* **52** (1988) 377.

3. G. C. Osbourn, *Phys. Rev.* **B27** (1983) 5126.

4. G. C. Osbourn, P. L. Gourley, I. J. Fritz, R. M. Biefeld, L. R. Dawson and T. E. Zipperian, in *Semiconductors and Semimetals*, ed. R. K. Willardson and A. C. Beer (Academic, New York, 1987) vol. 24, chap. 8.

5. K. J. Moore, G. Duggan, A. Raukema, and K. Woodbridge, *Phys. Rev.* **B42** (1990) 1326.

6. P. B. Kirby, J. A. Constable, and R. S. Smith, *Phys. Rev.* **B40** (1989) 3013.

7. N. G. Anderson, W. D. Laidig, and Y. F. Lin, *J. Elec. Mat.* **14** (1985) 187.

8. P. K. Bhattacharya, V. Das, F. Y. Juang, Y. Nashimoto, and S. Dhar, *Solid State Elect.* **29** (1986) 261.

9. I. J. Fritz, L. R. Dawson, T. J. Drummond, J. E. Schirber, and R. M. Biefeld, *Appl. Phys. Lett.* **48** (1986) 139.

10. M. J. Ludowise, W. T. Dietze, C. R. Lewis, M. D. Camras, N. Holonyak, B. K. Fuller, and M. A. Nixon, *Appl. Phys. Lett.* **42** (1983) 487.

11. D. R. Meyers, T. E. Zipperian, R. M. Biefeld, J. J Wiczer, *International Electronic Device Meeting 83* (1983) 700.

12. S. M. J. Liu, M. B. Dax, C. K. Peng, J. Klem, T. S. Henderson, W. F. Kopp, and H. Morkoc, *IEEE Trans. Electron Devices* **ED-33** (1986) 576.

13. Y. J. Yang, K. Y. Hsieh, and R. M. Kolbas, *Appl. Phys. Lett.* **51** (1987) 215.

14. J. E. Schirber, I. J. Fritz, and L. R. Dawson, *Appl. Phys. Lett.* **46** (1985) 187.

15. G. C. Osbourn, *Superlattices and Microstructures* **1** (1985) 223.

16. J. E. Schirber, I. J. Fritz, and L. R. Dawson, *Appl. Phys. Lett.* **46** (1985) 187.

17. I. J. Fritz, T. J. Drummond, G. C. Osbourn, J. E. Schirber, and E. D. Jones, *Appl. Phys. Lett.* **48** (1986) 1678.

18. D. C. Reynolds, K. R. Evans, C. E. Stutz, and P. W. Yu, *Phys. Rev* **B43** (1991) 4244.

19. D. C. Bertolet, J-K. Hsu, S. H. Jones, and K. M. Lau, *Appl. Phys. Lett.* **52** (1988) 293.

20. M. A. Herman and H. Sitter, (Springer-Verlog, New York, 1989).

21. K. R. Evans, C. E. Stutz, E. N. Taylor, and J. E. Ehret, *J. Vac. Sci. Technol.* **B9** (1991) 4.

22. J.-Y. Marzin, M. N. Charasse, and B. Sermage, *Phys. Rev.* **B31** (1985) 8298.

23. G. Ji, U. K. Reddy, D. Huang, T. S. Henderson, and H. Morkoc, *J. Appl. Phys.* **62** (1987) 3366.

24. J. Menendez, A. Pinczuk, D. J. Werder, S. K. Sputz, R. C. Miller, D. L. Sivco, and A. Y. Cho, *Phys. Rev.* **B36** (1987) 8165.

25. S. H. Pan, H. Shen, Z. Hang, F. H. Pollak, W. Zhuang, Q. Xu, A. P. Roth, R. A. Masut, C. Lacelle, and D. Morris, *Phys. Rev.* **B38** (1988) 3375.

26. M. J. Joyce, M. J. Johnson, M. Gal, and B. F. Usher, *Phys. Rev.* **B38** (1988) 10978.

27. A. Ksendzov, H. Shen, F. H. Pollak, and D. P. Bour, *Proceedings of the 4th International Conference on Modulated Semiconductor Structures*, ed. D. Tsui, L. L. Chang, and R. Merlin, (Ann Arbor, 1989) *Surf. Sci* **228** (1990) 326.

28. D. J. Arent, K. Deneffe, C. Van Hoof, J. De Boeck, and C. Borghs, *J. Appl. Phys.* **66** (1989) 1739.

29. J.-P. Reithmaier, R. Hoger, H. Riechert, A. Heberle, G. Abstreiter, and G. Weimann, *Appl. Phys. Lett.* **56** (1990) 536.

30. J.-P. Reithmaier, R. Hoger, and H. Riecher, *Appl. Phys. Lett.* **57** (1990) 957.

31. Y. Zou, P. Grodzinski, E. P. Menu, W. G. Jeong, P. D. Dapkus, J. J. Alwan, and J. J. Coleman, *Appl. Phys. Lett.* **58** (1991) 601.

32. C. G. Van de Walle and R. M. Martin, *Phys. Rev.* **B35** (1987) 8154.

33. C. G. Van de Walle, *Phys. Rev.* **B39** (1989) 1871.

34. N. E. Christensen, *Phys. Rev.* **B37** (1988) 4528.

35. S. P. Kowalczyk, W. J. Schaffer, E. A. Kraut, and R. W. Grant, *J. Vac. Sci. Technol.* **20** (1982) 705.

36. H. Kroemer, W-Y. Chien, J. S. Harris, and D. D. Edwall, *Appl. Phys. Lett.* **36** (1980) 295.

37. S. R. Forrest and O. K. Kim, *J. Appl. Phys.* **52** (1981) 5838.

38. K. Kazmierski, P. Philippe, P. Poulain, and B. de Cremoux, *J. Appl. Phys.* **61** (1987) 1941.

39. R. People, K. W. Wecht, K. Alavi, and A. Y. Cho, *Appl. Phys. Lett.* **43** (1983) 118.

40. A. K. Srivastava, J. L. Zyskind, R. M. Lum, B. V. Dutt, and J. K. Klingert, *Appl. Phys. Lett.* **49** (1986) 41.

41. B. Jogai, *Appl. Phys. Lett.*, to be published.

42. J. M. Gerard and J. Y. Marzin, *Phys. Rev.* **B40** (1989) 6450.

43. D. C. Reynolds, K. R. Evans, C. E. Stutz, B. Jogai, C. R. Wie, and P. W. Yu, to be published.

44. K. J. Moore, G. Duggan, K. Woodbridge, and C. Roberts, *Phys. Rev.* **B41** (1990) 1090.

45. B. Jogai and P. W. Yu, *Phys. Rev.* **B41** (1990) 12650.

46. H. L. Stormer, A. Pinczuk, A. C. Gossard, and W. Wiegmann, *Appl. Phys. Lett.* **38** (1981) 692.

47. L. C. Witkowski, T. J. Drummond, C. M. Stanchak, and H. Morkoc, *Appl. Phys. Lett.* **37** (1980) 1033.

48. T. J. Drummond, H. Morkoc, S. L. Su, R. Fischer, and A. Y. Cho, *Electron. Lett.* **17** (1981) 870.

49. C. Weisbuch, R. Dingle, A. C. Gossard, and W. Wiegmann, in *Gallium Arsenide and Related Compounds*, ed. H. W. Thim, (Briston, 1981) Inst. Phys. Conf. Ser. 56, Ch. 9 p.711.

50. L. Goldstein, Y. Horokoshi, S. Tarucha, and H. Okamoto, *Jpn. J. Appl. Phys.* **22** (1983) 1489.

51. D. C. Reynolds, K. K. Bajaj, C. W. Litton, P. W. Yu, S. Singh, W. T. Masselink, R. Fischer, and H. Morkoc, *Appl. Phys. Lett.* **46** (1985) 51.

52. D. C. Reynolds, K. K. Bajaj, G. Peters, C. Leak, W. M. Theis, P. W. Yu, H. Morkoc, K. Alavi, C. Colvard and I. Shidlovsky, *Superlattics and Microstructures* **4** (1988) 723.

53. R. L. S. Devine and W. T. Moore, *J. Appl. Phys.* **62** (1984) 3999.

54. B. V. Shanabrook and J. Comas, *Surf. Sci.* **142** (1984) 504.

55. D. C. Reynolds, K. K. Bajaj, C. W. Litton, P. W. Yu, W. T. Masselink, R. Fischer, and H. Morkoc, *Phys. Rev.* **B29** (1984) 7038.

56. Y. Nomura, K. Shinozaki, and M. Ishii, *J. Appl. Phys.* **58** (1985) 1864.

57. S. Charbonneau, T. Steiner, M. L. W. Thewalt, E. S. Koteless, J. Y. Chi, and B. Elman, *Phys. Rev.* **B38** (1988) 3583.

58. D. C. Reynolds, K. R. Evans, C. E. Stutz, and P. W. Yu, *Phys. Rev.* **B44**, (1991) 1839.

59. H. Venghaus, *J. Lumin.* **16** (1978) 331.

60. J. R. Haynes, *Phys. Rev. Lett.* **4** (1960) 4361.

61. D. C. Reynolds, C. E. Leak, K. K. Bajaj, C. E. Stutz, R. L. Jones, K. R. Evans, P. W. Yu, and W. M. Theis, *Phys. Rev.* **B40** (1989) 6210.

62. X. Rashba and Y. Gurgenishvili, *Fiz. Tverd. Tela. Leningrad* **4**, (1962) 1029 [*Sov. Phys. Solid State* **4**, (1962) 795].

63. D. C. Reynolds, K. R. Evans, K. G. Merkel, C. E. Stutz, and P. W. Yu, *Phys. Rev.* **B43** (1991) 9087.

64. J. Kusano, Y. Segawa, Y. O. Aoyagi, S. Namba, and H. Okamoto, *Phys. Rev.* **B40** (1989) 1685.

65. M. Altarelli and N. O. Lipari, *Phys. Rev.* **B9** (1974) 1733.

66. J. C. Maan, G. Belle, A. Fasolino, M. Altarelli, and K. Ploog, *Phys. Rev.* **B30**, (1984) 2253.

67. S. Tarucha, H. Okamoto, Y. Isawa, and N. Miura, *Solid State Commun.* **52** (1984) 815.

68. N. Miuram, Y. Iwasa, S. Tarucha, and H. Okamoto, in *Proceedings of the Seventeenth Conference on the Physics of Semiconductors*, ed. James D. Chadi and W. A. Harrison (Springer-Verlag, New York, 1985) p. 359.

69. W. Ossau, B. Jakel, E. Bangert, G. Landwehr, and G. Weimann, *Surf. Sci.* **174** (1986) 188.

70. H. Sakaki, Y. Arakawa, M. Nishioka, J. Yoshino, H. Okamoto, and N. Miura, *Appl. Phys. Lett.* **46** (1985) 83.

71. G. Belle, J. C. Mann, and G. Weimann, *Surf. Sci.* **170** (1986) 611.

72. T. Duffield, R. Bhat, M. Kozu, F. DeRosa, D. M. Hwang, P. Grabble, and S. J. Allen, *Phys. Rev. Lett.* **56** (1986) 2724.

73. N. S. Pulsford, J. Singleton, R. J. Nicholas, and C. T. B. Foxon, *J. Phys. (Paris) Colloq.* **48** (1987) C5-231.

74. G. Platero and M. Altarelli, *Phys. Rev.* **B39** (1989) 3758.

75. D. C. Reynolds, K. K. Bajaj, C. W. Litton, R. L. Greene, P. W. Yu, C. K. Peng, and H. Morkoc, *Phys. Rev.* **B35** (1987) 4515.

76. D. C. Reynolds, K. R. Evans, C. E. Stutz, and P. W. Yu, *Phys. Rev.* **B43** (1991) 4244.

77. F. Willmann, S. Suga, W. Dreybrodt, and K. Cho, *Solid State Commun.* **14** (1974) 783.

78. S. B. Naam, D. C. Reynolds, C. W. Litton, R. J. Almassy, T. C. Collins, and C. M. Wolfe, *Phys. Rev.* **B13** (1976) 761.

79. C. W. Weisbuch, R. C. Miller, R. Dingle, A. C. Gossard, and W. Wiegmann, *Solid State Commun.* **38** (1981) 709.

80. C. Weisbuch, R. Dingle, P. M. Petroft, A. C. Gossard, and W. Wiegmann, *Appl. Phys. Lett.* **38** (1981) 840.

81. L. Goldstein, Y. Horikoshi, S. Tarucha, and H. Okamato, *Jpn. J. Appl. Phys.* **22** (1983) 1489.

82. B. Deveaud, J. Y. Emery, A. Chomette, B. Lambert and M. Baudet, *Appl. Phys. Lett.* **45** (1984) 1078.

83. B. Deveaud, A. Regreny, J. Y. Emery and A. Chomette, *J. Appl. Phys.* **59** (1986) 1633.

84. H. Sakaki, M. Tanaka, and J. Yoshino, *Jpn. J. Appl. Phys.* **24** (1985) 1417.

85. T. Hayakawa, T. Suyama, K. Takahashi, M. Kondo, S. Yamamoto, S. Yano, and T. Hiijikata, *Appl. Phys. Lett.* **47** (1985) 952.

86. R. C. Miller, C. W. Tu, S. K. Sputz, and R. F. Kopf, *Appl. Phys. Lett.* **49** (1986) 1245.

87. D. Bimberg, D. Mars, J. N. Miller, R. Bauer, and D. Oertel, *J. Vac. Sci. Technol.* **B4** (1986) 1014.

88. D. C. Reynolds, K. K. Bajaj, C. W. Litton, J. Singh, P. W. Yu, P. Pearah, J. Klem, and H. Morkoc, *Phys. Rev.* **B33** (1986) 5931.

89. D. C. Reynolds, K. K. Bajaj, G. Peters, C. Leak, W. M. Theis, P. W. Yu, and H. Morkoc, *Phys. Rev.* **B37** (1988) 3117.

90. K. J. Moore, G. Duggan, K. Woodbridge, and C. Roberts, *Phys. Rev.* **B41** (1990) 1095.

91. L. Pavesi, E. Tuncel, B. Zimmermann, and F. K. Reinhart, *Phys. Rev.* **B39** (1989) 7788.

92. R. Lang and K. Nishi, *Appl. Phys. Lett.* **45** (1984) 98.

93. R. K. Littleton and R. E. Camley, *J. Appl. Phys.* **59** (1986) 2817.

94. K. J. Moore, G. Dugan, K. Woodbridge, and C. Roberts, *Phys. Rev.* **B42** (1990) 3024.

95. B. Jogai, Unpublished.

96. D. C. Reynolds, K. R. Evans, K. K. Bajaj, B. Jogai, C. E. Stutz, and P. W. Yu, *Phys. Rev.* **B43** (1991) 1871.

Growth and studies of antimony based III-V Compounds by Magnetron Sputter Epitaxy using Metalorganic and Solid Elemental Sources

James B.Webb and R.Rousina
National Research Council Canada
Institute for Microstructural Science
M-50, Montreal Rd., Ottawa, Ontario
Canada K1A 0R6

ABSTRACT

This paper reviews two forms of magnetron sputtering, magnetron sputter epitaxy (MSE) and metalorganic magnetron sputtering (MOMS), as applied to the epitaxial deposition of antimony based III-V semiconductors. In MOMS, the reactions between the metalorganic group III component with the sputtered group V antimony flux are discussed with reference to the non-intentional introduction of carbon impurities. The structural and opto-electronic properties of MOMS and MSE deposited InSb, GaSb and ternaries of $In_{1-x}Ga_xSb$ and $In_{1-x}Al_xSb$ are presented.

1. Introduction

The relatively recent advances in artificially structured materials, particularly in the area of semiconductor device applications, have resulted in a worldwide effort to develop and refine epitaxial deposition techniques having stable and reproducible control at the atomic level. Two techniques have demonstrated this high degree of control, namely metalorganic vapour phase epitaxy (MOVPE) and molecular beam epitaxy (MBE). The use of these techniques in depositing a wide variety of III-V and II-VI compounds has been well documented in the literature and continues to be the subject of new and innovative investigations into structures having not only bulk characteristics, but characteristics unique to two and even one dimensional systems. However, not all material systems can be easily accessed by these techniques and this has lead to a number of modifications to the basic growth processes. For instance, the use of ion beam sources to enhance dopant incorporation, laser irradiation during growth to enhance the chemical reactions at the growth interface in an attempt to lower epitaxial growth temperatures, the use of lasers to ablate targets to provide a highly controllable (mono-layer/pulse) vapour source and the use of gaseous sources in MBE to circumvent the difficulties associated with solid sources such as phosphorous. The latter has lead to "new" techniques based on a marriage between MBE and MOVPE. These "new" techniques have been termed metalorganic molecular beam epitaxy (MOMBE) or chemical beam epitaxy (CBE) depending on the nature of the sources. Clearly, no single growth technique that presently exists, can hope to achieve optimal results with all materials systems. Because of this, investigations into techniques which combine the advantages of various growth technologies are being pursued. One such technique, which will be discussed in this review, is based on the use of magnetron sputter sources (magnetron sputter epitaxy [MSE]), and/or various gaseous sources including metalorganic compounds (metalorganic magnetron sputtering [MOMS]).

Magnetron sputtering has been and is used extensively by the industrial sector for a wide variety of thin film coating applications such as tribological coatings, metallisation for integrated circuits, piezoelectric materials for surface acoustic wave devices and transparent conducting coatings for liquid crystal displays. A number of reviews covering these applications have been published previously, but few articles have dealt with some of the more recent advances in semiconductor epitaxy. Perhaps due to 1) the high voltages applied to the sputter cathode (particularly for non-magnetron sputtering), 2) the past difficulties in controlling the plasma which resulted in undesired sputtering of surfaces other than the target, or 3) perhaps because of the successful application of sputtering for various optical and wear coatings, magnetron sputtering has been perceived as a technique that would likely be unsuitable for the deposition of high quality single crystal epilayers of semiconducting compounds. Despite this perception, magnetron sputtering has been used to deposit high quality epitaxial layers of

III-V semiconductors. This article will review some of these recent advances and it is hoped that a much clearer picture of the process and its future potential will be realised.

2. The Magnetron Sputter Cathode and Epitaxy System

The design and I-V characteristics of the magnetron cathode have been discussed in numerous articles and will not be dealt with in this review. See for instance references [1],[2]. However, in answering the question " Why do we wish to use magnetron sputter sources?" we will point out some of the basic characteristics and advantages of this deposition source.

Diode sputtering is a purely physical process in which the material to be deposited is removed from a target through collisions involving high energy, non-reactive and relatively massive ions. Typically, argon ions formed in a plasma discharge between the cathode (target) and the anode (substrate-chamber), are accelerated to the cathode where they "knock out" or sputter away the target material. This material then travels outwards to the growth surface (substrate).

Unlike simple diode sputtering, magnetron sputtering confines the plasma using a toroidal magnetic field on the target surface. This results in an increase in ionisation efficiency and thus an increase in sputter rate. In addition, high energy secondary electrons created during the sputter process are prevented from reaching the substrate, thus allowing deposition on uncooled substrates at ambient temperature. Since sputtering can be carried out using either direct current or radio frequency power, virtually any material can be sputtered, including insulators such as teflon and highly refractory materials such as titanium. Using multi-targets and/or reactive gases, an extremely wide range of compounds can be deposited with excellent compositional control. Unlike thermal evaporation sources, the sputtered vapour has the same composition as the target material. This is an inherent characteristic of non-reactive sputtering and arises from the fact that the component or components that may be preferentially sputtered, become deficient on the target surface. This reduction in concentration continues until the sputter removal rate of all components approaches a stable equilibrium value identical to the target composition. However, disadvantages can include defect formation from energetic sputtered species and unwanted sputtering of components due to poor plasma confinement. However, with proper design considerations, these effects can be minimised.

The magnetron sputter epitaxy chamber requires the same design considerations as those used for an MBE system. In real terms, this means an MBE chamber able to accept a magnetron source. Fortunately, metal sealed and fully bakeable magnetron sources are now readily available. In addition to these requirements, a gas handling system similar to that used in MOVPE is necessary both for the introduction of high purity sputter gas and gaseous metalorganic

sources used in MOMS. A schematic representation of a MOMS/MSE system[3] which has been used to deposit epilayers of InSb, $In_{1-x}Ga_xSb$ and $In_{1-x}Al_xSb$ is shown in Figure 1. The substrate holder is a radiantly heated molybdenum block that can be rotated about a central axis to ensure both temperature and deposition uniformity from one, or simultaneously from all three concentrically mounted 5 cm diameter magnetron cathodes. All sputter

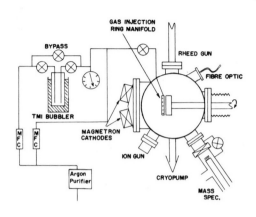

Figure 1. Schematic diagram of the MOMS/MSE deposition system. From T.Sudersena Rao et.al.[3].

sources are operated in the D.C. mode using computer interfaced, switching supplies capable of maintaining stable operating powers from less than a watt to 1KW. The sputter gas and metalorganic carrier gas is ultrahigh purity argon which is passed through a getter purifier before entering the metalorganic bubblers and/or sputter chamber. All gas flows are individually controlled using electronic mass flow controllers. Substrates are bonded to the molybdenum holder using indium solder. In the case of MOMS, the pressure in the bubblers are maintained at atmospheric pressure using a needle valve at the entrance to the chamber. This ensures stable operation of the bubbler and of the transfer of the metalorganic. By proper design of the injection manifold and of the magnetron cathode/operating profile, the targets can be effectively decoupled from the reactive components of the injected gas. This means that in MOMS, reactions are heterogeneous and occur at the substrate surface. The low sputter pressure (<3mTorr), ensures negligible gas phase reactions.

3. Epitaxy of III-V Compounds by Magnetron Sputtering

3.1 Metalorganic Magnetron Sputtering

The use of a metalorganic source in combination with a magnetron sputter source was first reported by our group at the National Research Council for the deposition of InSb for use in infrared detector applications[4]. An epilayer of indium antimonide was formed on a GaAs(100) substrate through a surface

reaction involving trimethylindium (TMI), which was injected near the substrate surface and antimony sputtered from a magnetron source (Eq.1).

$$In(CH_3)_3 + Sb \longrightarrow InSb + 3(CH_3) \qquad (1)$$

For temperatures in the range of 350 → 500°C no deposition of elemental antimony occurred in the absence of TMI. Under these conditions the growth rate is controlled by the arrival rate of TMI which reacts at the growth surface, releasing an indium atom which immediately combines with the arriving antimony. In the absence of the antimony beam, the epitaxial growth temperature is sufficiently high for antimony desorption to occur, which results in surface degradation and the formation of indium balls. The growth rate as a function of reciprocal temperature is shown in Figure 2. Also shown in the Figure are the growth rates for GaSb prepared by MOMS[5] and InSb prepared by MSE. This data will be discussed later. The dependence for InSb prepared by MOMS is similar to that observed for metalorganic vapour phase epitaxy where for low growth temperatures the rate is thermally activated, (a kinetically limited growth region), with an activation energy indicative of the energy required for the thermal breakup of the metalorganic[6]. In MOMS, an activation energy as low as 0.24 eV has been observed[3] and is considerably lower than the 1.77 eV metal-carbon dissociation energy. This indicates a substantial contribution to the breakup from the plasma, particularly at low growth temperatures. More effective decoupling of the plasma from the injected metalorganic increases the observed activation energy to 0.9 eV as shown in the Figure. In this regime, the surface morphology of the layers changes rapidly from a highly textured surface at 350°C (Figure 3a), to an almost basket-weave surface texture at 375°C (Figure 3b).

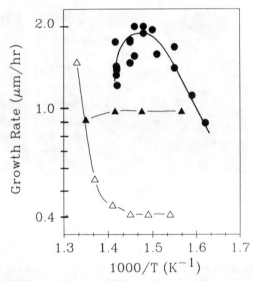

Figure 2. Growth rate as a function of reciprocal temperature for InSb on (100)GaAs.
(●)MOMS InSb, (▲)MSE InSb, (△)MOMS GaSb.

At higher temperatures a plateau region is observed in MOVPE and is

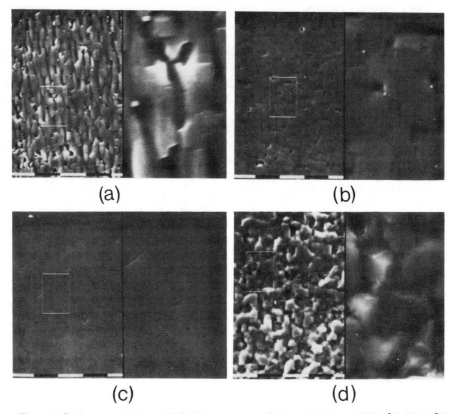

Figure 3. Surface morphology of InSb films grown on GaAs substrates at (a)350°C, (b)375°C, (c)420°C and (d)475°C. From J.B.Webb et.al.[10].

characteristic of a growth rate governed by the diffusion rate of metalorganic reactants across a stagnant boundary layer to the substrate surface[6]. In MOMS deposition of InSb this region is absent due to the much lower growth pressures (ie: MOMS operates in a near molecular flow regime similar to MOMBE). Instead, the rate begins to decrease from what is believed to be the result of an activated etching process, since faceting of the epilayer surface is observed (Figure 3d)[3]. Between these limits and centred approximately 30°C on either side of the temperature at which maximum growth rate is observed (420°C), lies a "window" where good surface morphology is obtained (Figure 3c). Layers grown in this temperature region exhibited optimum crystallinity and electronic properties.

The deposition characteristics for MOMS are obviously dependent upon the metalorganic source (ie:metal-carbon bond strength) and is clearly shown by

the growth rate dependence for GaSb[5]. Using trimethylgallium (TMG) which has a much higher energy associated with the removal of the first methyl group (Equ.2)[7], the growth rate was found to show only a weak temperature dependence at low temperatures (incomplete dissociation of the TMG).

$$Ga(CH_3)_3 + Sb \longrightarrow GaSb + 3(CH_3) \qquad (2)$$

At higher temperatures however, the growth rate increased markedly with an observed activation energy of 1.8 eV(42kcal/mol) as compared to 59.5 kcal/mol for the removal of the first methyl group (see Figure 2).

The large differences in thermal dissociation characteristics of TMG and TMI has been shown to introduce problems in the growth of the ternary $In_{1-x}Ga_xSb$[5]. At the growth temperature suitable for InSb, the breakup for TMG is incomplete. Increasing the growth temperature does not solve the problem since this causes a deterioration in surface morphology and moves one further away from the optimal growth temperature for both GaSb and InSb. Clearly the metalorganics must be chosen carefully to be compatible with the material to be grown and it's optimal epitaxy temperature.

The large difference in dissociation characteristics of TMI and TMG introduces another effect that has not been discussed so far, and that is, what happens to the carbon containing species freed during the growth?

3.2 Carbon Incorporation in MOMS

Unlike low or atmospheric pressure metalorganic vapour phase epitaxy, MOMS is carried out at much lower pressures, close to the molecular flow regime experienced in MOMBE or CBE. As pointed out previously, this essentially eliminates any possibility of gas phase reactions. Also, the group V species cannot provide atomic hydrogen as in MOVPE of GaAs using arsine (AsH_3). These factors limit the chemistry of growth to the substrate surface and limit the pathways available for carbon removal. As shown in Equations (1)and(2), the removal of carbon bearing species is essential, particularly if, as in the case of GaAs, carbon is electrically active.

Studies of the dissociation fragments of TMI, have been carried out using mass spectrometry for the MOMS deposition of InSb[8],[9],[10]. The effects of plasma power, added hydrogen and growth temperature on the evolved hydrocarbons were investigated in relation to the carbon content of the layers and to their electrical characteristics. With only thermal decomposition (ie: no plasma or H_2), the primary species observed were CH_n (n=1,3), methane (CH_4) and hydrocarbon radicals of the form C_2H_n (n=2 and 4). Adding H_2 or increasing the substrate temperature results in an increase in methane production (see Figure 4), a corresponding decrease in background carrier density and a

decrease in carbon content as determined from SIMS. This indicates that methane is not responsible for carbon incorporation. A similar observation was made for the pyrolysis of TMG in MOVPE.

The concentration dependence for the C_2H_n species was found to follow a similar pattern. It was therefore concluded that the more reactive radicals associated with CH_n (n=1,3) were likely responsible for the carbon found in the films. Keuch and Veuhoff[11] concluded that in MOVPE, carbon bearing radicals such as CH_3^+ adsorbed on the growth surface or as a complex CH_3-Ga, (or in this case as CH_3-In) will become incorporated in the layer if complete removal is inhibited. Typically atomic hydrogen, from a group V hydride such as arsine, combines with these radicals forming methane which is pumped away. Adding molecular H_2 during MOMS epitaxy would be expected to have a similar effect if the plasma were to ionise the molecular hydrogen. The observed increase in CH_4 with added hydrogen and sputter power shown in Figure 4 would

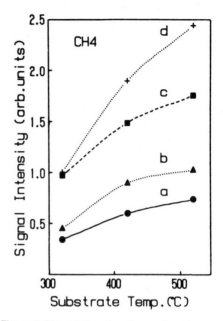

Figure 4. The relative abundance of methane at different temperatures and sputter ambients:(a)TMI only,(b)TMI + plasma, (c)TMI + H_2, and (d)TMI + H_2 + plasma. From T.Sudersena Rao et.al.[8].

support this interpretation. Alternatively, removal of the CH_3 radicals in the absence of any added hydrogen, can occur through a multi-step process[12] of the form;

$$CH_3 + CH_3 \longrightarrow CH_4 + CH_2 \qquad (3)$$

$$CH_3 + CH_3 \longrightarrow C_2H_4 + H_2 \qquad (4)$$

These processes would be consistent with the hydrocarbon species observed in the absence of H_2. In either case the metal-CH_3 bond strength will be an important factor in determining the desorption rate of the hydrocarbon species. This is confirmed by the increase in carbon content of $In_{1-x}Ga_xSb$ epilayers prepared by MOMS using increasing amounts of TMG[5]. Other metalorganic precursors such as triethylindium (TEI) and/or triethylgallium can be used. These

precursors generally show less carbon incorporation due to the formation of more stable and volatile species upon dissociation[6],[13], however the vapour pressures and/or dissociation energies may not be compatible with the optimal growth temperature of the desired material. A detailed discussion of these aspects are given in [14].

3.3 Magnetron Sputter Epitaxy

Another variation of magnetron sputtering, termed magnetron Sputter epitaxy (MSE)[15], employs only solid sources and in many respects is more closely related to conventional MBE. The operating parameters are similar to those employed in MOMS, however both the group III and group V elements are independently supplied from magnetron sputter sources. The major advantages of this technique are the absence of carbon contamination from the source material and a much simplified gas injection manifold, since only the argon sputter gas need be supplied.

A similar technique,(using conventional non-magnetron sources), was used more than 12 years ago to deposit superlattices of InSb/GaSb[16]. The technique (termed multi-target sputtering) employed targets of InSb and GaSb and substrates of cleaved BaF_2 or NaCl. Since no additional target for the group V element was used, the growth was carried out at temperatures near 320°C where the desorption of Sb was low. A growth rate of 0.17A°/s was found to give single crystal superlattice structures for layer thicknesses from 12 to 70A°. Following this study, Webb et al.[17] reported the use of a magnetron source to deposit epitaxial layers of InSb on (100)InSb and GaAs. Similar to the work reported earlier, they employed an InSb target and limited the temperature for growth to less than 400°C. The problem of antimony desorption at the optimum growth temperature of 420°C was not addressed in these studies.

More recently MSE[15] has been employed to deposit InSb homoepitaxial layers for infrared diode detectors and $InSb/In_{1-x}Al_xSb$ strained layer superlattices[18]. In these studies, magnetron sources were used to sputter targets of InSb, Al and Sb. Since the group V element was independently controlled, higher growth temperatures could be used. It was found that the growth rate as a function of growth temperature for InSb on (100)InSb followed a very different dependence than that of MOMS. As shown in Figure 2, the growth rate is virtually independent of temperature from 340°C -> 440°C. A similar behaviour was observed for growth on GaAs provided an initial low temperature (375°C) InSb buffer layer was deposited first. In both cases, 2-D growth was observed. Above 440°C, a small decrease in rate was observed, however the relative insensitivity of the growth rate to a wide growth temperature window, makes this technique particularly attractive for the growth of superlattices where precise thickness control is highly desirable.

4. Properties of InSb, $In_{1-x}Ga_xSb$ and $In_{1-x}Al_xSb$, InSb/InSb, InSb/GaAs

The growth of narrow gap materials on substrates such as GaAs and Si has been the subject of numerous investigations due to the potential for integration of detector array and read-out circuitry on a single chip. The epitaxy of InSb on GaAs presents numerous problems associated with the large 14.6% lattice mismatch between substrate and epilayer. Despite this, single crystal epilayers exhibiting a very high degree of crystalline perfection have been deposited by various growth techniques including MOMS and MSE. Figure 5 shows the dependence of X-ray full width half maximum (FWHM) of the $(004)K_B$ reflection for InSb layers grown on (001)GaAs as a function of layer thickness. As shown in the figure, the crystal quality improves dramatically with increasing thickness, with the observed FWHM decreasing from 650 arc sec. at 0.5 μm to 250 arc sec. for 6 μm thick layers.

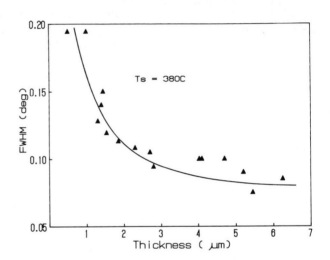

Figure 5. FWHM as a function of epilayer thickness for the $(004)K_B$ line of InSb. From T.Sudersena Rao et.al [3].

A similar effect has been observed in MBE grown material[19] and has been shown to be associated with the high density of misfit dislocations generated at the substrate/epilayer interface. Cross-sectional TEM studies on epilayers prepared by MOMS give defect densities in excess of 10^{11} cm^{-2} at the interface, decreasing to approx. 10^9 cm^{-2}, 3 μm from the interface. The defects observed were predominantly threading dislocations, stacking faults and twins. A high resolution lattice image showing the microtwins observed at the interface is shown in Figure 6[3]. Resonance Raman scattering from epilayers of InSb prepared by MOMS have also been reported[20]. Figure 7 shows the Raman spectra for a 1.1 μm thick epilayer on InSb The first order LO phonon peak at 193 cm^{-1} is clearly shown and has a FWHM of <10 cm^{-1}. The FWHM was found to decrease to <6 cm^{-1} for a layer thickness of 1.67 μm. This was found to be narrower than that observed for bulk wafers. For thin layers, a weak feature

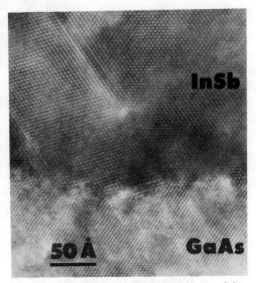

Figure 6. High resolution TEM lattice image of the InSb/GaAs interface. From J.B.Webb et.al.[10].

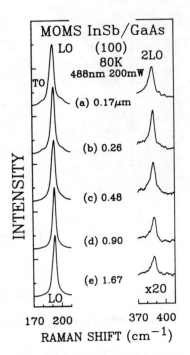

Figure 7. Raman Spectra at 80K as a function of layer thickness for epilayers of InSb grown on GaAs by MOMS. From Z.C.Feng et.al.[20].

appeared at 182 cm^{-1}, related to the forbidden TO phonon mode. The appearance of this feature is indicative of increased crystalline disorder for the thinner layers, in agreement with the TEM and X-ray data.

The electrical properties for both MOMS[8] and MSE[15] grown layers have also been investigated in detail. For the MOMS grown layers, the effect of carbon contamination was evident at low deposition temperatures, however at the optimum temperature for growth and with added hydrogen, the layers showed reduced background carrier densities at 300K of n= 5 x 10^{16} cm^{-3} and mobilities of >10^4 cm^2/V.s. Depending on the growth conditions, p-type conduction was also observed. It was concluded that the layers exhibited not only excess donor levels, but significant compensation as a result of impurities other than carbon.

The MSE grown layers showed carrier densities and mobilities similar in value to those found for the MOMS samples. For these layers the absence of carbon from metalorganic sources clearly indicated contamination from alternate sources. This was confirmed by SIMS analysis of both target and film which showed significant doping from iron introduced from the "high purity" antimony target. All layers showed n-type conduction at 300K except for those deposited

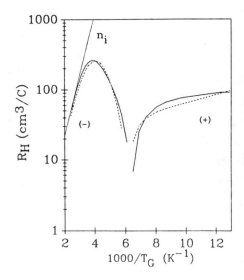

Figure 8. Hall coefficient as a function of reciprocal temperature for an epilayer of InSb on GaAs prepared by MSE. From R.Rousina et.al.[15].

above 475°C which were p-type at 300K. Maximum electron mobilities of 1.5×10^4 cm^2/V.s were observed at 300K. For the p-type layers (p-type at 77K), abackground carrier density of $p = 3 \times 10^{17}$ cm^{-3} and hole mobilities up to 500 cm^2/V.s were observed. The Hall coefficient as a function of reciprocal temperature for one of these layers is shown in Figure 8. The dashed curve is a first order fit to the data based on an acceptor concentration of 3×10^{17} cm^{-3}, mobility ratio of 37 and acceptor activation energy of 11.3 meV. The activation energy for Fe in InSb is 13 meV.

These layers have been used to fabricate homo-junction diodes (p-type epilayer on n-type substrate). The I-V characteristics of these diodes are shown in Figure 9. The forward and reverse characteristics followed the diode equation expressed as[21]

$$J = J_s[\exp(qV/nkT)-1] \qquad (5)$$

where n=2 when recombination currents, or n=1 when diffusion currents dominate. For the characteristics shown in the Figure, at low temperatures and applied voltages, generation-recombination dominates while at higher temperatures n approaches unity indicating a switch to diffusion limited currents. In contrast, the reverse characteristics are dominated by generation-recombination throughout the temperature range studied[15].

4.1 $In_{1-x}Ga_xSb$ on GaAs

A number of studies of the optical and electrical properties of bulk crystals of InSb/GaSb[22] alloys have been reported previously, however little information is available with regards to epilayer growth[23]. Recently, epilayers of $In_{1-x}Ga_xSb$ spanning the entire compositional range (0<x<1) have been grown on (100)GaAs by MOMS[5]. The optical absorption characteristics of these layers as a function of composition have been measured[21] and are shown in Figure 10. The absorption curves were characteristic of a direct gap semiconductor,

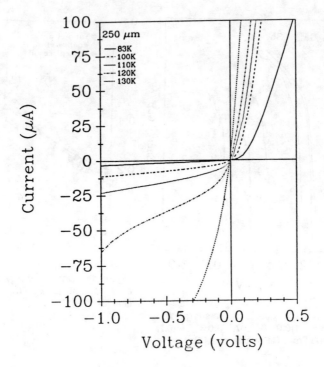

Figure 9. I-V characteristics of a 250 micron diameter InSb homojunction diode grown on (100)InSb. From R.Rousina et.al.[15].

and plots of $(\alpha h\nu)^{1/2}$ versus $h\nu$ were used to determine the band gaps of the alloys. Figure 11 gives a plot of the optical band gap as a function of composition. Also shown in the figure are the results reported for bulk[22] and those of Yano et al[23] for MBE grown material. The bowing parameter deduced from these measurements are similar to earlier values obtained from bulk measurements. The crystalline quality of these layers for all compositions was not optimal due to the fixed growth temperature used for all compositions. This is shown both in the X-ray diffraction data and in the Raman spectroscopy measurements. Figure 12 shows the Raman spectra for layers spanning the compositional range from InSb to GaSb[24]. Both InSb-like and GaSb-like phonons are present and as shown in the Figure, the appearance and strengthening of the forbidden TO mode is observed as the gallium content is increased, indicating increased crystalline disorder despite the reduced lattice mismatch for the higher gallium content layers. The overall crystallinity of the layers is quite good since the LO linewidths for all compositions is <10 cm^{-1}.

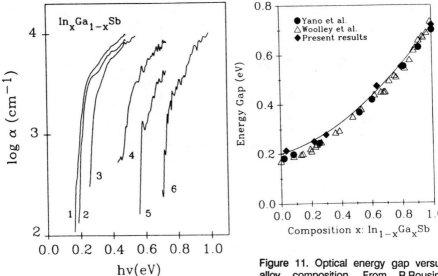

Figure 10. Absorption coefficient versus photon energy for x = (1)0.0, (2)0.07, (3)0.36, (4)0.7, (5)0.78, (6)0.97. From R.Rousina et.al.[5].

Figure 11. Optical energy gap versus alloy composition. From R.Rousina et.al.[3].

As reported earlier, alloys with increased gallium showed an increase in acceptor concentration. This has been associated with the increased amount of TMG used during growth and the resulting increase in carbon incorporated into the epilayers. Acceptor concentrations as high as 10^{20} cm^{-3} were observed for the GaSb epilayers. Although the doping effect of carbon could be minimised in the case of InSb prepared with TMI as the group III precursor, this could not be satisfactorily accomplished with GaSb using TMG. As a result the background impurity concentration for the gallium containing layers was limited by carbon contamination and not target impurities as was found for InSb prepared by MOMS or MSE.

4.2 $In_{1-x}Al_xSb$ Strained Layer Superlattices

Recently, MSE has been used to prepare superlattices of $In_{1-x}Al_xSb$ on (001)InSb[18]. Three magnetron sources are operated simultaneously to sputter targets of InSb, aluminum and antimony. Using this arrangement, 20 period unrelaxed superlattices have been grown using InSb and InSb with 20-40% aluminum. The layer thicknesses have been varied between 35 A° and 78 A°.

Initial studies based on TEM and x-ray diffraction indicate highly uniform layers with excellent crystallinity. X-ray linewidths of 86 arcsec have been obtained for the system $InSb/In_{0.6}Al_{0.4}Sb$. An X-ray diffraction scan for one of the superlattices is shown in Figure 13. The period of the superlattice is nominally 111A° with an average composition of 0.24. The lower solid curve shows the simulated X-ray spectrum based on the above parameters.

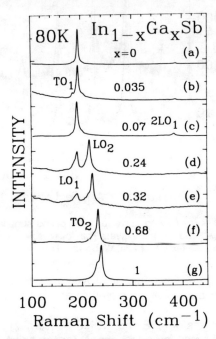

Figure 12. Raman spectra for MOMS grown $In_{1-x}Ga_xSb$ with x= (a)0.0, (b)0.035, (c)0.07, (d)0.24, (e)0.32, (f)0.68 and (g)1.0. From Z.C.Feng et.al.[24].

5. Conclusions:

It has been clearly demonstrated over the past number of years that both Metalorganic Magnetron Sputtering and Magnetron Sputter Epitaxy can be used to deposit high quality single crystal epilayers of antimony based III-V compounds. Both techniques show a high degree of control in layer thickness and in composition. The electrical properties of the deposited layers have indicated that the residual carrier density is limited by the source material purity,

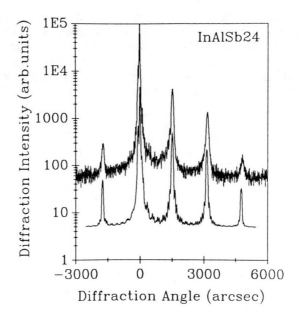

Figure 13. X-ray rocking curve for a 20 period superlattice of InSb/In$_{0.6}$Al$_{0.4}$Sb (Upper curve). The lower solid curve is a simulation (see text). From J.B.Webb et.al.[18].

particularly the group V antimony. Improvements in performance can be expected using higher purity starting materials and better cryoshrouding during growth.

The unique features of MSE and MOMS concerning use of a wide variety of target materials will be an important aspect for the epitaxial growth of metals, insulators and semiconductors for a wide variety of future device applications.

REFERENCES

[1] R.K.Waits in "Thin Film Processes", eds.J.L.Vossen and W.Kern, (Academic Press, New York 1978), p131.
[2] J.B.Webb in "Thin Films From Free Atoms and Particles", ed.K.Klabunde (Academic Press, Orlando 1985), p257.
[3] T.Sudersena Rao, C.Halpin, J.B.Webb, J.P.Noad and J.McCaffrey, J.Appl.Phys.65,(2),(1989),585.
[4] J.B.Webb and C.Halpin, Appl.Phys.Lett.47,(1985),831.
[5] R.Rousina, C.Halpin and J.B.Webb, J.Appl.Phys.68(5),(1990),2181.
[6] G.B.Stringfellow in "Organometallic Vapor-Phase Epitaxy", ed.G.B.Stringfellow,(Academic Press,Boston 1989), p252.
[7] Ibid.p21
[8] T.Sudersena Rao, J.B.Webb, Y.Beaulieu, J.L.Brebner, J.P.Noad and J.Jackman,J.Vac.Sci. Technol.A7(3),(1989),1215.
[9] T.Sudersena Rao, J.B.Webb, J.P.Noad and J.Jackman, Can.J.Phys.67,(1989),298.
[10] J.B.Webb, T.Sudersena Rao, C.Halpin and J.P.Noad in "Mechanisms of Reactions of Organometallics with Surfaces", eds.D.J.Cole-Hamilton and J.O.Williams, (Plenum 1989),p279.
[11] T.F.Kuech and E.Veuhoff,J.Cryst.Growth,68,(1984),148.
[12] M.G.Jacko and S.J.W.Price, Can.J.Chem.42,(1964),1198.
[13] N.Putz, H.Heinecke, M.Heyen, P.Balk, M.Weyers and H.Luth,J.Cryst.Growth 74,(1986),292.
[14] A.Brauers, M.Weyers and P.Balk,Chemtronics,4,(1989),8.
[15] R.Rousina and J.B.Webb,NATO Workshop on Narrow Gap Semiconductors,(Oslo, Norway 1991) in press.
[16] A.H.Eltouky and J.E.Greene,J.Appl.Phys.50(1),(1979),505.
[17] J.B.Webb, C.Halpin, J.Ehrismann and J.P.Noad, Can.J.Phys 65(8),(1987),872.
[18] J.B.Webb, G.H.Yousefi and R.Rousina, Appl.Phys.Lett. in press.
[19] A.J.Noreika, M.H.Francombe and C.E.C.Wood, J.Appl.Phys. 52,(1981),7416.
[20] Z.C.Feng, S.Perkowitz, T.S.Rao and J.B.Webb, Mat.Res.Soc.Symp.Proc. V160,(1990),739.
[21] S.M.Sze,"Physics of Semiconductor Devices",Wiley-Interscience (New York 1969)p.104.
[22] J.C.Wooley, J.A.Evans and C.M.Gillett, Proc.Phys.Soc.74,(1959),244.
[23] M.Yano, T.Takase and M.Kimata, Jpn.J.Appl.Phys.18,(1979),387.
[24] Z.C.Feng, S.Perkowitz, R.Rousina and J.B.Webb, Can.J.Phys.69,N3,(1991),4.

PROPERTIES OF CdMnTe FILMS AND CdMnTe-CdTe SUPERLATTICES GROWN BY PULSED LASER EVAPORATION AND EPITAXY

J. M. Wrobel
Department of Physics, University of Missouri
Kansas City, MO 64110, USA

and

J. J. Dubowski
Institute for Microstructural Sciences,
National Research Council of Canada
Ottawa, Ontario K1A 0R6, Canada

ABSTRACT

The research that is discussed in this review is concerned with the structural and optical properties of $Cd_{1-x}Mn_xTe$ films and $CdTe-Cd_{1-x}Mn_xTe$ superlattices grown by Pulsed Laser Evaporation and Epitaxy (PLEE): a technique that has been especially developed for the deposition of thin films, quantum wells and superlattices of semiconductor materials.

1. INTRODUCTION

This chapter is devoted to the properties of CdMnTe-based thin films, quantum wells and superlattices grown by Pulsed Laser Evaporation and Epitaxy (PLEE). It summarizes the last several years of research on the use of PLEE for the growth of microstructures and heterojunctions. After a short discussion of the significance of $Cd_{1-x}Mn_xTe$, the characteristics of the material obtained by PLEE are described. These include crystal structure, surface morphology, chemical composition, film orientation, presence of impurities, and some optical properties.

$Cd_{1-x}Mn_xTe$ alloys first attracted the attention of the scientific community at the end of the seventies. These compounds became the first and most studied members of a class of materials known either as Diluted Magnetic Semiconductors [1-7], or as Semimagnetic Semiconductors. Today

hundreds of papers have been written on $Cd_{1-x}Mn_xTe$. It is believed that the presence of paramagnetic ions leads to the strong Faraday rotation, as well as causing the large Zeeman splitting of electronic levels, the formation of the spin glass phase, the formation of magnetic polarons, magnetic field induced metal-insulator transitions, etc. The band gap, lattice constant, and phonon structure are determined by the chemical composition of these materials. Over a wide range of composition these dependencies are linear functions of the manganese content.

$Cd_{1-x}Mn_xTe$ forms single crystals of cubic, zinc-blende structure where up to 77 percent of the Cd cations are substituted by manganese Mn^{+2} ions [8]. Both cadmium and manganese contribute their two s-electrons to the six p-electrons of tellurium to form a tetrahedral s-p^3 bonding. The upper limits for cubic structure are dictated by the fact that binary bulk MnTe does not crystallize in this structure. However, it should be noted that epitaxial thin films of MnTe have recently been obtained in a cubic phase [9,10], thereby shifting the limit upwards from 77 percent.

The most frequently used methods for producing epitaxial thin films of $Cd_{1-x}Mn_xTe$ include Molecular Beam Epitaxy (MBE) [11-15], Metalorganic Chemical Vapor Deposition (MOCVD) [16-18], Pulsed Laser Evaporation and Epitaxy [19-21], Atomic Layer Epitaxy [22,23], and Ionized Cluster Beam Deposition [24]. These methods are also the most widely recognized techniques for the creation of CdTe-CdMnTe superlattices [25] where control of the atomic scale growth is essential [26-33].

2. PULSED LASER EVAPORATION AND EPITAXY

Pulsed Laser Evaporation and Epitaxy (PLEE) is a high vacuum growth technique which has been developed at the National Research Council of Canada particularly for the growth of semiconductor layers [34,35] . The growth chamber is very similar to that of MBE except that instead of using Knudsen effusion cells as molecular beam sources, suitable pulsed lasers are used for the vaporization of solid targets [35,36]. Recently, this technique, which is also known as Pulsed Laser Deposition (PLD), has achieved broader recognition because of its use in the production of high T_c superconductors [37]. Deposition from pulsed laser induced vapors is also known as Laser-Molecular Beam Epitaxy.

The mechanisms involved in laser ablation may be due to thermal or photo-processes, or a combination of both. The advantage of the use of a laser for vaporization of solid targets is that it makes possible local vaporization from a small spot where a focused laser beam strikes the target. If laser power is kept at its minimum level to maintain such a local vaporization, the rest of the target will remain at a near ambient temperature. This will result in greatly reduced contamination from high temperature devices that are normally applied in conventional vacuum deposition systems. Consequently, pulsed lasers, with a time between pulses allowing for the dissipation of excessive heat, are of particular interest for deposition.

The properties of the material pulses (train of pulses) ejected from an ablated target are determined by laser light wavelength, the angle of incidence, pulse duration, the energy density at the target, and laser fluence [38-48]. Most experiments with pulsed laser evaporation demonstrate that congruent evaporation takes place, preserving the chemical composition of the target. This feature has been successfully utilized for the growth of high T_c superconductors, and it is also a basis for the growth of the ternary compounds discussed in this chapter.

The ions and energetic neutrals produced by laser evaporation interact with the substrate through several mechanisms depending on their energy [49]. These processes include: (1) desorption of weakly bound atoms from the substrate surface, (2) sputtering, (3) surface diffusion (migration) of atoms, (4) displacement of lattice atoms, (5) adsorption of atoms arriving on the surface, (6) implantation, (7) electronic excitation.

Due to its nature, two different parameters characterize deposition in PLEE: pulsed growth rate, R_p, and mean growth rate, R_m. The first is defined as the average thickness of the film deposited in one pulse, and the second as the thickness of the film deposited in a time unit. Typical values in epitaxial deposition are R_p=0.02-0.5 Å/pulse and R_m=0.3-5.0 μm/h [35].

With the above mentioned deposition rate, it takes several laser pulses to deposit one atomic layer. Therefore, PLEE provides ideal control of abrupt atomic interfaces while maintaining a relatively high mean growth rate. Obviously, other processes at interfaces, particularly interdiffusion, counteract this control, and interface abruptness can be lost. Another opportunity provided by the typical pulsed growth rate is the possibility of precise mixing of several laser produced fluxes, resulting in the production of structures whose chemical composition can be arbitrarily designed. Such a feature is difficult to match by other methods of film deposition.

3. EPITAXIAL LAYERS AND MICROSTRUCTURES OF $Cd_{1-x}Mn_xTe$

The layers are routinely characterized with Reflection High-Energy Electron Diffraction (RHEED), Scanning Electron Microscopy (SEM), and Energy-Dispersive X-ray Spectroscopy (EDX), in order to accurately evaluate their quality and suitability for manufacturing devices. Secondary Ion Mass Spectroscopy (SIMS) is used for detection of impurities and their distribution. It also enables us to study the uniformity of chemical composition in the depths of the samples. Optical properties of the films are studied with Raman spectroscopy, photoluminescence, modulated reflectivity, and the photovoltaic effect.

3.1. substrate preparation

The attraction of GaAs as a substrate for II-VI semiconductors is due to the current development of the device and material technology of this

material. GaAs wafers are less expensive than any of the II-VI semiconductors, and it is possible to achieve epitaxy of II-VI materials on such wafers. One can imagine a whole series of II-VI semiconductor devices built of layers grown on GaAs containing circuits supporting these devices.

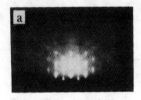 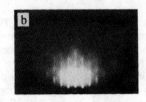 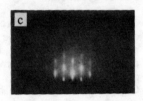

Fig. 1. RHEED patterns of substrates: a) annealed (001) GaAs surface ([1$\bar{1}$0] azimuth); b) ion cleaned (001) GaAs surface (same azimuth); c) ion cleaned (001) InSb surface ([110] azimuth) (from ref. [36,50]).

Special care has to be taken to prepare the substrates for successful epitaxy of $Cd_{1-x}Mn_xTe$. For GaAs substrates, the standard procedure consists of degreasing in chloroform, acetone and alcohol, then etching in $H_2SO_4:H_2O_2:H_2O$ (7:1:1). After this treatment, the substrates can be introduced into the PLEE system to be further treated in order to remove oxides from the surface. This can be accomplished by a 2 minute annealing at 580°C possibly under overpressure of As in order to prevent loss of As from the surface of GaAs. This stage is crucial for the properties of the film, and has to be monitored in-situ by RHEED. A typical RHEED pattern obtained from an annealed (001) GaAs surface is shown in figure 1a. Alternatively or additionally, other techniques can be used for the final preparation of the substrate. Ion bombardment, for example, can also be used to prepare the substrate surface of GaAs for epitaxy [36]. An oxide free GaAs surface prepared in the PLEE system reveals a sharp RHEED pattern with clear Kikuchi bands as seen in figure 1b.

Because its lattice constant is almost equal to that of CdTe, InSb is a very attractive substrate for the epitaxy of some II-VI semiconductors. In the preparation of InSb substrates, the standard degreasing procedure is applied. A brief chemical etch in concentrated HCl is followed by rinsing in deionized water. Finally, in the vacuum chamber, the substrates undergo ion-cleaning by bombardment with Ne^+ ions at 300°C [50]. In-situ monitoring with the RHEED apparatus insures satisfactory preparation of the surface. For (001) InSb, the pattern observed at the [110] azimuth typically exhibits 2×4 reconstruction as shown in figure 1c.

Successful epitaxy on CdTe or CdZnTe requires a different procedure for substrate preparation. After degreasing, the substrates are etched in a solution of Br in CH_3OH (1%) and annealed for about 15 minutes at 350°C [32]. The vacuum annealing leads to crystallographically dependent sublimation [51]. For example at 350°C the rates of sublimation from the [111] tellurium face and [001] surfaces of CdTe are 70 Å/min and 15 Å/min, respectively. As a rule we

find that the substrate preparation procedures which are suitable for MBE are also suitable for PLEE.

3.2. composition

The chemical composition of the $Cd_{1-x}Mn_xTe$ layer produced in PLEE can differ from that of the target. This variation is caused by the different vapor pressures and sticking coefficients of the constituent elements. The films discussed here were grown in a temperature range of 180°C - 320°C. High quality epitaxial layers, with a chemical composition equal to that of the $Cd_{1-x}Mn_xTe$ targets, can be obtained by supplementing the laser generated flux of vapors with cadmium from another source [19,20,35]. Deposition, using the XeCl laser to evaporate CdMnTe, without additional Cd flux, systematically resulted in layers with higher Mn concentration. For example, a (111) $Cd_{1-x}Mn_xTe$ epitaxial film with x=0.7 was successfully obtained from a $Cd_{0.44}Mn_{0.56}Te$ target. The manganese concentration in the film depends on the substrate temperature during growth [19,20,35]. Generally, the lower the growth temperature, the closer the Mn content in the film approximates that in the target.

Computer control of the lasers used in PLEE allows instantaneous changes in the vaporization rates of CdTe and $Cd_{1-x}Mn_xTe$ targets, which has an almost immediate effect on the composition of the beam. This degree of control makes possible the growth of quantum wells and superlattice structures. Figure 2 shows a typical SIMS profile of a CdTe-CdMnTe multiple quantum well (MQW) structure. This particular structure was grown on a CdZnTe substrate with a 0.5 μm thick CdMnTe buffer layer. Considering both the composition and thickness of the layers, the SIMS data indicates a very uniform structure with sharp interfaces.

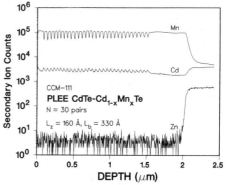

Fig. 2. Typical SIMS profiles of a PLEE-grown $CdTe-Cd_{0.9}Mn_{0.1}Te$ MQW structure (from ref. [32]).

In fact, the actual quality of the interfaces is probably even better than seen in the figure, as the data is affected by the resolution of the SIMS technique.

Heteroepitaxy of CdTe and $Cd_{1-x}Mn_xTe$ on GaAs and InSb resulted in the presence of group III elements in the layers [52,53]. The distribution of the elements is nonuniform. SIMS mapping, seen in figure 3, shows a high surface concentration of Ga. The area with increased Ga concentration forms islands ranging in size from 2 μm to 10 μm in diameter. On the average,

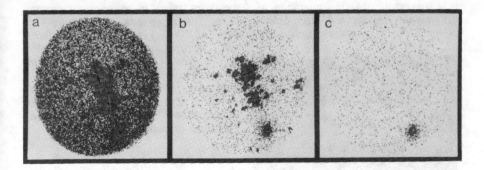

Fig. 3. The Ga ion images of a 1.85 μm thick (111) $Cd_{0.92}Mn_{0.08}Te$ layer; a) at the surface; b) at a depth of about 0.07 μm; c) 0.8 μm deep (from ref. [52]).

10^4-10^6 islands can be found on a 1 cm^2 area of the film surface. Increased average Ga (In) concentration of 10^{18} - 10^{19} cm^{-3} at the surface decreases rapidly below the SIMS detection limit of 5×10^{15} cm^{-3} within 0.2 μm from the surface. Beneath the islands there are often some channels where increased Ga (In) concentration extended across the layer to the interface with the substrate. Near the interface a 0.3 - 0.4 μm thick region also exhibits higher Ga (In) concentration, as is indicated in figure 4. The distribution of the group III elements and their concentration can be altered thermally. Annealing of the layer for just two hours at 400°C increases the concentration of the impurity inside the layer by a few orders.

It appears that the effect of the outdiffusion of these impurities is not dependent on the growth temperature. The presence of Ga from the substrate was observed in approximately the same concentration even at the surface of a CdTe layer grown at 25°C. Qualitatively, the distribution of the impurity depends on the film orientation. Gallium was evenly spread at the (111) CdTe film surface, while islands of high concentration exist at the (001) surface.

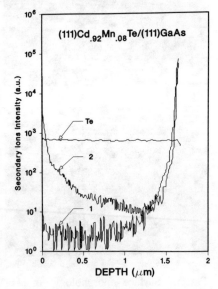

Fig. 4. SIMS Ga profiles for the (1) as grown (111) $Cd_{0.92}Mn_{0.08}Te$ layer, and (2) after a 2 hour annealing at 400°C (from ref. [52]).

It is suggested that the migration of the element takes place primarily during the initial growth, and that defects in the film drive this migration. The presence of migration channels, precipitation of Ga (In) on the grown surface and incorporation in the film can cause some technological problems in devices built in layers grown on III-V substrates. These impurities can be optically active [54] and lead to the deterioration of the optical and electrical properties of devices.

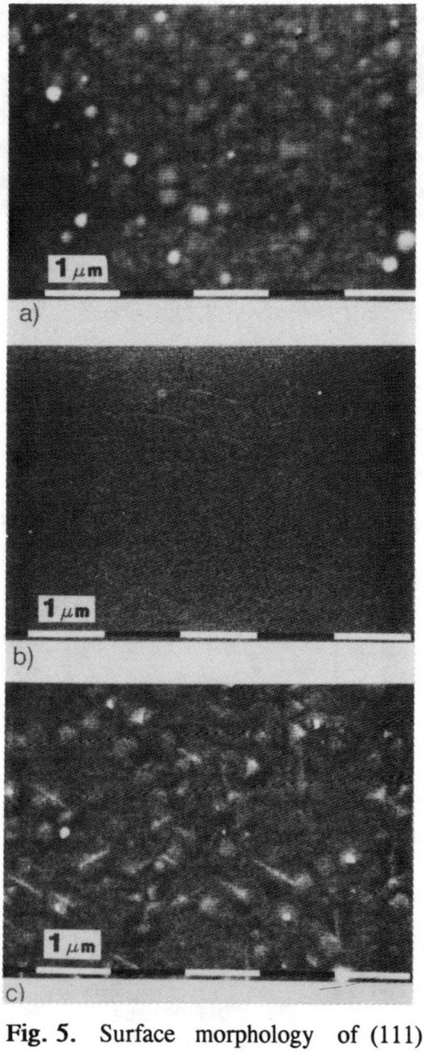

3.3. surface morphology and crystal structure

Other physical properties of the film, besides composition, are affected by the substrate temperature, particularly the crystal structure and surface morphology. Figure 5 indicates the effect of an additional Cd flux on the quality of $Cd_{1-x}Mn_xTe$ epitaxial layers grown at 300°C on (111) GaAs [19]. Either a deficiency or an excess of Cd atoms will have a negative effect on the crystal structure and surface morphology of the layers grown at this temperature. Only the appropriate Cd flux of about $(2 - 4) \times 10^{15}$ atoms/cm^{2}s, produces a smooth surface morphology as seen in figure 5b and results in the lattice producing the sharp electron channeling patterns shown in figure 6. In the presence of the Cd flux and the appropriate surface temperature, SEM micrographs revealed a featureless surface morphology with microparticles, 2-3 μm in diameter, occasionally embedded in the film. With insufficient Cd flux, the density of such particles is much higher. With excessive Cd flux, an additional structure, related to the orientation of the layer, appears. The morphology has the triple symmetry characteristic for the (111) direction. In fact, the

Fig. 5. Surface morphology of (111) $Cd_{1-x}Mn_xTe$ films grown on (111) GaAs at 300°C; a) no additional Cd flux; b) optimal Cd flux; c) excessive Cd flux (from ref. [19]).

surface quality imposes much narrower limits on the partial pressure of Cd vapors than does the composition.

A smooth surface morphology can also be obtained if deposition is carried out in the absence of an additional flux of Cd, on substrates kept at a temperature of 240°C or below [20,33,55]. For example, the smooth surface morphology was characteristic for high quality structures with $x=0.1$, which were grown from a $Cd_{0.93}Mn_{0.07}Te$ target [55]. Surface morphology was very strongly correlated with the quality of the crystal structure. Additionally, the scattering of electrons by features observed on the surface significantly decreased the sharpness of the electron channeling pattern.

Fig. 6. Electron channeling pattern for a 1.7 μm thick (111) $Cd_{0.95}Mn_{0.05}Te$ layer obtained with optimal Cd flux (from ref. [19]).

3.4. film orientation

An example of the RHEED pattern is shown in figure 7a together with the SEM image for a (001) $Cd_{0.91}Mn_{0.09}Te$ layer obtained on a (001) CdZnTe substrate. A flat surface is evidenced by the presence of streaks and Kikuchi bands. The surface morphology and RHEED pattern of a (111) $Cd_{1-x}Mn_xTe$ layer grown on (111) GaAs are shown in figure 7b. The SEM image indicates a surface of similar quality to that of the previous example, but a slight modulation of the intensity of the RHEED rods is an indication of less smooth morphology. In both cases, weaker lines between the main pattern indicate

a)

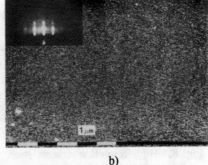
b)

Fig. 7. Surface morphology and associated RHEED pattern for: a) (001) $Cd_{0.91}Mn_{0.09}Te$ layer (from ref. [33]); b) (111) $Cd_{0.917}Mn_{0.083}Te$ layer (from ref. [20]).

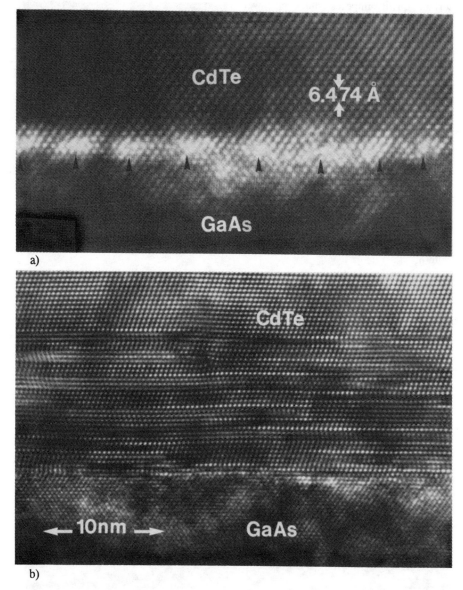

Fig. 8. High resolution transmission electron microscope image of PLEE a) (001)CdTe/(001)GaAs, and b) (111)CdTe/(001)GaAs interfaces. The arrows indicate positions of the misfit dislocations (from ref. [56]).

surface reconstruction of the film.

The pregrowth preparation process influences the properties of the substrate surface, which in turn have a direct influence on the orientation of the epitaxial layer [36]. Annealing of the (001) GaAs substrate without overpressure of As in the PLEE chamber leads to a rough surface with no reconstruction. The RHEED pattern of a substrate prepared in such a way (figure 1a) indicates volume diffraction with no trace of surface reconstruction. This kind of surface always leads to parallel epitaxy of CdTe. Ne ion sputtering of the (001) GaAs surface produces a smoother, reconstructed surface, indicated by a RHEED pattern with weaker lines between the main streaks (figure 1b). In this case the CdTe layer grows in the (111) orientation with azimuth [1$\bar{1}$0] CdTe parallel to the [110] GaAs direction.

Both materials, CdTe and GaAs, crystalize in the same zinc-blende structure but their lattice constants differ by approximately 14 percent. Figure 8a illustrates the interface between PLEE grown (001) CdTe and a (001) GaAs substrate obtained with a high-resolution transmission electron microscope (TEM) [56]. The stress at the interface is released by numerous mismatch dislocations which, on the average, occur every seven lattice constants. The arrows in the figure indicate the position of the dislocations. The nonparallel epitaxy of (111) CdTe on (001) GaAs produces lower stress and a lower number of mismatch dislocations, as the atomic layer spacing in the [$\bar{1}$12] direction of CdTe multiplied by a factor of 3 is almost equal to that in the [$\bar{1}$10] direction in GaAs multiplied by a factor of 2 [57,58]. A TEM image of the (111)CdTe/(001)GaAs interface is presented in figure 8b.

4. OPTICAL PROPERTIES

4.1. photoluminescence

Information about the electronic structure of semiconductors, particularly energy levels within the band gap, can be obtained from photoluminescence studies. The band gap luminescence in $Cd_{1-x}Mn_xTe$ compounds is divided into three regions: excitonic, near edge emission, and defect band emission.

The highest energy excitonic region usually exhibits two transitions called L1 and L2 [59]. The first, attributed to recombination of excitons bound to neutral acceptors, can show a structure in some samples. The second, with energy very close to the free exciton energy, is attributed to an exciton trapped via magnetic interaction with Mn^+ ions [59]. The energies at which these transitions are observed in PLEE layers [19,20,55,60] agree with findings for bulk crystals [61], which at a temperature of 4 K are:

$$E_{L2} = 1604.7 + 1397x \text{ (meV)} \text{ for } 0.05 \leq x \leq 0.2 \tag{1}$$

$$E_{L2} = 1575 + 1536x \text{ (meV)} \text{ for } 0.2 \leq x \leq 0.4 \tag{2}$$

$$E_{L1} = 1588.8 + 1440x \text{ (meV)} \text{ for } 0 < x \leq 0.1 \tag{3}$$

Sometimes band-to-band or free-to-bound transitions are observed in the excitonic region.

Near edge emission includes free-to-bound and donor-acceptor-pair recombination transitions. This region is particularly important in the investigation of levels introduced by various impurities. The energies at which the lines occur are related to the energies of the impurity levels in the following way [62]:

$$E_{FB} = E_g - E_{A(D)} + \tfrac{1}{2}kT \qquad (4)$$

$$E_{DAP} = E_g - E_A - E_D + \frac{e^2}{4\pi\epsilon_0\epsilon R} \qquad (5)$$

where E_{FB} and E_{DAP} are the energies of the photons emitted in free-to-bound and donor-acceptor-pair transitions, E_g is the energy gap, E_A and E_D are the energies of the acceptor and the donor levels with respect to the top of the valence and the bottom of the conduction bands, T is the temperature of the electrons in the system, and R is the distance between the elements of the donor-acceptor-pair.

The mechanisms which produce the defect band are identical to those which are responsible for near edge emission. However, the near edge emission is caused by impurities in the material, and the defect band is caused by optically active structural defects of the crystal. These are most likely vacancies in the cation lattice which are always present in $Cd_{1-x}Mn_xTe$.

In figure 9 we present an example of a PL spectrum for a bulk material (figure 9a) and compare it with that of a (001) layer (figure 9b) of similar composition grown on (001) InSb [63]. Both spectra are dominated by a sharp line, L2 [59]. For a given chemical composition, the width of the L2 line is a measure of the quality of the material [12,64], and is dependent on both the uniformity of the composition and the concentration of defects. Both spectra indicate that the

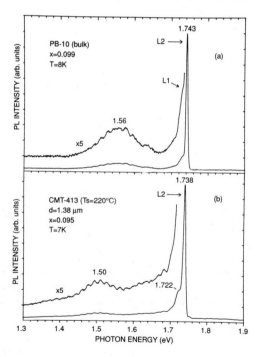

Fig. 9. Photoluminescence spectra of $Cd_{0.9}Mn_{0.1}Te$; a) bulk sample; b) PLEE grown epitaxial layer (from ref. [63]).

samples are of very good quality, although the layer exhibits a slightly broader line L2 (8 meV) as compared with the bulk (6 meV). The chemical composition of the bulk material and the layer can be determined from the energy dependence of the L2 transition (eq.2). In the studied case, the composition of the bulk was $x=0.099$, and that of the film was $x=0.095$. The low energy shoulder, L1, in the excitonic region is due to an exciton bound to an acceptor [59]. The broad band observed at the lowest energies of the spectrum has the properties of a donor-acceptor-pair recombination. Photoluminescence spectra obtained for other layers exhibit characteristics consistent with the published data for bulk material confirming the high performance of PLEE [19,20,35,55,60], although some incorporation of the group III element from the substrate is observed, as was discussed in paragraph 3.2.

The electronic confinement in superlattices alters the electron energy levels as compared to the bulk crystals. This alteration strongly depends on the geometrical configuration of the superlattice (uniformity of the well thickness within one layer and from layer to layer). Conclusions related to the superlattice geometry can be drawn from the photoluminescence experiments.

Line X shown in figure 10 is caused by a weakly bound E_1-HH_1 excitonic recombination in a multiple quantum well structure grown by PLEE [65]. Line X_{hh} represents the photoluminescence excitation spectrum for the same transition. The 88 Å thick CdTe wells are separated, in this case, by 225 Å thick $Cd_{0.9}Mn_{0.1}Te$ barriers. The energy difference between the X and X_{hh} lines, the so called Stokes shift, is caused by the trapping of the free excitons on the interface defects [66]. A large Stokes shift indicates rough interfaces. For PLEE grown superlattices, the values of the shift are comparable, and in some cases even smaller than for those grown by other methods [25,67].

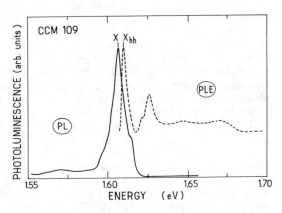

Fig. 10. Photoluminescence (PL) and photoluminescence excitation (PLE) spectra of a 22 MQW CdTe-$Cd_{0.9}Mn_{0.1}Te$ structure (from ref. [65]).

The width of the photoluminescence line is related to the geometrical uniformity of the superlattices. Well width fluctuations lead to a dispersion of the confinement energies of the electron and hole fundamental states, thus to a dispersion of the excitonic recombination energy. The average width of the excitonic lines in the PLEE superlattices is about 8 meV at half maximum, which is comparable to the values obtained in samples grown by other

methods [67,68].

The energy of the E_1-HH_1 transition is mainly determined by the well width. Strains and band offsets only weakly affect this energy. Therefore it is possible to approximate the thickness of the wells in relation to the transition energy. Such optically measured thickness has satisfactory accuracy for narrow wells not exceeding 150 Å. Above this value, the analysis worsens significantly, because the energy of the E_1-HH_1 transition becomes only weakly dependent on the well width.

4.2. Raman spectra

Raman scattering is a good tool for determining the quality of the crystal structure, the composition, and the presence of various compounds in the layer. The intensity and width of the observed lines can be directly related to the quality of the structure. The formation of other compounds within the layer can be indicated by additional lines in the spectrum.

$Cd_{1-x}Mn_xTe$ is a material which produces two separate "CdTe-like" and "MnTe-like" modes of vibration. Their frequencies are dependent on the Mn concentration x, and are well known [69-71]. The spectrum presented in figure 11 was obtained at 14 K for a sample containing 56 percent manganese [72]. The high order Raman lines seen in this spectrum are in outgoing resonance with E_g electronic transition, which enhances the LO mode intensity. All the lines in the spectra presented in the figure can be identified as $Cd_{1-x}Mn_xTe$ modes. The number of lines observed for the epitaxial layers exceeds that observed for bulk crystals. This is also true of Raman spectra obtained at 80 K as well [60]. More than four overtones were never reported for bulk $Cd_{1-x}Mn_xTe$ material. Table I assigns the combinations of phonons to the Raman features seen at 80 K in $Cd_{0.51}Mn_{0.49}Te$ [60].

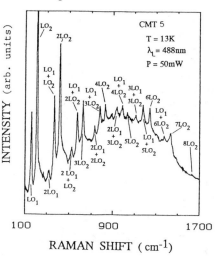

Fig. 11. Raman spectrum at 13 K of a (111) $Cd_{0.44}Mn_{0.56}Te$ layer grown by PLEE on (111) GaAs (from ref. [72]).

In PLEE $Cd_{1-x}Mn_xTe$ layers grown on GaAs and InSb substrates, one can expect the occurrence of Raman lines associated with the migration of the group III element from the substrate, the formation of interfacial Te layers and some compounds in the layer. For example deposition of CdTe on InSb can

Table I. Raman transitions observed in $Cd_{0.51}Mn_{0.49}Te$ at 80 K (from ref. [60]).

Group	Assignment	Energy (cm^{-1})
1LO:	LO_1	160
	LO_2	208
2LO:	$2LO_1$	320
	LO_1+LO_2	368
	$2LO_2$	416
3LO:	$2LO_1+LO_2$	526
	LO_1+2LO_2	576
	$3LO_2$	623
4LO:	$3LO_1+LO_2$	683
	$2LO_1+2LO_2$	736
	LO_1+3LO_2	785
	$4LO_2$	833
5LO:	$2LO_1+3LO_2$	943
	LO_1+4LO_2	993
	$5LO_2$	1034
6LO:	$6LO_1$	961
	$5LO_1+LO_2$	1006
	$4LO_1+2LO_2$	1051
	$3LO_1+3LO_2$	1097
	$2LO_1+4LO_2$	1148
	LO_1+5LO_2	1202
	$6LO_2$	1249
7LO:	$7LO_1$	1121
	$4LO_1+3LO_2$	1265
	$2LO_1+5LO_2$	1363
	LO_1+6LO_2	1400
	$7LO_2$	1453
8LO:	$8LO_1$	1289
	$6LO_1+2LO_2$	1375
	$5LO_1+3LO_2$	1429
	$3LO_1+5LO_2$	1536
	$2LO_1+6LO_2$	1564
	LO_1+7LO_2	1618

form an undesirable compound with the metal from the substrate, such as In_2Te_3 [73,74]. Modes of Sb have been reported in CdTe films grown on InSb by MBE [73,74]. Free Te is also readily detected [75]. No presence of In_2Te_3 at the $Cd_{1-x}Mn_xTe/InSb$ interface was found in PLEE grown samples. However, there were some indications of the presence of free Te and Sb [76].

The effect of composition on the energies of phonons observed in PLEE films is nearly identical to its effect in bulk material, and follows the linear dependence:

$$\omega_{LO1} = 171 - 22x \quad \text{(CdTe-like)} \tag{6}$$

$$\omega_{LO2} = 194 + 27x \quad \text{(MnTe-like)} \tag{7}$$

in units of cm^{-1} and at 80 K. The plot in figure 12 compares the results obtained for PLEE thin films with data obtained for bulk material, in both Raman [69] and infrared spectroscopy [70,71]. As the manganese concentration rises, the energy of the CdTe-like phonon drops and that of the MnTe-like increases. The line intensities for the films are also affected by the composition just as they are in bulk form.

The width at half maximum of photoluminescence and Raman lines is often used as a measure of the crystalline quality of epitaxial films. As different compositions are compared, this procedure can be misleading. $Cd_{1-x}Mn_xTe$ may be considered as a mixture of CdTe and MnTe crystals, each having its own vibration mode. As the latter compound is introduced into pure CdTe, it broadens the CdTe-like line by disturbing the original lattice. In general, the broadening of the CdTe-like line is proportional to the manganese content. Similarly, the introduction of CdTe into the MnTe lattice results in the broadening of the MnTe-like line. Indeed, such a tendency, presented in figure 13, is observed in the $Cd_{1-x}Mn_xTe$ film grown by PLEE.

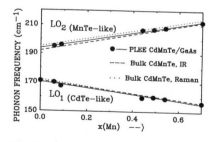

Fig. 12. Composition dependencies of LO phonon energies in $Cd_{1-x}Mn_xTe$ films grown by PLEE compared to bulk crystals (from ref. [60]).

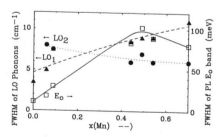

Fig. 13. The full width at half maximum of Raman and photoluminescence lines of PLEE-grown $Cd_{1-x}Mn_xTe/GaAs$ structures. Measurements taken at 80 K (from ref. [60]).

If a single spectral line is produced in a spectrum, as takes place for example in photoluminescence, the explanation of the broadening is similar. The line should be narrower for binaries than for an alloyed ternary compound. In the case of a single line there should be a composition at which the width reaches the maximum value. Figure 13 indicates that at about 50 percent of manganese the width of the E_0 photoluminescence line has the greatest value.

In multiple quantum well and superlattice structures, weak Raman lines, which are caused by the zone folding along one direction, occur at very low energy [77,78]. However, the Raman spectra are dominated by features characteristic to both well and barrier materials, and provide information about individual layers. Figure 14 represents a Raman spectrum of a PLEE grown CdTe-CdMnTe superlattice. As seen in the figure, both barriers and wells leave their signatures in the spectrum. Additionally, it is possible to observe an IF line due to the so called interface mode [25,78]. Because the electronic gap is increased by electron confinement in a superlattice, the onset of resonance for multiple quantum wells shifts towards the high energy side with respect to bulk material or single epitaxial layers.

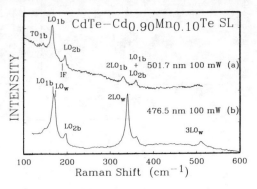

Fig. 14. Raman spectra of a CdTe-$Cd_{0.9}Mn_{0.1}$Te superlattice structure grown by PLEE; a) non-resonant excitation; b) resonant excitation (from ref. [65]).

4.3. phototreflectance

Figure 15 shows photoreflectance spectra obtained at 2 K for bulk $Cd_{0.9}Mn_{0.1}$Te, a PLEE-grown $Cd_{0.9}Mn_{0.1}$Te layer [63], and a MQW consisting of 20 CdTe-$Cd_{0.9}Mn_{0.1}$Te periods [79]. The theory of photoreflectance [80,81] links the spectra with sample quality [64] and composition [82,83] through the broadening parameter Γ and the energy of the observed features respectively. The conclusions concerning the characteristics of the film and the target are very consistent with the results obtained from photoluminescence. Free and bound excitons are seen in the spectrum at low temperature. However when the temperature is increased, the bound exciton feature disappears. Studies of the samples grown on InSb substrates by both techniques, photoreflectance and photoluminescence, show that the quality of PLEE layers improves with the thickness. Layers which are thinner than 0.5 μm lack excitonic features in the photoluminescence, and in the photoreflectance spectra the broadening

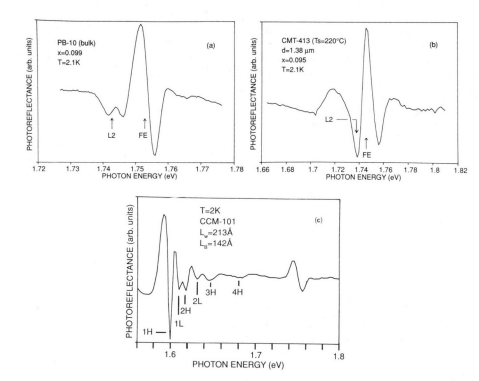

Fig. 15. Photoreflectance spectra at 2 K; a) bulk $Cd_{0.9}Mn_{0.1}Te$; b) PLEE-grown $Cd_{0.9}Mn_{0.1}Te$ epilayer (from ref. [63]); c) CdTe-$Cd_{0.9}Mn_{0.1}Te$ MQW structure (from ref. [79]).

parameter resumes higher values. We feel that the high concentration of indium that migrates from the substrate and the dislocations in the thin layers are the causes of this behavior.

The photoreflectance spectrum from the MQW structure, containing 20 periods of 213 Å thick (001) CdTe wells separated by 142 Å thick (001) $Cd_{0.9}Mn_{0.1}Te$ barriers, covered by a 300 Å thick $Cd_{0.9}Mn_{0.1}Te$ cap layer, is dominated by a strong 1H feature near 1.600 eV. This feature corresponds to a free exciton transition between the heavy hole level and the lowest electron state. The accompanied structure observed at 1.610 eV (1L), 1.618 eV (2H), 1.631 eV (2L), and 1.646 eV (3H) is related to the transitions between the two hole levels with confined free exciton levels. The feature at 1.750 eV is attributed to the free exciton recombination in the cap layer.

4.4. photovoltaic effect

The photovoltaic effect in $Cd_{1-x}Mn_xTe$ films grown by PLEE exhibits the characteristic of the fundamental absorption edge in the bulk material [83]. In figure 16, the spectra represent the data for two $Cd_{0.9}Mn_{0.1}Te$ layers grown on GaAs with and without a CdTe buffer layer [63]. The spectra were obtained using the frequency modulated light beam of the interferometer. Under those experimental conditions, the spectrum shown in figure 16a is similar to that of the absorption spectrum of bulk $Cd_{1-x}Mn_xTe$ with the same manganese content. Such a spectrum exhibits a sharp rise in absorption below the band edge and a peak in absorption at free exciton energy. The differences between the absorption and photovoltaic spectra are attributed to the variation in the carrier lifetime and light penetration depth, with photon energies involved in the measurements [63].

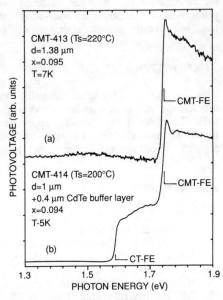

Fig. 16. Low temperature photovoltaic spectra of PLEE $Cd_{0.9}Mn_{0.1}Te$ epitaxial layers; a) grown on InSb; b) grown on CdTe buffer layer (from ref. [63]).

5. CONCLUSIONS

The structural and optical properties of $Cd_{1-x}Mn_xTe$ films and CdTe-$Cd_{1-x}Mn_xTe$ quantum wells and superlattices grown by Pulsed Laser Evaporation and Epitaxy (PLEE) have been discussed in this review. The structures have the characteristics of the best currently available material. This is indicated by the sharp interfaces observed with high resolution transmission electron microscopy, the sharp excitonic lines in photoluminescence, narrow double crystal X-ray diffraction spectra, and the small Stokes shifts observed in the superlattices with photoexcitation luminescence spectroscopy. Some features, such as the high order (n=8) optical phonon Raman scattering have been observed in $Cd_{1-x}Mn_xTe$ for the first time. The success of PLEE in the fabrication of high quality semiconductor structures is, at least partly, a result of 'low-temperature' epitaxy carried out in a clean vacuum environment. Under such conditions, transfer of impurities other than those originally embedded in the ablated target into the deposited structures is practically excluded. The potential of PLEE in the growth of advanced semiconductor structures has only been revealed recently. Further progress in this area requires detailed study of

the mechanisms of laser ablation in order to better understand basic phenomena inherent to laser-solid interaction, and to provide adequate feedback for the atomic scale growth process.

ACKNOWLEDGEMENTS

We would like to recognize the contribution of the coauthors of our articles quoted in this review. The work of Dr. P. Becla who made all the bulk CdMnTe material is especially appreciated. We thank Mrs. T. Wrobel for help in editing this chapter. Finally, we would like to acknowledge the constant support of our wives, Theresa and Barbara.

REFERENCES

1. R.R. Galazka, in the Proceedings of the 14th International Conference on the Physics of Semiconductors, Edinburgh, 1978, edited by B.L.H. Wilson; Inst. of Phys. Conf. Ser. **43**, 133 (1978).
2. J.A. Gaj, J. Phys. Soc. Jpn. Suppl. A **49**, 797 (1980).
3. J.K. Furdyna, J. Appl. Phys. **53**, 7637 (1982).
4. N.B. Brandt and V.V. Moshchalkov, Adv. Phys. **33**, 193 (1984).
5. J.K. Furdyna, J. Appl. Phys. **64**, R29 (1988).
6. "Diluted Magnetic (Semimagnetic) Semiconductors", Mat. Res. Soc. Symp. Proc. **89**, (1987); edited by R.L. Aggarwal, J.K. Furdyna and S. von Molnar; published by the Material Research Society.
7. "Diluted Magnetic Semiconductors", Semiconductors and Semimetals, R.K. Willardson and A.C. Beer, vol **25** (1988); edited by J.K. Furdyna and J. Kossut; published by Academic Press.
8. A. Pajaczkowska, Prog. Cryst. Growth Charact. **1**, 289 (1978).
9. S.M. Durbin, J. Han, Sungki O., M. Kobayashi, D.R. Menke, R.L. Gunshor, Q. Fu, N. Pelekanos, A.V. Nurmikko, D. Li, J. Gonsalves, and N. Otsuka, Appl. Phys. Lett. **55**, 2087 (1989).
10. H. Anno, T. Koyanagi, and K. Matsubara, 5th International Conference on II-VI Compounds, Tamano, Japan, 1991; J. Cryst. Growth (to be published).
11. L.A. Kolodziejski, T. Sakamoto, R.L. Gunshor, and S. Datta, Appl. Phys. Lett. **44**, 799 (1984).
12. D.E. Ashenford, B. Lunn, J.J. Davies, J.H.C. Hogg, J.E. Nicholls, C.G. Scott, D. Shaw, H.H. Sutherland, C.P. Hilton, T.J. Gregory, D. Johnston, B.C. Cavenett, G.K. Johnson, and M. Haines, J. Cryst. Growth **95**, 557 (1989).
13. D.E. Ashenford, J.H.C. Hogg, D. Johnston, B. Lunn, C.G. Scott, and D. Staudte, J. Cryst. Growth **101**, 157 (1990).
14. R.N. Bicknell-Tassius, A. Waag, Y.S. Wu, T.A. Kuhn, and W. Ossau, J. Cryst. Growth **101**, 33 (1990).

15. R.N. Bicknell, R.W. Yanka, N.C. Giles-Taylor, D.K. Blanks, E.L. Buckland, and J.F. Schetzina, Appl. Phys. Lett. **45**, 92 (1984).
16. A. Nouhi and R.J. Stirn, Appl. Phys. Lett. **51**, 2251 (1987).
17. Z.C. Feng, R. Sudharsanan, S. Perkowitz, A. Erbil, T.K. Pollard, and A. Rohatgi, J. Appl. Phys. **64**, 6861 (1988).
18. G.N. Pain, N. Bharatula, G.I. Christiansz, M.H. Kibel, M.S. Kwietniak, C. Sandford, T. Warminski, R.S. Dickson, R.S. Rowe, K. McGregor, G.B. Deacon, B.O. West, S.R. Glanvill, D.G. Hay, C.J. Rossouw, and A.W. Stevenson, J. Cryst. Growth **101**, 208 (1990).
19. J.M. Wrobel and J.J. Dubowski, Appl. Phys. Lett. **55**, 469 (1989).
20. J.J. Dubowski, J. Cryst. Growth. **101**, 105 (1990).
21. X.L. Zheng, C.A. Huber, P. Becla, M. Shih, and D. Heiman, Mat. Res. Soc. Symp. Proc. **161**, 465 (1990).
22. M. Pessa and O. Jylhä, Appl. Phys. Lett. **45**, 646 (1984).
23. M.A. Herman, O. Jylhä, and M. Pessa, J. Cryst. Growth **66**, 480 (1984).
24. T. Koyanagi, K. Matsubara, H. Takaoka, and T. Tagaki, J. Appl. Phys, **61**, 3020 (1987).
25. L.L. Chang, Superlattices and Microstructures **6**, 39 (1989).
26. L.A. Kolodziejski, T.C. Bonsett, R.L. Gunshor, S. Datta, R.B. Bylsma, W.M. Becker and N. Otsuka, Appl. Phys. Lett. **45**, 440 (1984).
27. L.A. Kolodziejski, R.L. Gunshor, S. Datta, T.C. Bonsett, M. Yamanishi, R. Frohne, T. Sakamoto, R.B. Bylsma, W.M. Becker, and N. Otsuka, J. Vac. Sci. Technol. **B3**, 714 (1985).
28. R.N. Bicknell, N.C. Giles-Taylor, J.F. Schetzina, N. G. Anderson, and W.D. Laidig, Appl. Phys. Lett. **46**, 238 (1985).
29. R.N. Bicknell, N.C. Giles-Taylor, D.K. Blanks, R.W. Yanka, E.L. Buckland, and J.F. Schetzina, J. Vac. Sci. Technol. B **3**, 709 (1985).
30. D.K. Blanks, R.N. Bicknell, N.C. Giles-Taylor, J.F. Schetzina, A. Petrou, and J. Warnock, J. Vac. Sci. Technol. A **4**, 2120 (1986).
31. T.J. Gregory, C.P. Hilton, J.E. Nicholls, W.E. Hagston, J.J. Davies, B. Lunn, and D.E. Ashenford, J. Cryst. Growth **101**, 594 (1990).
32. J.J. Dubowski, A.P. Roth, Z.R. Wasilewski, and S.J. Rolfe, Appl. Phys. Lett. **59**, 1591 (1991).
33. J.J. Dubowski, J.R. Thompson, S.J. Rolfe, and J.P. McCaffrey, Superlattices and Microstructures **9**, 327 (1991).
34. J.J. Dubowski, Chemtronics **3**, 66 (1988).
35. J.J. Dubowski, Acta Physica Polonica A **80**, 221 (1991).
36. J.J. Dubowski, D.F. Williams, J.M. Wrobel, P.B. Sewell, J. LeGeyt, C. Halpin, and D. Todd, Can. J. Phys. **67**, 343 (1989).
37. "Laser Ablation for Materials Synthesis", Mat. Res. Symp. Soc. Proc. **191** (1990); edited by D.C. Paine and J.C. Bravman; published by the Material Research Society.
38. J.T. Cheung, Mat. Res. Soc. Symp. Proc. **29**, 301 (1984).
39. G.P. Schwartz, V.E. Bondybey, J.H. English, and G.J. Gaultieri, Appl. Phys. Lett. **42**, 952 (1983).
40. T. Nakayama, Surf. Sci. **133**, 101 (1983).
41. V.S. Ban and B.E. Knox, Int. J. Mass Spectrom. Ion Phys. **3**, 131, (1969).

42. A.D. Akhsakhalyan, Yu.A. Bityurin, S.V. Gaponov, A.A. Gubkov, V.I. Luchin, Zh. Tekh. Fiz. 52, 1584 (1982); (Sov. Phys. Tech. Phys. 27, 969 (1982)).
43. J.J. Dubowski, P.K. Bhat, D.F. Williams, and P. Becla, J. Vac, Sci. Technol. A 4, 1879 (1986).
44. A. Namiki, K. Watabe, H. Fukamo, S. Nishigaki, and T. Noda, J. Appl. Phys. 54, 3443 (1983).
45. A. Namiki, K. Watabe, H. Fukamo, S. Nishigaki, and T. Noda, Surf. Sci. 128, L243 (1983).
46. A. Namiki, T. Kawai, and K. Ichige, Surf. Sci. 166, 129 (1986).
47. P.K. Bhat, J.J. Dubowski, and D.F. Williams, Chemtronics 1, 82 (1986).
48. P.K. Bhat, J.J. Dubowski, and D.F. Williams, Phys. Stat. Sol. (a) 96, K9 (1986).
49. T. Tagaki, J. Vac, Sci. Technol. A 2, 382 (1984).
50. J.R. Thompson, J.J. Dubowski, and D.J. Northcott, Can. J. Phys. 69, 274 (1991).
51. J.J. Dubowski, J.M. Wrobel, and D.F. Williams, Appl. Phys. Lett. 53, 660 (1988).
52. J.J. Dubowski, J.M. Wrobel, S. Rolfe, J.A. Jackmann, J.H. Mazur, and J. Noad, Mat. Res. Soc. Symp. Proc. 161, 443 (1990).
53. J.J. Dubowski, J.M. Wrobel. J.A. Jackman, and P. Becla, Mat. Res. Soc. Symp. Proc. 131, 143 (1989).
54. J.M. Wrobel, J.J. Dubowski, and P. Becla, J. Vac. Sci. Technol. A 7, 338 (1989).
55. J.J. Dubowski, J.R Thompson, R. Benzaquen, A.P. Roth, Z. Wasilewski, Can. J. Phys. 69, 270 (1991).
56. J.H. Mazur, unpublished results.
57. B. Ortner and G. Bauer, J. Cryst. Growth 92, 69 (1988).
58. S. Tatarenko, J. Cibert, Y. Gobil, G. Feuillet, K. Saminadayar, A.C. Chami, and E. Ligeon, Appl. Surface Sci. 41/42, 470 (1989).
59. A. Golnik, J. Ginter, and J.A. Gaj, J. Phys. C: Solid State Phys. 16, 6073 (1983).
60. Z.C. Feng, S. Perkowitz, and J.J. Dubowski, Appl. Phys. Lett. 69, 7782 (1991).
61. D. Heiman, P. Becla, R. Kershaw. D. Ridgley, K. Dwight, A. Wold, and R.R. Galazka, Phys. Rev. B 34, 3961 (1986).
62. "Luminescence Spectroscopy"; edited by M.D. Lumb; published by Academic Press, (1978).
63. D. Labrie, C. Qiu, X. Wang, and J.J. Dubowski, J. Appl. Phys. Phys. 71, 2811 (1992).
64. K.Y. Lay, H. Neff, and K.J. Bachmann, Phys. Stat. Sol. (a) 92, 567 (1985).
65. J.J. Dubowski, A.P. Roth, E. Deleporte, G. Peter, Z.C. Feng, and S. Perkowitz, 5th International Conference on II-VI Compounds, Tamano, Japan, 1991; J. Cryst. Growth 117, 862 (1992).
66. G. Bastard, C. Delalande, M.H. Meynadier, P.M. Frijlink, and M. Voos, Phys. Rev. B 29, 7042 (1984).

67. L. Viña, L.L.Chang, and J. Yoshino, Journal de Physique, Colloque C5, 48, 317 (1987).
68. X.-C. Zhang, S.-K. Chang, A.V.Nurmikko, L.A.Kolodziejski, R.L. Gunshor, and S. Datta, Phys. Rev. B 31, 4056 (1985).
69. S. Venugopalan, A. Petrou, R.R. Galazka, A.K.Ramdas, and S. Rodrigues, Phys. Rev. B 25, 2681 (1982).
70. W. Gebicki and W. Nazarewicz, Phys. Stat. Sol. (b) 86, K135 (1978).
71. J.M. Wrobel, B.P. Clayman, P. Becla, R. Sudharsanan, and S. Perkowitz, J. Appl. Phys 64, 310 (1988).
72. W.J. Keeler, H. Huang, and J.J. Dubowski, Phys. Rev. B 42, 11355 (1990).
73. D.R.T.Zahn, R.H.Williams, T.D.Golding, J.H. Dinan, K.J. Mackey, J. Geurts, and W. Richter, Appl. Phys. Lett. 53, 2409 (1988).
74. D.R.T. Zahn, W. Richter, T. Eickhoff, J. Geurts, T.D. Golding, J.H. Dinan, K.J. Mackey, and R.H. Williams, Appl. Surf. Sci. 41/42, 497 (1989).
75. P.M. Armitharaj and F.H. Pollak, Appl. Phys. Lett. 45, 789 (1984).
76. W.J. Keeler and J.J. Dubowski, Can. J. Phys. 69, 255 (1991).
77. S. Venugopalan, L.A.Kolodziejski, R.L. Gunshor, and A.K.Ramdas, Appl. Phys. Lett. 45, 974 (1984).
78. E.-K Suh, D.U.Bartholomew, A.K.Ramdas, S. Rodriguez, S. Venugopalan, L.A.Kolodziejski, and R.L. Gunshor, Phys. Rev. B. 36, 4316 (1987).
79. X. Wang, C. Qiu, D. Labrie, and J.J. Dubowski, Thin Solid Films (to be published).
80. D.E. Aspnes and J.E. Rowe, Sol. State Commun. 8, 1145 (1970).
81. D.E. Aspnes, Surf. Sci. 37, 418 (1973).
82. Y.R. Lee, A.K.Ramdas, and A.L. Aggarwal, Phys. Rev. B 38, 10600 (1988).
83. N. Bottka, J. Stankiewicz, and W. Giriat, J. Appl. Phys. 52, 4189 (1981).

$Zn_{1-y}Cd_ySe_{1-x}Te_x$ QUATERNARY II-VI WIDE BANDGAP ALLOYS AND HETEROSTRUCTURES

R.E. Nahory, M.J.S.P. Brasil and M.C.Tamargo
Bell Communications Research
Red Bank, NJ 07701

ABSTRACT

We discuss the $Zn_{1-y}Cd_ySe_{1-x}Te_x$ quaternary II-VI alloy, which we have grown by MBE techniques on GaAs or InP substrates. We model the bandgap vs composition (x,y) for this system and find the results to be in good agreement with measured values of bandgap and lattice parameter. Photoluminescence in the quaternary, while very efficient, is strongly affected by relatively deep levels, especially for low Te concentrations, probably due to Te-clustering effects. We also discuss band-offset engineering possibilities with respect to ZnSe for the system, which has regions of type I and type II allignments. Growth of the quaternary on InAs is shown to be potentially useful for lattice-matched double heterostructures which could produce visible-light lasers in the red.

1. Introduction

Interest in wide-bandgap semiconductor injection lasers operating in the visible region of the spectrum has stimulated considerable effort in studies of II-VI compounds. This interest has been enhanced by the operation of the first blue-green injection laser[1] which was demonstrated in 1991 using a separate confinement heterostructure design consisting of the materials ZnCdSe/ZnSeS/ZnSe and nitrogen doped p-type ZnSe.[2] This advance represents a culmination of considerable research work on two basic issues: pn-junctions and heterostructures in II-VI alloys. The material n-type ZnSe has been available for some time, while the development of p-type ZnSe is a recent accomplishment in which the use of nitrogen as acceptors has been advanced.[2] This latter progress was accomplished with the use of a nitrogen radical source in MBE. In addition to pn-junctions, however, suitable carrier and optical confinement are also required for efficient semiconductor laser operation. Confinement requires the use of appropriate heterostructures with proper band-offsets and indices of refraction. This latter issue has not yet been resolved adequately.

In this report we discuss some of the heterostructure issues. We report the growth by molecular beam epitaxy (MBE) techniques of the quaternary $Zn_{1-y}Cd_ySe_{1-x}Te_x$ along with selected optical properties such as bandgap versus composition and lattice parameter. We also address the topic of band-offsets for the ZnCdSeTe/ZnSe system, which can be type I for some range of quaternary compositions and type II for other ranges. The point is made that this available range of offsets in fact allows for band- offset engineering, and optimization, in this system through adjustment of composition.

2. Ternary $Zn_{1-y}Cd_ySe$ and $ZnSe_{1-x}Te_x$

We begin by briefly addressing properties of the ternary II-VI compounds $Zn_{1-y}Cd_ySe$ and $ZnSe_{1-x}Te_x$. In Fig. 1 we show the measured low temperature bandgaps vs composition for these two alloy materials. Both span regions of the visible spectrum but show striking differences in curvature, or bowing parameter. Table I lists several mathematical relations for these materials and summarizes bandstructure parameters of these ternaries.

One of these ternaries, $Zn_{1-y}Cd_ySe$, appears well-behaved with only the usual, rather small bowing parameter. It is readily prepared over the entire range of compositions, having been grown by MBE and reported by Samarth et al,[3] and used for the first

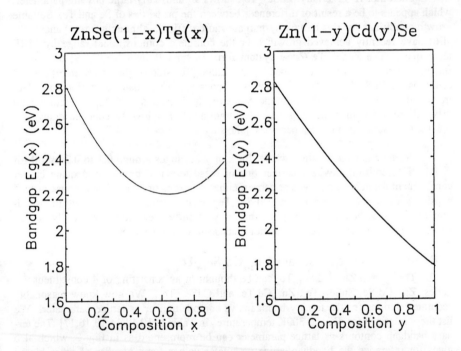

Figure 1. Bandgap vs composition for two ternary II-VI alloys. a) $ZnSe_{1-x}Te_x$, b) $Zn_{1-y}Cd_ySe$

injection lasers by Haase et al.[1] This system shows relatively narrow shallow bound exciton spectra in photoluminescence measurements, consistent with well-behaved materials expectations. However for compositions up to y = 0.2, with manageable strains on ZnSe, the estimated valence band offset for this ternary with respect to ZnSe is very small. It has been estimated to be only in the range of 15 meV for unstrained cases, and

somewhat larger in the range of 50 meV for strained layers.[4] Both of these values are rather small for practical devices, which must operate above room temperature and still provide adequate confinement for both electrons and holes. For example, for the well-known $Al_xGa_{1-x}As/GaAs$ lasers a value x = 0.3 is typically used, which gives $\Delta E_C = 220 meV$ and $\Delta E_V = 150 meV$, taking a 60%/40% offset rule.[5] To maintain adequate electron and hole confinement the band offsets in the II-VI lasers should be similar in magnitude to these AlGaAs/GaAs values. Otherwise thresholds increase and efficiency drops due to carrier leakage over the barriers, especially at temperatures above room temperature.

The other II-VI ternary, $ZnSe_{1-x}Te_x$, shows an unusually large bowing parameter, which appears to be a result of differences between the properties of Se and Te. Suitable growth conditions for this ternary system are more difficult, although Turco-Sandroff et al[6] have recently achieved growth over the complete compositional range by MBE techniques. The $ZnSe_{1-x}Te_x/ZnSe$ system forms a type II interface,[7,8,9] tending to separate carriers such that holes occupy the $ZnSe_{1-x}Te_x$ side of the hetero-interface and electrons the ZnSe side. This situation is, of course, not adequate to confine both electrons and holes together. Photoluminescence results indicated that the conduction band offset decreases with decreasing Te concentration.[9] For low Te concentrations, x < 0.2, the conduction band offset becomes quite small.

We thus have a situation where, for x or y in ranges around 0.1 to 0.2 useful for visible light emitting devices, neither one of these ternaries provides adequate carrier confinement for at least one type of carrier, either electrons or holes. We propose instead consideration of the quaternary $Zn_{1-y}Cd_ySe_{1-x}Te_x$ as an alternative to deal with this situation, as discussed below. This material has only recently been grown epitaxially[10,11], while a limited range of bulk compositions has been reported earlier.[12]

3. Bandgap of the quaternary $Zn_{1-y}Cd_ySe_{1-x}Te_x$

The system $Zn_{1-y}Cd_ySe_{1-x}Te_x$ can be thought of as consisting of 4 constituent ternaries: $Zn_{1-y}Cd_ySe$, $ZnSe_{1-x}Te_x$, $Zn_{1-y}Cd_yTe$, and $CdSe_{1-x}Te_x$. We now know reasonably accurate relations for the ternary bandgaps vs composition or vs lattice parameter. We list these relations for 4K and 300K temperatures in Table I.[10,13,14,15,16,17] The ternary bandgap contours vs lattice parameter can be represented as in Fig. 2, which also gives for reference the bandgap/lattice parameter values for a number of other binary semiconductors. The quaternary ZnCdSeTe occupies the area inside the curves drawn in the figure. It is interesting to note how the ternaries ZnCdSe and ZnSeTe overlap for ZnSe-rich compositions. In fact, this overlap occurs for all visible wavelengths shorter than red. Thus any quaternary composition in this range has an approximately fixed bandgap at a particular lattice parameter. For example, at a fixed lattice parameter of 5.8 Angstroms, the bandgap is about 2.35 eV (530 nm, green) at low temperature for the complete range of quaternary compositions, which lies between the ternary endpoints $ZnSe_{.68}Te_{.32}$ and $Zn_{.64}Cd_{.36}Se$. However, as will be discussed further below, the band-

Table I. (a) Bandgap at 4K and lattice parameter (at 300 K) as a function of composition for the quaternary $Zn_{1-y}Cd_ySe_{1-x}Te_x$ and its four constituent ternary compounds.

$Zn_{1-y}Cd_ySe_{1-x}Te_x$:

$$E_g(x,y) = -0.752x^2y - 0.184xy^2 + 1.507x^2 \\ + 0.35y^2 + 1.182xy - 1.935x - 1.39y + 2.82 \quad [10]$$

$$a(x,y) = 5.6676 + 0.4334x + 0.3824y - 0.0064xy$$

$ZnSe_{1-x}Te_x$:

$$E_g(x) = 2.820 - 1.935x + 1.507x^2 \quad [17]$$
$$a(y) = 5.6676 + 0.4334x$$
$$E_g(a) = 285.836 - 95.407a + 8.023a^2$$

$Zn_{1-y}Cd_ySe$:

$$E_g(y) = 2.820 - 1.39y + 0.35y^2 \quad [16]$$
$$a(y) = 5.6676 + 0.3824y$$
$$E_g(a) = 100.304 - 30.766a + 2.393a^2$$

$Zn_{1-y}Cd_yTe$:

$$E_g(y) = 2.392 - 0.960y + 0.166y^2 \quad [15]$$
$$a(y) = 6.101 + 0.376y$$
$$E_g(a) = 61.674 - 16.881 + 1.174a^2$$

$CdSe_{1-x}Te_x$:

$$E_g(x) = 1.78 - 0.937x + 0.755x^2 \quad [14]$$
$$a(x) = 6.05 + 0.427x$$
$$E_g(a) = 166.558 - 52.278 + 4.139a^2$$

(b) Bandgap at 300K as a function of composition for the quaternary $Zn_{1-y}Cd_ySe_{1-x}Te_x$ and its four constituent ternary compounds.

$Zn_{1-y}Cd_ySe_{1-x}Te_x$:

$$E_g(x,y) = -0.841x^2y - 0.222xy^2 + 1.621x^2 \\ + 0.35y^2 + 1.360xy - 2.101x - 1.388y + 2.73$$

$ZnSe_{1-x}Te_x$:

$$E_g(x) = 2.730 - 2.101x + 1.621x^2$$

$Zn_{1-y}Cd_ySe$:

$$E_g(y) = 2.730 - 1.388y + 0.35y^2$$

$Zn_{1-y}Cd_yTe$:

$$E_g(y) = 2.250 - 0.869y + 0.128y^2$$

$CdSe_{1-x}Te_x$:

$$E_g(x) = 1.692 - 0.962x + 0.78x^2$$

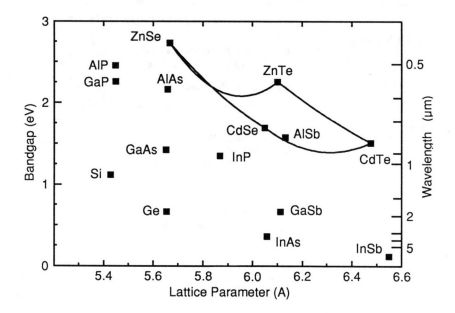

Figure 2. Bandgap vs lattice parameter at 300K for the 4 ternary II-VI alloys $ZnSe_{1-x}Te_x$, $Zn_{1-y}Cd_ySe$, $CdSe_{1-x}Te_x$ and $Zn_{1-y}Cd_yTe$ and various binary compounds.

offsets with ZnSe vary in useful ways over this range of quaternary compositions.

We take a different look at this situation in Fig. 3, which shows the bandgap of the quaternary as a function of composition x,y in a 3-dimensional style. The four corners represent the constituent binaries, as labeled. The strong bowing in the x-direction (Se-Te variation) is readily apparent in the figure. For the present purposes we have grown ZnCdSeTe materials with compositions in the vicinity of the ZnSe corner of this diagram, where each wafer grown for this work is indicated by a solid dot in the figure. The surface follows the equation given in Table I. This surface, a cubic, is geometrically the simplest which matches the known ternary boundaries, and is found experimentally for samples grown in this work to give a good approximation of the actual bandgap surface. For the III-V quaternary InGaAsP this manner of analysis also gives reasonably accurate estimates of the bandgap vs composition.

4. Growth and characterization of $Zn_{1-y}Cd_ySe_{1-x}Te_x$ epilayers

The $Zn_{1-y}Cd_ySe_{1-x}Te_x$ epilayers were grown by MBE [10,11] using elemental sources of 6-N Zn, Se and Te and 7-N Cd in conventional effusion cells. Substrates were

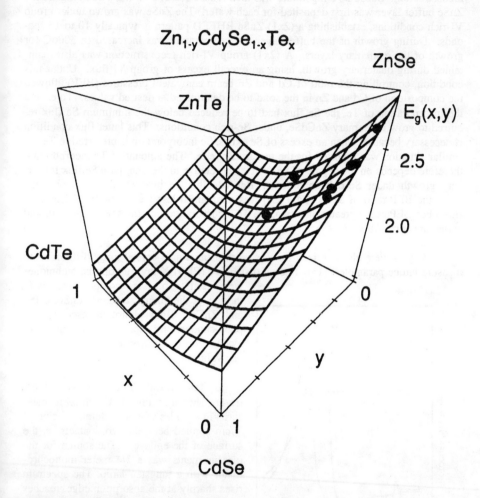

Figure 3. Bandgap surface vs composition (x,y). The four corners represent the four constituent binaries, as labeled. The solid dots represent measured values for the grown epitaxial layers.

<100> GaAs. Oxide desorption, monitored by reflection high energy electron diffraction (RHEED), was established by observing the transition from (2x4) to (3x1) GaAs surface reconstruction. Substrate temperature was then reduced to 270° C and a 500 Angstrom ZnSe buffer layer was first deposited for each wafer. The ZnSe was grown under group VI-rich conditions, establishing a (2x1) ZnSe RHEED pattern in typically 10 to 15 seconds. During growth of the buffer layer, the temperature was increased to 300oC for growth of the quaternary layers. A (2x1) group VI-rich reconstruction was also maintained during quaternary growth, using an overall excess of group VI flux. Under this condition, controlling the ratio of Cd and Zn beam equivalent pressures (BEP) allowed the composition of Cd and Zn in the solid to be determined at desired values. To ensure the incorporation of Te, the Se flux had to be reduced below the minimum Se flux required to grow the ternary ZnCdSe, under Se-rich conditions. This latter flux condition is necessary because, when an excess of Se exists, Se incorporation is preferred to Te. A similar situation was reported for the ternary ZnSeTe. The amount of Te incorporation therefore depends on the degree of Se deficiency relative to the minimum Se flux for ternary growth under Se-rich conditions. Reproducible Te incorporation was obtained when the BEP ratio of Se to [Cd + Zn] was maintained less than 2, with sufficient Te flux. For BEP ratios greater than two, the Te incorporation into the solid was small and varied uncontrollably.

Layers were characterized for this work using single crystal x-ray diffraction to measure lattice parameters.[11] Energy-dispersive (EDS) X-ray microprobe techniques were used to measure compositions, with the use of the binary standards ZnSe, CdSe and ZnTe. Both photoluminescence and photoconductivity spectra were used to evaluate optical properties and obtain bandgap values.

Two typical photoconductivity spectra are shown in Fig. 4, which were measured in samples with an external bias of 12 V/cm applied between two contacts on the surface of the epilayer. The source for the measurements was a 1/4 meter monochromator with a tungsten lamp. The spectrum rises sharply at the absorption edge, reaches a peak, and then falls as surface recombination becomes dominant for high absorption and generation of carriers very close to the surface. The peak gives a good estimate of the bandgap, within an uncertainty estimated

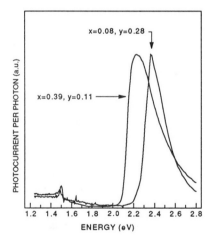

Figure 4. Photoconductivity spectra at 5K for typical $Zn_{1-y}Cd_ySe_{1-x}Te_x$ layers.

to be better than 20 meV. The compositions given in the figure are measured values using X-ray microprobe techniques.

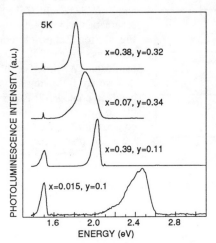

Figure 5. Photoluminescence spectra at 5K for typical layers of $Zn_{1-y}Cd_ySe_{1-x}Te_x$.

Photoluminescence (PL) spectra were also used to characterize these materials, using the 375 nm or 488 nm output of an Ar ion laser as pump, and a 3/4 meter spectrometer with photomultiplier detector. Figure 5 shows typical PL spectra for 4 different epilayers. The PL is relatively efficient, but narrow excitonic lines are not observed. The quaternary layers exhibit a single, rather broad band, 70 to 230 meV in width. The width is greater for $Zn_{1-y}Cd_ySe_{1-x}Te_x$ compositions with small x, a behavior similar to that observed in $ZnSe_{1-x}Te_x$.[6] As found for the case of the ternary $ZnSe_{1-x}Te_x$, the photoluminescence is dominated by a fairly deep defect or impurity level, about 200 to 350 meV below the bandgap, where the depth varies with composition. The nature of these deep levels is not understood, but it is believed that they are related to Te clustering effects, which give a strong localization of carriers.[17,18] The photoconductivity spectra thus give a more precise measure of the bandgap. Table II gives detailed values measured in this work for composition (x,y), bandgap energy (obtained experimentally from photoconductivity spectra, or calculated using the equation given in Table I), photo-luminescence spectral peak energy, and lattice parameter (measured using X-ray diffraction, or calculated using the measured composition and the equation in Table I). According to these results, the bandgap may be calculated using the equation in Table I within better than 50 meV from a given composition, while lattice parameter may be calculated within better than 0.02 Angstroms.

From the kind of picture already presented in Fig. 3, showing the location of the epilayers grown in the present work near the ZnSe corner of the bandgap surface, it is easy to envision an iso-bandgap path linking the ternaries ZnSeTe and ZnCdSe. We will discuss band-offsets and their expected variation along such paths below. A two-dimensional contour map locating these grown quaternaries is shown in Fig. 6. In our program, we have emphasized growth in the region near ZnSe where the contours largely overlap. One epilayer was grown outside of this region near the lattice parameter 6.2 Angstroms as a test case, as shown.

5. Discussion of band-offsets

The estimated band-offsets for ZnSeTe, ZnCdSe and ZnCdSeTe with ZnSe are schematically depicted in Fig. 7 for ZnSe-rich compositions all with the same bandgap. We imagine here that we are following an iso-bandgap contour. It is possible to adjust the quaternary composition along this iso-bandgap path between the two ternaries so that only the band-offset changes with composition. For ZnCdSe/ZnSe the valence band offset

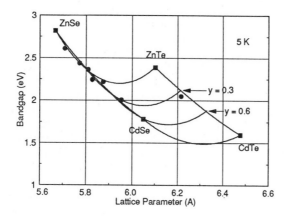

Figure 6. Bandgap energy contours (lines) calculated for $Zn_{1-y}Cd_ySe_{1-x}Te_x$ and data points (dots).

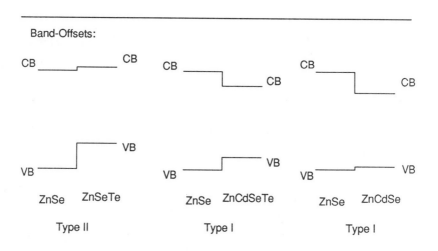

Figure 7. Schematic of the band-offsets for the three systems ZnSe/ $Zn_{1-x}Se_xTe$, ZnSe/$Zn_{1-y}Cd_ySe_{1-x}Te_x$, and ZnSe/$Zn_{1-y}Cd_ySe$ for ZnSe-rich alloys.

is rather small, while for the ZnSeTe/ZnSe case the offset is of type II but with small conduction band offsets for x < 0.2. For the quaternary ZnCdSeTe/ZnSe offsets along this iso-bandgap path, there exists a region where they are type I and lie in between those of the two ternaries, as depicted in Fig. 7. The band offsets should vary from small values in the conduction band, of type II, for $ZnSe_{1-x}Te_x$, to small values in the valence band, of type I, for $Zn_{1-y}Cd_ySe$, with substantial values for both band offsets for the quaternary $Zn_{1-y}Cd_ySe_{1-x}Te_x$ compounds along the iso-bandgap contour. Measurements of these offsets have not yet been made.

6. $Zn_{1-y}Cd_ySe_{1-x}Te_x$ on InAs or InP substrates

It is interesting to compare the ZnCdSeTe quaternary system with well-known binary III-V compounds. As seen in Fig. 2, interesting binary III-V substrates for use with the II-VI system are GaAs, InP, InAs or GaSb. So far heterostructures grown on GaAs have been predominant. We have already found experimentally that InP is a useful

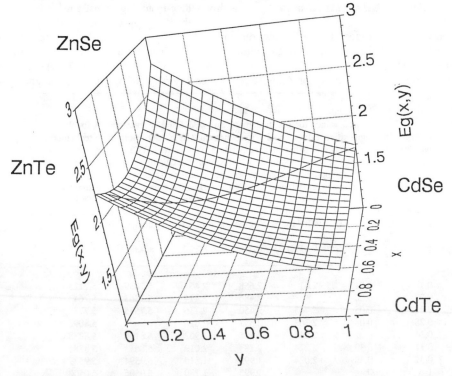

Figure 8. $Zn_{1-y}Cd_ySe_{1-x}Te_x$ bandgap surface as a function of composition (x,y). The line indicates the contour for lattice-matched growth on InAs.

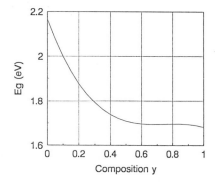

Figure 9. $Zn_{1-y}Cd_ySe_{1-x}Te_x$ bandgap energy vs composition y for quaternaries lattice-matched to InAs substrates.

substrate for growth of either ZnSeTe or ZnCdSe. Also of importance is the recent development of $In_xGa_{1-x}As$ bulk materials grown by LEC techniques. Taking this bulk ternary with $x = 0.04$, a precise lattice match to ZnSe is obtained. [19]

We conclude with a discussion of ZnCdSeTe bandgap parameters for lattice-matched growth on either InAs or InP substrates. For the former, we choose a fixed lattice parameter for the quaternary of 6.058 Angstroms, equal to that of InAs. Figures 8 and 9 show graphically the range of quaternary bandgaps obtainable under this lattice-matching constraint. Fig. 8 shows this bandgap contour on the quaternary bandgap surface. Figure 9 shows directly the bandgap vs composition y, where x is determined by the lattice matching condition. The room temperature bandgap ranges for these conditions from near 1.7 eV (about 730 nm in the infrared, for $CdSe_{.98}Te_{.02}$) to 2.15 eV (about 570 nm, yellow, for $ZnSe_{.1}Te_{.9}$). This energy range of nearly 0.5 eV might al-

Table II: Measured values for the quaternary $Zn_{1-y}Cd_ySe_{1-x}Te_x$ are shown for 9 epitaxial layers. The composition values x and y were measured by X-Ray Microprobe techniques, the bandgap E_g was measured using low temperature photocurrent (PC) spectra, the calculated bandgap energy E_g(calc) was determined for T=5K using the equation from Table I, E(PL) gives the measured low temperature photoluminescence peak energy, a(X-ray) gives the room temperature lattice parameter obtained from X-ray diffraction measurements, and a(calc) gives the calculated lattice parameter evaluated using an extended form of Vegard's law for the quaternary. For the latter, we used the equation of Table I, which assumed that the usual linear form of Vegard's law is satisfied for each of the constituent ternary compounds.

y	x	E_g(PC)	E_g(cal)	E(PL)	a(Xray)	a(calc)
0.1	0.015	2.61	2.658	2.46	5.7039	5.7123
0.23	0.02		2.486	2.214	5.7363	5.7642
0.22	0.05	2.443	2.450	2.136	5.7575	5.7733
0.28	0.08	2.362	2.337	2.016		5.8092
0.34	0.07	2.244	2.285	1.902	5.8156	5.8278
0.11	0.39	2.224	2.183	2.019	5.865	5.8784
0.32	0.38	2.024	1.995	1.815	5.9594	5.9539
1	0		1.995	1.780	6.0506	6.0500
0.31	1	2.049	2.110	2.048	6.2046	6.2176

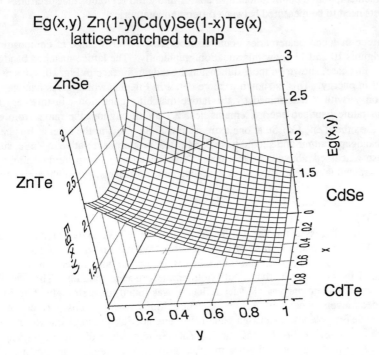

Figure 10. $Zn_{1-y}Cd_ySe_{1-x}Te_x$ bandgap surface as a function of composition (x,y). The line indicates the contour for lattice-matched growth on InP.

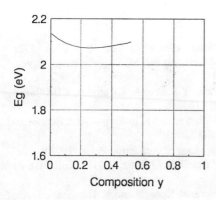

Figure 11. $Zn_{1-y}Cd_ySe_{1-x}Te_x$ bandgap energy vs composition y for quaternaries lattice-matched to InP substrates.

low the design of a lattice-matched double hetereostructure, with adequate electron and hole confinement, which could function as a laser in the red region of the spectrum. The band-offsets need to be measured between different quaterneries.

Lattice-matched quaternaries could also be grown over a range of compositions on InP. Figures 10 and 11 show this situation graphically. The lattice-matched bandgap contour for this case, shown on the 3-dimensional bandgap surface in Fig. 10, stays nearly constant in energy. This is shown in more detail in Fig. 11, which gives bandgap vs composition y, with x determined by the lattice-matching condition. In this case the room temperature bandgap energy remains near 2.1 eV throughout the range from one ternary to the other ($Zn_{.47}Cd_{.53}Se$ at one extreme and $ZnSe_{.54}Te_{.46}$ at the other). It is nearly an iso-bandgap contour and should have band-offset properties, relative to ZnSe, similar to those discussed above. Furthermore, these quaternary compositions might be useful for strained layer heterostructures with tunable band-offsets, with light emission in the red.

7. Summary

We have discussed the $Zn_{1-y}Cd_ySe_{1-x}Te_x$ quaternary II-VI alloy, which we have grown by MBE techniques on GaAs or InP substrates. We have modeled the bandgap vs composition (x,y) for this system and found the result to be in good agreement with values measured by photoconductivity and photoluminescence techniques. The photoluminescence in the quaternary, while very efficient, was found to be strongly affected by relatively deep levels, especially for low Te levels, similar to the proposed Te-clustering effects in ZnSeTe. We also discussed band-offset engineering possibilities with respect to ZnSe for the system ZnSeTe - ZnCdSeTe - ZnCdSe to produce different confinement options for electrons and holes. Lattice-matched growth of the quaternary on InAs, for example, was shown to be potentially useful for double heterostructures which could produce visible-light lasers in the red.

8. References

1. M.A.Haase, J.Qui, J.M.DePuydt and H.Cheng, Appl. Phys. Lett. 95, (1991) 1272.
2. R.M.Park, M.B.Troffer, C.L.Rouleau, J.M.DePuydt and M.A.Haase, Appl. Phys. Lett. **57** (1990) 2127.
3. N.Samarth, H.Luo, J.K.Furdyna, R.G.Alonso, Y.R.Lee, A.K.Ramdas, S.B.Quadri and N.Otsuka, Appl. Phys. Lett. 56 (1990) 1163.
4. H.Jeon, J.Ding, A.V.Nurmikko, H.Luo, N.Samarth, J.K.Furdyna, W.A.Bonner and R.E.Nahory, Appl. Phys. Lett. **57** (1990) 2413.
5. H.C.Casey and M.B.Panish, Heterostructure Lasers, Academic Press (1978).
6. F.S.Turco-Sandroff, R.E.Nahory, M.J.S.P.Brasil, R.J.Martin and H.L.Gilchrist, Appl. Phys. Lett. **58**, (1991) 1611.
7. M.Kobayashi, N.Mino, H.Katagiri, R.Kimura, M.Konigai and K.Takahashi, J. Appl. Phys. **60** (1986) 773.

8. Y.Rajakarunanayake, R.H.Miles, G.Y.Wu and T.C.McGill, J.Vac. Sci. Technol. B **6**, (1988) 1354.
9. M.J.S.P.Brasil, R.E.Nahory, F.S.Turco and R.J.Martin, *Proc. of the 1990 Fall Meeting of the Materials Research Society*, Materials Research Society, Pittsburgh, PA.
10. M.J.S.P.Brasil, M.C.Tamargo, R.E.Nahory, H.L.Gilchrist and R.J.Martin, Appl.Phys.Lett. **59** (1991) 1206.
11. M.C.Tamargo, M.J.S.P.Brasil, R.E.Nahory, R.J.Martin, A.L.Weaver and H.L.Gilchrist, Semicond. Sci. Technol. **6** (1991) A8.
12. G.W.Iseler and A.J.Strauss, J. Luminescence **3** (1970) 1.
13. The compositional dependence of the bandgap for each of the ternaries in Table I is taken from the literature, references 14 - 17. The relations have been slightly adjusted to ensure continuity at the binary connecting points. The relation for the quaternary has been reported in reference 10.
14. I.B.Ermolovich, A.M.Pavelets and I.N.Khanat, Thin Solid Films **143** (1986) 225.
15. A.G.Areshkin, G.S.Pecar, G.N.Polissit, T.B.Popova, L.G.Suslina and D.L.Fedorov, Sov. Phys. Solid State **28** (1986) 2109.
16. D.J.Olego, J.P.Faurie, S.Sivananthan and P.M.Raccah, Appl. Phys. Lett. **47** (1985) 1172.
17. M.J.S.P. Brasil, R.E,.Nahory, F.S.Turco-Sandroff, H.L.Gilchrist and R.J.Martin, Appl. Phys. Lett. **58** (1991) 2509.
18. T.Yao, M.Kato, J.J.Davies and H.Tanino, J.Cryst. Growth **86** (1988) 552.
19. W.A.Bonner, B.J.Skromme, E.Berry, H.L.Gilchrist and R.E.Nahory, Gallium Arsenide and Related Compounds 1988, Inst. Phys. Conf. Ser. No. 96, London: IOP Pub. Ltd., (1988) 337.

Intersubband Transitions in SiGe/Si Quantum Structures

R. P. G. Karunasiri, K. L. Wang, and J. S. Park
66-127G, Engineering IV
Department of Electrical Engineering
University of California, Los Angeles
Los Angeles, CA 90024

ABSTRACT

Intersubband transitions in quantum wells and superlattices have created a great deal of interest because of their potential applications in infrared detection and imaging. To date, most of the studies are based on the III-V based heterostructures. The Si-based devices have the advantage of monolithic integration with the conventional silicon signal processing electronics. With the current advances in silicon molecular beam epitaxy (Si-MBE) and other low temperature epitaxial techniques, strained SiGe layers can be grown on Si in a controlled manner. Recently, quantum size effects such as resonant tunneling and intersubband transitions have been demonstrated using SiGe/Si heterostructures. In this chapter, we will begin with a brief introduction of the current status of the SiGe material system which will be followed by the physics of intersubband transition in quantum wells and superlattices. The experimental observation of intersubband transition using SiGe/Si quantum wells and δ-doped layers in Si will be discussed. The importance of many-body effects in understanding the experimental observations of δ-doped layers, will also be discussed. Finally, the application of SiGe/Si multiple quantum well structures for the fabrication of an infrared detector will be illustrated.

1. Introduction

The advances of epitaxial growth techniques make possible the synthesis of artificial quantum well and superlattice structures with novel optical and electrical properties which do not exist in the bulk constituents. Initially, much of the effort focused on the lattice matched materials (e.g. GaAs/AlGaAs and InP/InGaAsP), which later propagated into lattice mismatch systems as the growth of strained layers come to be understood. Among the lattice mismatch systems, Si/Ge heterostructures have created an added interest due to their potential of being integrated with the conventional silicon VLSI technology.

The growth of $Si_{1-x}Ge_x$ had been attempted earlier by Kasper et al. [1] using MBE above 750 °C and also by others using high temperature CVD techniques [2]; however,

in these approaches three dimensional growth and islanding were observed. It was not until the early 1980's when Bean et al. [3] first reported the low temperature commensurate growth of $Si_{1-x}Ge_x$ on Si at 580 °C and attained smooth high quality films. The most important feature of the latter work and all other recent works [4, 5, 6] is the low temperature virtue which enables the films to be grown pseudomorphically. Due to the lattice mismatch of 4.17% between Ge and Si at room temperature, the two dimensional epitaxial growth of $Si_{1-x}Ge_x$ proceeds until the thickness of the film exceeds a value such that the strain energy in the layer surpasses the energy for the generation of dislocations. In this case, misfit dislocations appear at the interface and the lattice relaxes, resuming the natural lattice constant as determined by Vegard's law. Several models are available for predicting the critical thickness for a given Ge content [7, 8, 9]. For multiple layer growth, the critical thickness can be estimated using the average Ge composition of a single period of the structure (i.e., the average Ge content of a period, $x_{Ge} = (x_1 d_1 + x_1 d_2)/(d_1 + d_2)$, where x and d are the Ge content and thickness of each constituent layer, respectively) [9]. If a relaxed SiGe buffer layer is used with a Ge fraction equal to the above x_{Ge}, strain symmetrization can be accomplished [10]. In this case, the net strain of the multiple layers on top of the buffer layer vanishes and the total thickness of the layers is no longer limited. The band structure of $Si_{1-x}Ge_x$ is also affected by the strain and depends on the substrate orientation. Van de Walle and Martin [11] have calculated the heterojunction band lineups at the Si/Ge interface by an *ab initio* pseudopotential method. Using their results and assuming the linear interpolation of the strain along with the empirical data of deformation potentials, we can obtain the band offsets for a substrate with any Ge composition. This appears to work well for the (100) growth direction [12]. To date there is a reasonable understanding of the growth of strained SiGe as well as the band alignments. This makes the demonstration of a wide variety of quantum devices possible [13].

In the following sections, we will discuss first the physics and the experimental observations of intersubband transitions in SiGe/Si quantum wells as well as δ-doped quantum well structures. Then the application of such structures for the fabrication of an infrared detector operating near 10 μm will be discussed.

2. Physics of Intersubband Transitions

Intersubband transitions occur between quantized states in a quantum well or in a superlattice between minibands as illustrated in Fig. 1. For semiconductor heterostructures, the quantum well can be formed either in the conduction band or in the valence band, depending on the band discontinuity. For example, in a SiGe/Si heterostructure grown pseudomorphically on Si, most of the band offset appears in the valence band, while for III-V heterostructures it appears in the conduction band. Most of the relevant features of the intersubband process can be understood in the effective mass framework. In the following, we will analyze the intersubband process

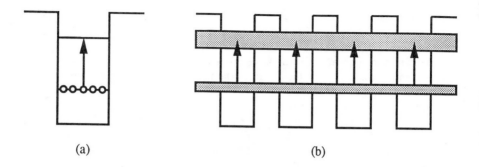

Figure 1: Intersubband (or intraband) transition in (a) a quantum well, (b) between minibands of a superlattice.

using a parabolic quantum well as an example. This not only provides an understanding of the fundamentals of the process, but also makes the analysis less complex compared to the more familiar square quantum well. For comparison, we will include the results for the square quantum well as well.

2.1 Transition in a Quantum Well

First, we consider the absorption between two states of an isolated parabolic well as shown in Fig. 2. To a good approximation we can describe the envelope states as those given by that of an infinite parabola. The Hamiltonian (H) of the system in the effective mass approximation (where the effect of the nearest bands are taken into account using an empirical effective mass m^*) can be written as,

$$H = \frac{p^2}{2m^*} + \frac{1}{2m^*}\Omega^2 z^2 \tag{1}$$

where z is the growth direction and p is the momentum of the electron (or hole) along the z direction. Ω is given in terms of the curvature (K) of the parabola by $\sqrt{K/m^*}$. Energy eigenvalues of the system are given by

$$E_n = (n+\frac{1}{2})\hbar\Omega + \frac{\hbar^2 k_\perp^2}{2m^*} \tag{2}$$

where $k_\perp$ is the wave vector of the electron (or hole) perpendicular to the growth direction and n = 0,1,2... .

The wave function $\phi_n(\vec{r})$ of the nth quantized state has the form:

$$\phi_n(\vec{r}) = u_n(\vec{r})\frac{e^{i\vec{k}_{i\perp}\cdot\vec{\rho}}}{\sqrt{A}}n(z) \tag{3}$$

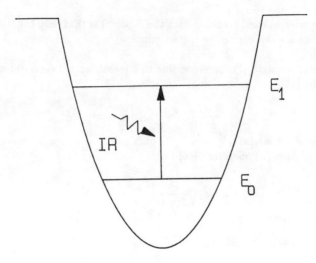

Figure 2: Schematic diagram of a parabolic quantum well showing the intersubband transition.

where A is the area of the well, $\vec{k}_\perp$ and $\vec{\rho}$ are the wave and the position vectors in the xy plane, u is the cell period function near the band extremum and n(z) is the envelope function due to the quantization along the z-direction.

The selection rules for the intersubband transition can be obtained by considering the momentum matrix elements. In the dipole approximation, the matrix element for the transition between states $|m>$ and $|n>$ can be expressed as [14]

$$<u_n n | \hat{\epsilon} \cdot \vec{p} | u_m m > \approx <u_n | \hat{\epsilon} \cdot \vec{p} | u_m >_{cell} <n|m>$$
$$+ <u_n | u_m >_{cell} <n| \epsilon_z z | m> \quad (4)$$

where $\hat{\epsilon}$ is the photon polarization vector, $\vec{p}$ is the momentum of the hole and z is the growth direction. The above expansion is possible because the spatial dependence of the envelope functions within a unit cell can be neglected. In the case of a transition between two quantum states in the same band (i.e. intersubband transition or $u_m = u_n$), the first term in Eq. 4 vanishes due to the orthogonality of the envelope functions. However, the second term is non-zero as long as $\epsilon_z \neq 0$ and $|n>$ and $|m>$ are of opposite parity (i.e. the difference of quantum numbers, Δn = odd). On the other hand, for transitions involving two different hole bands (i.e. intraband transition), the second term of Eq. 4 vanishes due to the orthogonality of the Bloch functions. The first term is non-zero only when $|n>$ and $|m>$ have the same quantum number. For example, the transition between the ground states of valence and conduction bands is allowed. In general, for a transition involving two different bands, the quantum numbers of the envelope states have to be the same (i.e. Δn = 0). One of the key

features of the intersubband transition (at the Γ-point) is that only the light incident parallel to the layers can excite the transition.

The oscillator strength (f) corresponding to a transition between states $|m>$ and $|n>$ is given by [15]

$$f_{m \to n} = \frac{2m_o}{\hbar^2} (E_n - E_m) \; |<n|z|m>|^2. \tag{5}$$

The matrix element $<n|z|m>$ can easily be obtained by writing z in terms of creation (a^+) and annihilation (a) operators [16] as

$$z = \sqrt{\frac{\hbar}{2m^*\Omega}}(a^+ + a) \tag{6}$$

where a^+ and a satisfy

$$a^+|n> = \sqrt{n+1}\,|n+1>, \text{ and } a|n> = \sqrt{n}\,|n-1>. \tag{7}$$

The matrix element can be evaluated readily using z from Eq. (6) as

$$<n|z|m> = \sqrt{\frac{\hbar}{2m^*\Omega}}\, [\,\sqrt{m+1}\,\delta_{n,m+1} + \sqrt{m}\,\delta_{n,m-1}]. \tag{8}$$

This implies that for a parabolic well, non-zero dipole matrix elements occur only between two adjacent states. For a square well, the dipole matrix element is non-zero between any two states with opposite parity as stated previously [15]. In the case of absorption, we need to consider only the first term of Eq. (8). Then the oscillator strength for the transition from n to n+1 is given by

$$f_{n \to n+1} = \frac{m_o}{m^*}(n+1). \tag{9}$$

This may be compared with the case of a square well with an infinite barrier, where the oscillator strength for the same transition is given by [15]

$$f_{n \to n+1} = \frac{64 m_o}{\pi^2 m^*} \frac{(n+1)^2(n+2)^2}{(2n+3)^3} \tag{10}$$

For the transition between the ground state and the first excited state, the oscillator strengths for the parabolic and the square wells are given by m_o/m^* and $0.96 m_o/m^*$, respectively. The strength of the absorption can be evaluated using the transition rate (W) which is given by [17],

$$W = \frac{2\pi}{\hbar} \sum_F |<\phi_F|V_p|\phi_I>|^2 \; \delta(E_F - E_I - \hbar\omega) \tag{11}$$

where E_I and E_F are the initial and final state energies, ω is the angular frequency of the incident photon, ϕ_I and ϕ_F are the initial and final state wave functions, and the photon interaction potential V_p is given by [17]

$$V_p = \frac{e}{m^*} \frac{\hbar}{2\epsilon_o n_r \omega c} \hat{\epsilon} \cdot \vec{p} \qquad (12)$$

where n_r is the refractive index and $\hat{\epsilon}$ and $\vec{p}$ are the polarization of the incident photon and the momentum of the electron (or hole), respectively. By evaluating the transition rate using Eq. (11) and integrating it over the occupied states and the incident photon energy at Brewster's angle, the fraction of energy absorbed per unit length is [18]

$$\frac{e^2 \hbar}{4\epsilon_o m^* c} \frac{\rho_s}{l} \frac{f}{n_r(n_r^2+1)} \qquad (13)$$

where ρ_s is the two dimensional density of electrons (or holes) in the well and l is the quantum well thickness.

2.2 Transition between Superlattice Minibands

Next, we consider an optical transition of a superlattice made of parabolic quantum wells. Fig. 3 shows the schematic diagram of the superlattice, where the ground states are isolated and the excited states form a miniband. This structure is particularly important in infrared detector applications since the isolated ground states keep the leakage current low under an external bias. In this case, the analysis of the transition process is very similar to that of an impurity-to-band transition where a bound electron (or hole) makes a transition to the conduction (or valence) band. Here we use the tight binding method [16] to find the final state wave function using the infinite parabolic well wave functions as our basis set. Under these conditions the initial and final state wave functions can be written as

$$\Phi_I = u_o \frac{e^{i\vec{k}_{i\perp} \cdot \vec{\rho}}}{\sqrt{A}} \phi_{om}(z) \qquad (14)$$

$$\Phi_F = u_o \frac{e^{i\vec{k}_{f\perp} \cdot \vec{\rho}}}{\sqrt{A}} \sum_n \frac{e^{iqnL}}{\sqrt{N}} \phi_{1n}(z) \qquad (15)$$

where A is the area of the well in xy plane, L is the periodicity, N is the number of wells in the superlattice (assumed very large) and ϕ_{om} and ϕ_{1n} are the ground state and the excited state envelope wave functions of an infinite parabolic well at positions rL and nL, respectively.

Taking into account only the nearest-well interactions, the dispersion relation for the first excited state can be written as [19]

$$E_f = E_1 + 2\Delta \cos(qL) + \frac{\hbar^2 k_\perp^2}{2m^*} \qquad (16)$$

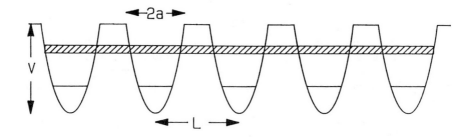

Figure 3: Parabolic well superlattice showing the isolated ground state and the miniband at the first excited state. Such a structure keeps the leakage current through ground states low when used for detector applications.

where E_1 is the first excited state energy of an isolated parabolic quantum well and 2Δ is the bandwidth of the miniband. Δ is given by given by

$$\Delta = 2m^*\Omega^2\sqrt{\frac{\beta^3}{\pi}} \int_{-l/2}^{+l/2} z(z-L)(z^2 - \frac{l^2}{4}) e^{-\beta[(z-L)^2 + z^2]/2} dz \quad (17)$$

where $\beta = m^*\Omega/\hbar$ and l is again the width of the parabolic well. For given parameters of the superlattice, Δ can easily be obtained from Eq. 17 by numerical integration. Using the initial and final state wave functions we can find the matrix element of the transition to be

$$<\Phi_F|\,\hat{\epsilon}\cdot\vec{p}\,|\Phi_I> = \sum_n \frac{e^{-iqnL}}{\sqrt{N}} <\phi_{1n}|\,\hat{\epsilon}\cdot\vec{p}\,|\phi_{om}>. \quad (18)$$

Since ϕ_{om} is localized at the rth well position, the major contribution to the matrix element occurs when n = r. In this case, it can easily be shown using Eq. (8) that Eq. (18) reduces to

$$|<\Phi_F|\,\hat{\epsilon}\cdot\vec{p}\,|\Phi_I>|^2 \approx \frac{m^*2\Omega^2}{N}\frac{\hbar}{2m^*\Omega}(\hat{\epsilon}\cdot\hat{e}_z)^2. \quad (19)$$

Substituting this into the expression for the transition rate, we find the absorption constant as a function of photon energy ($\hbar\omega$) to be

$$\alpha = \frac{e^2\hbar}{2m^*c\epsilon_o n_r}\frac{\rho_s}{L}\frac{\Omega}{\omega}\frac{(\hat{\epsilon}\cdot\hat{e}_z)^2}{\sqrt{(E_{max} - \hbar\omega)(\hbar\omega - E_{min})}} \quad (20)$$

where $E_{max} = E_1 - E_o + 2\Delta$ and $E_{min} = E_1 - E_o - 2\Delta$.

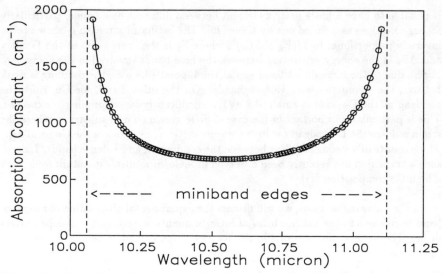

Figure 4: Absorption constant as a function of wavelength. Here we assume IR incident at Brewster's angle. Quantum well parameters are chosen to get a response near 10 μm with a 10 meV bandwidth.

Figure 4 shows the absorption constant as a function of wavelength when the device parameters are chosen so as to obtain a response near 10 μm. The parabolic quantum well is assumed to contain 2×10^{18} holes/cm^3. The absorption constant shows a singular behavior at the band edges due to one-dimensional van Hove singularities of the joint density of states. This gives a relatively high absorption coefficient near the band edges as seen in Fig. 4. The origin of this behavior is due to the exact parallelism between the in-plane dispersion relations of the initial and final subbands. In reality, the non- parabolicity and other effects which alter this parallelism, would soften the singular behavior resulting in a finite absorption constant near the band edges.

2.3 Intervalence Band Transition

In the above sections we have discussed the transition in quantum wells or in superlattices due to coupling of the photon electric fields between the envelope state wave functions. However, under certain conditions, it is possible to have optical transitions between different hole bands (for example, between the heavy and light hole bands) which also lie in the same wavelength range as in the intersubband transition. The key advantage to such a transition is that it can provide normal incident detection because of the dipole matrix element which occurs between the Bloch states. In general, the matrix elements between p-like Bloch states in the valence band vanish due to the odd parity of both the initial and final wave functions. If there is a significant coupling of the s-like Γ conduction band Bloch state with that of the valence band, it

is possible to have a finite matrix element between different hole bands, particularly at large k-values as pointed out by Kane [20]. The oscillator strength is shown to be inversely proportional to $1/(E_g + \Delta E_v)^2$, where E_g is the energy gap at the Γ-valley and ΔE_v is the energy separation between the hole bands involved in the transition. For Si, due to the large direct band gap at the Γ-point (3.5 eV), the coupling is weak between the conduction and valence bands. On the other hand, for Ge, since the bandgap at the Γ−point is small (0.8 eV), a significantly large coupling is expected. This is particularly important in the case of SiGe grown on a Si substrate, where the strain reduces the Γ-bandgap (at high Ge compositions, even below the Ge band gap) which results in a strong coupling between the conduction and valence bands. Indeed, such a transition has recently been observed in SiGe/Si multiple quantum wells with a high Ge composition [21].

In the following sections, we will discuss the experimental observation of intersubband transitions in the valence band of SiGe/Si quantum wells and of δ-doped layers in Si. The general features of the results will be explained in terms of the above discussion.

3. Intersubband Transition in SiGe/Si Quantum Wells

The intersubband transitions in SiGe/Si quantum wells were first observed for holes [22] using a multiple quantum well structure grown pseudomorphically on a Si substrate. For pseudomorphic structures grown on Si, most of the band offset appears at the valence band. More recently, electron intersubband absorption has been observed [23, 24] using a multiple quantum well structure grown on a relaxed SiGe buffer layer in order to obtain appreciable band offset at the conduction band. In this section, we discuss the experimental observation of hole intersubband infrared absorption in $Si_{1-x}Ge_x$/Si multiple quantum wells grown on Si substrates [22].

All of the samples described here are grown in a molecular beam epitaxy (Si-MBE) system on high resistivity (100 Ω-cm) Si(100) wafers. This reduces the infrared absorption due to free carriers in the substrate. The substrate is cleaned by a standard procedure [25] before it is loaded into the chamber and the protective oxide layer is then removed by heating the substrate to 900°C for 20 minutes. The substrate temperature is kept in the 540 − 600°C range depending on the Ge composition of the quantum well structure.

First, an undoped 3000 Å Si buffer is grown on a Si substrate. The multiple quantum well structure consists of 10 periods of 40 Å $Si_{0.6}Ge_{0.4}$ wells and 300 Å Si barriers. The center 30 Å region of the quantum wells is p-doped to about 1×10^{19} cm^{-3}. The p-type doping is obtained using a thermal boron source [26]. The quantum well structure is capped with a 1000 Å Si layer. The thick barrier layer keeps the average Ge composition of the entire structure sufficiently low even though the total thickness

Figure 5: TEM picture of the $Si_{0.6}Ge_{0.4}$ multiple quantum well structure.

of the $Si_{0.6}Ge_{0.4}$ layers is well above the critical thickness for a 40% Ge content [8]. A TEM picture of the quantum well structure is shown in Fig. 5. Within the field of the TEM, there were no dislocations visible in the structure.

The transmission spectrum of the sample is taken at room temperature using a Fourier transform infrared (FTIR) spectrometer. A waveguide structure (5 mm long and 0.5 mm thick) is employed (see inset of Fig. 6) for the measurement in order to enhance the absorption since the absorption strength is small due to the relatively large mass of the heavy hole (0.26 m_o). A beam condenser is used to focus the infrared beam on the waveguide. An infrared polarizer is placed in the path of the infrared beam in order to probe the polarization dependence of the absorption process.

The measured absorption spectra as a function of photon energy using the waveguide structure is shown in Fig. 6. The set of curves show different polarization angles of the infrared beam. Zero degrees corresponds to the polarization of the beam along the growth direction. A peak at 153 meV (8.1 μm) with a long tail towards the short wavelength region is observed. The small features superimposed on the spectra are mainly due to strong infrared absorption near 136 meV by SiO_2 and water bands which are in the spectral range of interest. The strength of the intersubband absorption decreases as the polarization of the infrared beam is rotated from the perpendicular direction towards the direction parallel to the plane of layers. Figure 7 illustrates the relative absorption strength as a function of the polarizer angle ϕ. The data follows the $\cos^2\phi$ dependence (dashed curve) very well in agreement with the selection rule of the intersubband transition as discussed in section 2. The absorption strength (I_A) due to multiple reflections in the waveguide can be found using Eq. 10

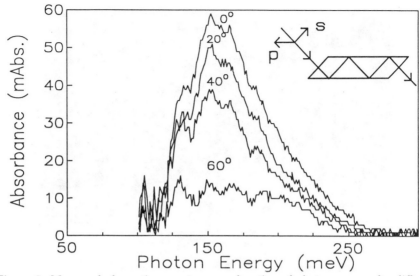

Figure 6: Measured absorption spectra as a function of photon energy for different polarizations of the incident infrared beam. The peak at 153 meV is due to intersubband absorption between the heavy hole ground state and the first excited state. The absorption strength at large polarization angles is shown to decrease in accordance with the selection rules of intersubband transition at Γ-point. The inset shows the waveguide structure used in the measurement.

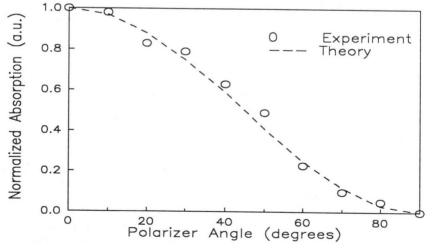

Figure 7: Normalized absorption strength as a function of the polarization of the incident infrared beam. Zero degrees corresponds to polarization along the growth direction of the structure. The dashed curve shows the theoretically expected $\cos^2\phi$ dependence.

to be
$$I_A = \rho_s N_T \frac{e^2 h}{4c\varepsilon_o n_r m^*} f \frac{\cos^2\theta}{\sin\theta} \cos^2\phi \qquad (21)$$

where N_T is the total number of quantum wells that the infrared beam passes through due to multiple reflections, ρ_s is the two dimensional density of holes in the well, I_A is the integrated absorption strength, θ is the angle of incidence extending from the normal to the plane, and the effective mass m^* is taken along (100) direction. Here, we ignore the non-parabolic effects on the intersubband transition due to the quantization and the strain present in the quantum well.

The oscillator strength of the transition (f) can be estimated using the Eq. 21. For the structure, we used $N_T = 100$, $m^* = 0.26$ m_o, $n_r = 3.4$, $\theta = 45°$, $\rho_s = 3\times 10^{12}$ cm^{-2} and the integrated absorption strength, $I_A = 5$ meV. These parameters give an oscillator strength of 0.98. This is in good agreement with the calculations as well as the values obtained for intersubband absorption in III-V based quantum well structures [15, 27] which are in the 0.5 to 1.2 range. It should be noted that this is an approximate estimate since the non-parabolicity of the hole band can be severe in the quantum well and the joint density-of-states effective mass can deviate significantly from the bulk value [28].

In assessing the origin of the observed transition, it is recognized that the presence of three hole bands (heavy, light and split-off) with energy separations close to the observed absorption peaks, makes the identification of the transitions difficult. However, the strong polarization dependence indicates that the transitions involving quantized states of two different hole bands is relatively small. As discussed in section 2.3, the intervalence band transition is possible when the Ge composition as well as the doping in the well are large [29]. This simplifies the deconvolution process. Figure 8 shows a detailed band diagram of the quantum wells with the light and heavy hole bands and their bound state energies. The bound state energies in the quantum well are calculated using an envelope function approximation with the band offsets under compressive strain obtained from a linear interpolation of those of Van der Walle and Martin [30]. The effective masses of the light and heavy holes along the (100) direction are obtained from Ref. [31], in which the effective masses are calculated using three band k.p and strain Hamiltonians [32, 33]. The band offsets for the light and heavy holes are 267 meV and 336 meV and the effective masses are 0.22m_o and 0.26m_o, respectively. There are two heavy hole bound states at 49 meV and 189 meV and two light hole states at 123 meV and 266 meV in the quantum well. All of the energies are measured relative to the heavy hole band edge in the well region.

The light and heavy hole bands are degenerate in the Si barrier regions and are split in the well. This is due to different mass quantizations and the strain in the Si$_{0.6}$Ge$_{0.4}$ layer [34]. In the quantum well, the light hole band lies about 70 meV above the heavy hole band, while the split-off hole band is about 290 meV above

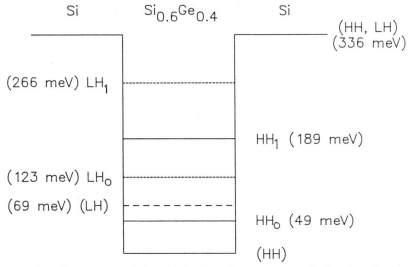

Figure 8: Band structure of the Si/Si$_{0.6}$Ge$_{0.4}$/Si quantum well showing the bound states due to both light and heavy hole bands. The heavy hole (HH) and light hole (LH) band edges are shown in solid and broken lines, respectively.

the heavy hole band. These energy separations are relatively large compared with kT at room temperature ($\sim$ 26 meV). At thermal equilibrium most of the holes occupy the heavy hole ground state. Thus, the intersubband absorption mainly occurs between quantized heavy hole states. The absorption peak at 153 meV (8.1 μm) is due to the transition from the heavy hole ground state to the first excited state. The absorption peak width is about 75 meV and is about seven times larger than the intersubband absorption peak widths observed in AlGaAs/GaAs quantum well structures [15, 27, 35]. This may be due to the strong non-parabolicity of the hole bands, particularly in the presence of a strain. The calculated peak position for the transition between the ground state and the first excited heavy hole state is 140 meV. This agrees reasonably well with the experimental value of about 153 meV.

4. Intersubband Transitions in δ-doped Quantum Wells

Next, we will discuss the use of δ-doped layers in Si for the study of intersubband transitions. The δ-doping in semiconductors can be viewed as an alternative way to achieve quantum well structures without heterojunctions. The potential profile associated with δ-doping closely resembles that of a parabolic quantum well due to the finite width of the doped layer [36]. The well thickness and the barrier height can be controlled by the thickness of the doped layer and the doping density. The potential profile and the energy level spectrum in the well are usually obtained by solving Schrödinger's and Poisson's equations self-consistently (Hartree approximation). The existence of quantized states in such a system in Si was probed using tunneling

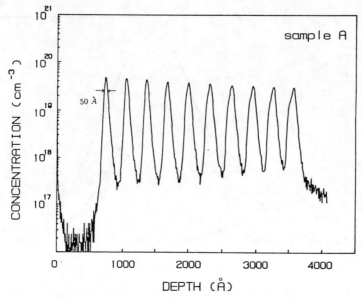

Figure 9: SIMS depth profile of sample A. It reveals 10 periods of boron δ-doped Si layers with a FWHM of about 50 Å.

spectroscopy [37] and optical transitions between quantized states [38, 39, 29]. For intersubband detector applications, the population of carriers in such structures are considerably larger than those typically used in heterojunction quantum wells. As a result, the many-body effects play an important role in determining the optical properties of δ-doped quantum wells. In the following section, we describe the intersubband absorption of δ-doped quantum wells which is then followed by a discussion of the effect of many-body interactions on the optical transitions.

4.1 Doping Dependence of Intersubband Absorption

A typical structure used in this study consists of an undoped Si buffer layer followed by 10 periods of 35 Å heavily boron-doped Si layers and 300 Å undoped Si spacers. Four samples (A, B, C, D) with doping concentrations of about 0.7, 2.7, 4.0 and 5.7×10^{20} cm^{-3}, respectively are prepared to study the dependence of the energy level separation on the population of holes in the subbands. The doping density is calibrated by secondary ion mass spectroscopy (SIMS). Figure 9 shows the (SIMS) depth profile for a typical δ-doped structure (sample A). A full width at half maximum (FWHM) of approximately 50 Å is obtained from the depth profile.

The absorption spectra of the samples are taken at room temperature using a Fourier transform infrared spectrometer (FTIR). To enhance the absorption strength, waveguide structures are used which are similar to that used for SiGe/Si multiple

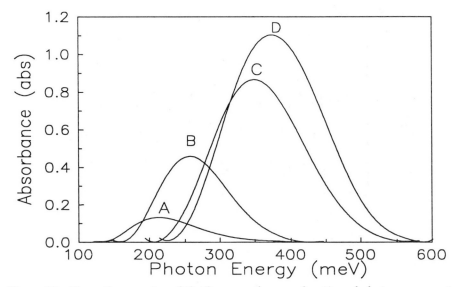

Figure 10: Absorption spectra of the four samples as a function of photon energy at 300 K. The set of curves are due to different doping concentrations.

quantum well structures discussed in section 3. The measured absorption spectra of the samples as a function of photon energy are shown in Fig. 10. Peak positions are found near 210, 259, 348, and 377 meV for samples A, B, C, and D, respectively. It can be clearly seen that the absorption spectra shift towards the high energy regime with increasing absorption as the doping density is increased. The shift of the absorption peak is mainly due to the increase of the potential well depth at high doping densities. The widths of the absorption peaks are more than an order of magnitude larger than those observed in GaAs/AlGaAs quantum well structures, which are typically about 10 meV. The non-parabolicity of hole bands plays an important role in determining the peak width. In particular, at high doping densities, the hole bands can be filled up to several hundreds of meV.

To further confirm the nature of the transition, Fig. 11 shows the relative absorption spectra of the sample C as a function of polarization angle ϕ, where 0° corresponds to the s-polarization and 90° for the p-polarization (see inset of Fig. 11). The reduction of absorption strengths with the polarization angle ϕ as shown in Fig. 11 is in agreement with the selection rule of the intersubband transitions as in the case of SiGe/Si multiple quantum wells discussed in section 3.

In order to understand the obtained data, particularly the variation of peak positions as a function of the 2D hole density in the wells, the subband energies are first estimated using a multi-band self-consistency calculation (Hartree approximation) [40]. In the calculation, we assume that the doping is uniformly distributed over a 35 Å layer of Si and that the light and heavy hole bands are degenerate at

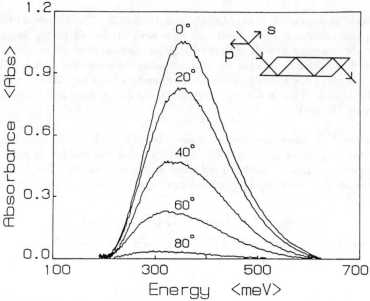

Figure 11: Polarization dependent absorption spectra of sample C at 300K. The decrease of the absorbance with increasing polarizer angle is due to the reduction of the component of photon polarization along the growth direction similar to that of SiGe/Si multiple quantum wells.

the Γ-point while the split-off band is 44 meV above the light and heavy hole bands. The light, heavy and split-off hole effective masses of 0.2, 0.30 and 0.24, respectively, along the Si (100) direction and density-of-states effective masses of 0.15, 0.40 and 0.24, respectively in the [100] plane are used for the subband energy calculation. The calculated peak positions for the samples A, B, C, and D are 62, 118, 139, and 159 meV, respectively. These values are considerably smaller than the measured peak positions. This discrepancy is mainly due to the failure of the single particle approximation in the calculation. The many-body effects discussed in the next section play an important role in determining the confining potential as well as the plasma shift of the peak position due to collective transition of the carriers [41, 42]. This is particularly important in the case of boron (B) doping. Because of its large solid solubility in Si, a considerably higher doping density can be achieved [26].

4.2 Many-body Effects

The importance of many-body effects has been recognized in the understanding of optical transitions in inversion layers [43, 44, 45, 46]. The optical transitions between quantized subbands in an inversion layer show a large discrepancy between the experimentally observed peak positions and the self-consistent calculations using the Hartree approximation [47, 48]. By incorporating many-body effects, such as exchange-correlation and depolarization the calculated values agree with the experimental results [49, 41, 42]. Recently, a similar behavior has been observed in

intersubband absorption in GaAs/AlGaAs quantum wells. The differences between observed and calculated energy positions have been attributed to the lowering of ground state energy as a result of the exchange interactions between electrons in the ground subband [50, 51]. In all of the previous cases, the amount of shift is relatively small (a few meV) due to the limitations of achieving a large population in the subbands. In the case of boron δ-doped layers in Si, the discrepancy as described above is more than 100 meV.

The many-body effects are usually incorporated in the calculation using the local-density-functional approximation [42]. In the following, we will briefly discuss the steps that are used in the multiband self-consistent calculation [40]. The Schrödinger equation, including the many-body effects, can be written as

$$[-\frac{\hbar^2}{2m^*}\nabla^2 + V_H(z) + V_{xc}(z)]\psi_n = E_n\psi_n \qquad (22)$$

where V_H and V_{xc} are the Hartree and exchange-correlation potentials, respectively. The Hartree potential is given by

$$V_H(z) = \frac{e}{\epsilon\epsilon_o}\int(z-z')\rho(z')dz' \qquad (23)$$

where $\rho(z) = e[N_D - n(z)]$, ϵ is the dielectric constant, N_D is the dopant density,

$$n(z) = n^{2D}\sum_n |\psi_n(z)|^2 \qquad (24)$$

is the carrier concentration at z and n^{2D} is the two dimensional density of carriers. The exchange-correlation potential in the local-density-functional approximation using the Hedin and Lundqvist parameterization [52] can be expressed as

$$V_{xc}(z) = -[1 + \frac{0.7734 r_s}{21}ln(1 + \frac{21}{r_s})](\frac{2}{\pi\alpha r_s})Ry^* \qquad (25)$$

where $r_s = [\frac{4}{3}\pi a^{*3}n(z)]^{-1/3}$, the effective Bohr radius $a^* = 4\pi\epsilon\hbar^2/m^*e^2$, $\alpha = [4/(9\pi)]^{1/3}$ and the effective Rydberg $Ry^* = e^2/(8\pi\epsilon a^*)$. In this calculation, the doping distribution as well as the three hole bands are taken into account in a manner similar to the Hartree approximation described in section 4.1. It is also known that the photon induced many-body contributions such as depolarization [41] and exciton-like interaction [42] between the ground and excited states shifts the calculated energy position by a substantial amount. These two effects come about as a result of the plasma frequency dependence on the charged sheet. For a two-dimensional gas, such effects are estimated by solving the time dependent Schrödinger equation self-consistently using the Hartree and exchange-correlation potentials. The peak position, including the shift due to plasma effects, is given by [48]

$$\tilde{E}_{10}^2 = E_{10}^2(1 + \alpha - \beta) \qquad (26)$$

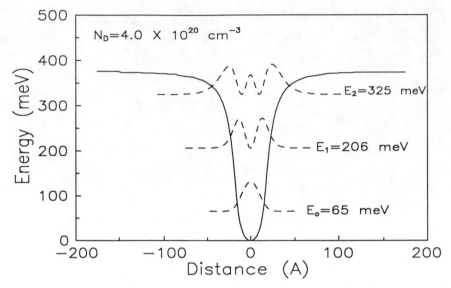

Figure 12: Calculated potential well and wave functions for the sample with 2.7×10^{20} cm^{-3} using a multiband Hartree approximation. Only the data for the heavy hole band is included in this figure.

where $\tilde{E}_{10}$ is the shifted energy and E_{10} is the difference $(E_1 - E_0)$ of the subband energy levels calculated by Eq. 22 self-consistently. The quantities α and β for the transition between the ground state and the first excited states (assuming only the ground state is occupied) are given by [42]

$$\alpha = \frac{2e^2 n_0^{2D}}{\epsilon \epsilon_o} \left(\frac{\hbar^2}{2m^* E_{10}}\right)^2 \int_{-\infty}^{+\infty} dz \left[\psi_1(z)\frac{d\psi_o(z)}{dz} - \psi_o(z)\frac{d\psi_1(z)}{dz}\right]^2 \quad (27)$$

and

$$\beta = \frac{2n_0^{2D}}{E_{10}} \int_{-\infty}^{+\infty} dz \psi_1(z)^2 \psi_o(z)^2 \frac{\partial V_{xc}[n(z)]}{\partial n(z)} . \quad (28)$$

In order to estimate the magnitude of the many-body effects, we first calculated the wave functions and the bound state energies self-consistently using Hartree and exchange-correlation potentials. The contribution due to depolarization and exciton-like effects are then estimated using Eqs. 27 and 28. The calculated potential well, eigenstates and wave functions which correspond to the heavy hole band for the sample with 2.7×10^{20} cm^{-3} doping density are shown in Fig. 12. Fig. 13 summarizes the experimental and calculated peak positions using different levels of approximations. The dashed curve with filled circles show the measured peak positions while the dashed curve represents the calculated peak positions without the plasma shifts. As mentioned before, these values are considerably smaller than the experimental values. The solid curve in the Fig. 13 shows the incorporation of plasma (i.e., depolarization and exciton-like) shifts to the calculated energy level separation. This

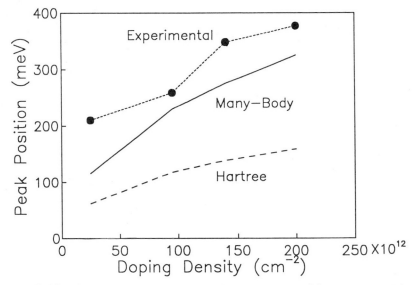

Figure 13: Subband separation as a function of doping density (a) experimental data (dashed curve with filled circles), (b) using the Hartree approximation (dashed curve), and (c) the Hartree approximation with exchange-correlation potential including depolariztion and exciton-like shifts (solid curve).

brings the calculated and experimental peak positions reasonably close to agreement. The slightly lower values obtained in the calculation may be due to the omission of the non-parabolic effects and the occupancy of the excited state in the analysis. We will now discuss the use of intersubband absorption in quantum wells for infrared detector applications.

5. Detector Applications

There is a considerable interest in the application of multiple quantum wells for infrared detectors. This is due to the flexibility of tuning the response for a desired wavelength as well as the availability of a number of material systems with advantageous properties. Most of the work to date is based on intersubband absorption of electrons of III-V based material systems [53, 54]. The fabrication of GaAs/AlGaAs multiple quantum well detector arrays with hybrid Si signal processing electronics have also been demonstrated [55, 56]. Recently, infrared absorption between quantized hole states in $Si_{1-x}Ge_x$/Si quantum wells have been observed using Fourier transform infrared spectroscopy (FTIR) [22]. The large valence band offset of $Si_{1-x}Ge_x$/Si as well as the small hole effective mass favor the hole intersubband absorption for mid infrared applications. In this section, we illustrate the fabrication of an infrared detector using $Si_{1-x}Ge_x$/Si multiple quantum wells. One of the key advantages of the Si- based infrared detectors is the possibility of monolithic integration

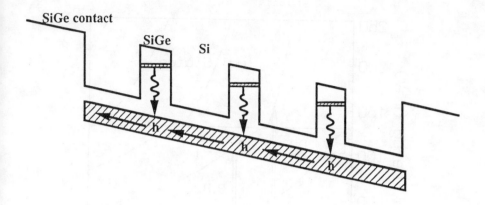

Figure 14: Schematic diagram of photoexcited hole transport via extended miniband states.

with the mature Si signal processing electronics for focal plane array applications.

For detector applications, quantum well structures with a single bound state and excited extended states close to the barrier are desirable [14, 57]. This allows the photoexcited carriers to travel without seeing barriers while maintaining a large absorption coefficient [14]. The carrier transport process in the extended miniband is schematically shown in Fig. 14. For design purposes, we have calculated the position of the bound and extended states as a function the well width using an effective mass approximation. In this calculation, the Ge composition in the well is taken to be 15%. The corresponding band offset is about 130 meV [30] and the heavy hole effective mass along the growth direction is taken as 0.28 m_o. The energy of the extended states is taken as the maximum of the transmission coefficient. The calculated values as shown in Fig. 15 indicate that a quantum well 30 Å thick can have absorption near 10 μm with the extended state lying above the Si barrier.

In this structure the temperature of the substrate during growth is kept at about 600 °C. This improves the epitaxial quality of the layers, in particular for Si barriers where most of the photoexcited carrier transport occurs. The quantum well structure consists of 50 periods of 30 Å $Si_{0.85}Ge_{0.15}$ wells (p-doped to $\sim 1 \times 10^{19}$ cm^{-3}) separated by 500 Å undoped Si barriers. The entire multiple quantum well structure is sandwiched between a doped (p = 1×10^{19} cm^{-3}) 1 μm bottom and a 0.5 μm top layers for electrical contacts. Figure 16 shows the room temperature absorption spectrum as a function of wavelength using a 45° multipass waveguide structure with dimensions 4 mm long and 0.5 mm wide. Here, we have subtracted the monotonically increasing background absorption (mainly due to free carrier absorption) in order to magnify the intersubband absorption. A peak near 8.6 μm is observed with a full width at half maximum of about 100 meV. This absorption is due to the transition

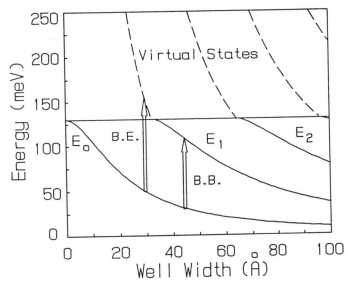

Figure 15: Bound and extended states of a $Si_{0.85}Ge_{0.15}$ quantum well as a function of well thickness. The barrier height is about 130 meV. The arrows show the bound to bound (B.B.) and bound to extended (B.E.) state transitions.

from the heavy hole ground state to the extended state since only the heavy hole ground state is confined in the quantum well. The splitting of the light and heavy hole bands [34] in the strained $Si_{0.85}Ge_{0.15}$ quantum well pushes the light hole ground state about 50 meV above the heavy hole ground state. For the doping density used in this study, the two dimensional Fermi energy is only about 20 meV above the ground state energy. This results in a negligible amount of holes occupying the light hole ground state. The absorption coefficient is comparable to that observed in a similar structure in the III-V system [58]. The spectrum is considerably broader than normally observed in III-V based quantum well structures ($\sim$ 40 meV). This is mainly due to the non-parabolicity of the hole band causing the transition energy to be partially dependent on the transverse momentum [40].

5.1 Detector Characterization

To study the photoresponse, mesa diodes 200 μm in diameter are fabricated with a 45° facet on the edge of the wafer. The measured leakage current and differential resistance at about 77 K are shown in Fig. 17 as a function of bias across the device. For the present non-optimized device, the leakage current at 2 V is an order of magnitude higher than for a typical GaAs/AlGaAs infrared detector [53]. This is mainly due to the higher doping (almost an order of magnitude larger) used in the present device to compensate for the reduction of absorption strength due to the larger hole effective mass compared to the electron effective mass in GaAs. The leakage current increases

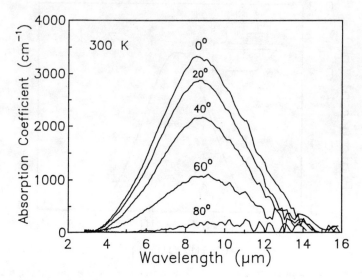

Figure 16: Measured absorption spectrum at 300 K as a function of wavelength using a waveguide structure.

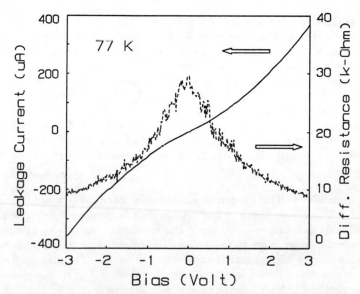

Figure 17: Current-voltage and differential resistance measured at 77 K. Positive bias means the top contact positive.

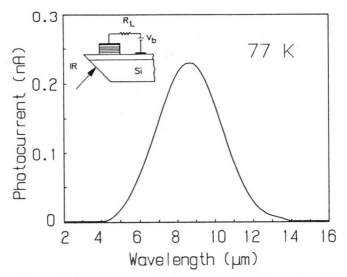

Figure 18: Measured photocurrent at 77 K as a function of wavelength for a 200 μm mesa diode with infrared incident from a 45° facet as shown in the inset. The bias across the device is 2 volts.

rapidly as the bias across the device is increased. This is due to the enhancement of tunneling and the thermionic components of the current. The spectral dependence of the photocurrent is measured using a glowbar-source and a grating monochromator. The light polarized in the plane of incidence is illuminated on the facet at the normal such that the incidence angle on the multiple quantum well structure is 45° which is shown in the inset of Fig. 18. Figure 18 shows the measured photocurrent as a function of wavelength using a 7 kΩ load resistor with 2 V bias across the device. A broad peak near 8.2 μm (150 meV) is observed in the photoresponse which is close to that observed in the absorption measurement using FTIR. The slight shift of the peak to the lower wavelength may be due to the low temperature used in the photocurrent measurement.

The responsivity (R) of the detector is measured using the Globar-monochromator setup at 9.5 μm wavelength. The photon flux is measured using a calibrated GaAs/AlGaAs quantum well detector. The measured responsivity as a function of bias across the device is shown in Fig. 19. The responsivity rapidly increases as the bias is increased to 2.5 V, where it approaches saturation. The maximum value of R for the present device (which is not optimized) is about 0.3 A/W, comparable to that typically observed in III-V based quantum well detectors.

To understand the dependence of photocurrent on the transport parameters, we calculate the photocurrent ($I_p(E)$) as a function of photo-generated carriers at a given quantum well multiplied by the probability of their reaching the contact:

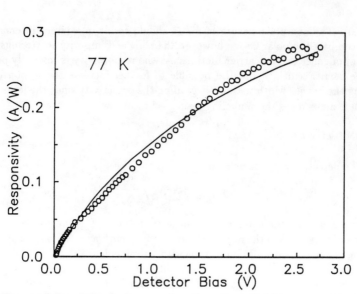

Figure 19: Responsivity of the device as a function of applied bias across the device at 77 K. The solid curve is calculated values using Eq. (31).

$$I_p(E) = \sum_{n=1}^{N} \frac{eP_o}{h\nu}(e^{-\alpha n l} + e^{-\alpha(2N-n)l})\alpha l e^{-nL/v(E)\tau}, \quad (29)$$

where P_o is the incident infrared power, $h\nu$ is the photon energy, α is the absorption coefficient, l is the well width, L is the period of the multiple quantum wells, N is the number of quantum wells, $v(E)$ is the drift velocity, and τ is the excited carrier lifetime. Since $\alpha l \ll 1$ the above equation can be simplified to

$$I_p(E) \approx \frac{2eP_o}{h\nu}\alpha l \sum_{n=1}^{N} e^{-nL/v(E)\tau}. \quad (30)$$

After evaluating the summation and assuming that for moderate electric fields $L/v(E)\tau < 1$, the responsivity $R(E)$ $(=I_p(E)/P_o)$ can be found using the drift velocity for holes $v(E) = \mu E/(1 + \mu E/v_s)$ [59] to be

$$R(E) \approx \frac{2e\alpha\tau}{h\nu}\frac{l}{L}\frac{\mu E}{1 + \mu E/v_s}, \quad (31)$$

where μ is the mobility, E is the electric field and v_s is the saturated drift velocity. In order to estimate the mobility and the excited carrier lifetime, the experimental data is fitted to Eq. 31 using the saturated drift velocity [59] of about 1×10^7 cm s^{-1} and the measured absorption coefficient at 9.5 μm (2800 cm^{-1}). The mobility of holes of 1000 cm^2/V s and the excited carrier lifetime of 15 picoseconds seems to fit the experimental data well as shown by the solid line of Fig. 19. The values of mobility and lifetime are comparable to those obtained for a typical GaAs/AlGaAs detector

[53, 55]. The non-polar nature of Si/Ge should result in a larger intersubband lifetime compared to polar GaAs; however, the alloy and impurity scatterings can have a strong influence on the carrier lifetime. As the responsivity is inversely proportional to the barrier width, it may be possible to further increase the responsivity by reducing the barrier width without degrading the detectivity since the large heavy hole effective mass keeps the tunneling leakage current low.

5.2 Detectivity

The detectivity, D*, is defined as

$$D^* = \frac{S/N}{P}\sqrt{A\Delta f}, \qquad (32)$$

where S and N are the signal and the noise in Amperes, P is the power received by the detector in watts, A is the area of the detector in cm^2, and Δf is the bandwidth. For the case of the quantum well detector with light incidence at an angle θ, the effective area becomes $A\cos\theta$ and the noise current (shot noise) is given by $\sqrt{4eI_d\Delta f}$, where I_d is the dark current. Using the effective area and the noise current, the detectivity can be written as

$$D^* = \eta(eA\cos\theta/I_d)^{1/2}/2h\nu \qquad (33)$$

where $\eta = 1\text{-}e^{-2\alpha l}$ is the quantum efficiency, A is the area of the device and $h\nu$ is the photon energy. The detectivity can be estimated using Eq. 33 with the measured leakage current and the quantum efficiency. For the present non-optimized device, the estimated value is about D*(9.5 μm) = 1×10^9 cm$\sqrt{Hz}$/W at 77 K. A detectivity an order of magnitude larger is observed in III-V based quantum well detectors by optimizing the device parameters. This indicates that by optimizing device parameters, the sensitivity of the SiGe/Si quantum well detector can be further improved.

6. Summary

In summary, the physics of intersubband transitions is discussed using a parabolic quantum well as an example. The demonstration of intersubband infrared absorption in Si$_{1-x}$Ge$_x$/Si multiple quantum well structures is discussed. The nature of the optical transitions is characterized by using the polarization dependence measurement and is in good agreement with the selection rule. Similarly, the use of $\delta-$doped Si layers for intersubband transitions is illustrated. The importance of many-body effects in the determination of optical transition energies is discussed. An intersubband infrared detector using Si$_{1-x}$Ge$_x$/Si multiple quantum wells is demonstrated. The broad photoresponse covers a large portion of the 8-14 μm atmospheric window in contrast to a narrow band response of III-V based detectors. The peak responsivity as well as the detectivity of the device can be further improved by optimizing the device parameters. This demonstration shows a potential of monolithic integration

of SiGe/Si multiple quantum well infrared detectors with Si signal processing circuits for focal plane applications.

Acknowledgements: The authors would like to thank Michael Chu and Martin Tanner for critical reading of the manuscript. This work was supported in part by Army Research Office (Dr. John Zavada) and the Air Force Office of Scientific Research (Dr. Gerald Witt).

References

[1] E. Kasper, and H. J. Herzog. *Thin Solid Films*, **44**, 357 (1977).

[2] H. M. Manasevit, I. S. Gergis, and A. B. Jones. *Appl. Phys. Lett.*, **41**, 464 (1980).

[3] J. C. Bean, L. C. Feldman, A. T. Fiory, S. Nakahara, and I. K. Robinson. *J. Vac. Sci. Technol*, **A2**, 436 (1984).

[4] S. S. Rhee, J. S. Park, R. P. G. Karunasiri, Q. Ye, and K. L. Wang. *Appl. Phys. Lett.*, **53**, 204 (1988).

[5] B. S. Meyerson, K. J. Uram, and F. K. Legoues. *Appl. Phys. Lett.*, **53**, 2555 (1988).

[6] J. S. Hoyt, C. A. King, D. B. Noble, C. M. Gronet, J. F. Gibbons, M. P. Scott, S. S. Laderman, S. J. Rosner, K. Nauka, J. Turner, and T. I. Kamins. *Thin Solid Films*, **183**, 93 (1989).

[7] J. H. Van de Merwe. *J. Appl. Phys*, **34**, 117 (1963).

[8] R. People, and J. C. Bean. *Appl. Phys. Lett*, **39**, 538 (1986).

[9] R. Hull, J. C. Bean, F. Cerdeira, A. T. Fiory, J. M. Gibson. *Appl. Phys. Lett.*, **48**, 56 (1986).

[10] E. Kasper. *Surf. Sci.*, **176**, 630 (1986).

[11] C. G. Van de Walle and R. M. Martin. *Phys. Rev.*, **B34**, 5621 (1986)

[12] Q. M. Ma and K. L. Wang. *Appl. Phys. Lett.*, **58**, 1184 (1991).

[13] R. P. G. Karunasiri and K. L. Wang. *J. Vac. Sci. Technol.*, **B9**, 2064 (1991).

[14] D. D. Coon, and R. P. G. Karunasiri. *Appl. Phys. Lett.*, **45**, 649 (1984).

[15] L. C. West, and S. J. Eglash. *Appl. Phys. Lett.*, **46**, 1156 (1985).

[16] S. Wieder. *The Foundation of Quantum Theory*. Academic Press, Inc., New York, 1973.

[17] B. K. Ridley. *Quantum Processes in Semiconductors*. Oxford University, London, 1982.

[18] R. P. G. Karunasiri and K. L. Wang. *Superlattices and Microstructures*, **4**, 661 (1988).

[19] P. Voisin, G. Bastard, and M. Voos. *Phys. Rev. B*, **29**, 935 (1984).

[20] E. O. Kane. *J. Phys. Chem. Solids.*, **1**, 82 (1956).

[21] J. S. Park, R. P. G. Karunasiri, and K. L. Wang. *unpublished*.

[22] R. P. G. Karunasiri, J. S. Park, Y. J. Mii and K. L. Wang. *Appl. Phys. Lett.*, **57**, 2585 (1990).

[23] H. Hertle, G. Schuberth, E. Gornik, G. Abstreiter, and F. Schaffler. *Appl. Phys. Lett.*, **59**, 2977 (1991).

[24] C. H. Lee and K. L. Wang. *presented in 11th MBE workshop, Sep. 16-18, Austin, TX (1991)*.

[25] A. Ishizaka and Y. Shiraki. *J. Electrochem. Soc.*, **133**, 666 (1986).

[26] S. S. Rhee, R. P. G. Karunasiri, C. H. Chern, J. S. Park and K. L. Wang. *J. Vac. Sci. Tech.*, **B7**, 327 (1989).

[27] B. F. Levine, R. J. Malik, J. Walker, K. K. Choi, C. G. Bethea, D. A. Kleinman, and J. M. Vandenberg. *Appl. Phys. Lett.*, **50**, 273 (1987).

[28] R. Wessel amd M. Altarelli. *Phys. Rev. B*, **40**, 12457 (1989).

[29] J. S. Park, R. P. G. Karunasiri, Y. J. Mii and K. L. Wang. *Appl. Phys. Lett.*, **58**, 1083 (1991).

[30] C. G. Van de Walle and R. Martin. *J. Vac Sci. Technol.*, **B3**, 1257 (1985).

[31] S. K. Chun and K. L. Wang. *Effective Mass and Mobility of Holes in Strained $Si_{1-x}Ge_x$ Layers on (001) $Si_{1-y}Ge_y$ Substrate*, unpublished.

[32] G. E. Pikus and G. L. Bir. *Soviet Phys. Solid State*, **1**, 1502 (1960).

[33] J. C. Hensel and G. Feher. *Phys. Rev.*, **129**, 1041 (1963).

[34] R. People. *Phys. Rev. B*, **32**, 1405 (1985).

[35] Y. J. Mii, K. L. Wang, R. P. G. Karunasiri, and P. F. Yuh. *Appl. Phys. Lett.*, **56**, 1046 (1990).

[36] A. Zenner, F. Koch, and K. Ploog. *Surf. Sci.*, **196**, 671 (1988).

[37] H. P. Zeindl, T. Wegehaupt, I. Eisele, H. Oppolzer, H. Reisinger, G. Tempel and F. Koch. *Appl. Phys. Lett.*, **50**, 1164 (1987).

[38] G. Tempel, N. Schwarz, F. Müller and F. Koch. *Thin Solid Films*, **184**, 171 (1990).

[39] F. Müller, G. Tempel and F. Koch. *Proc. ICPS*, **II**, 1254 (1990).

[40] R. P. G. Karunasiri, J. S. Park, K. L. Wang, and S. S. Rhee. *unpublished*.

[41] S. J. Allen, D. C. Tsui, and B. Vinter. *Solid State Commun.*, **20**, 425 (1976).

[42] T. Ando. *Z. Phys. B*, **26**, 263 (1977).

[43] R. G. Wheeler and H. S. Goldberg. *Phys. Rev. Lett.*, **27**, 925 (1971).

[44] A. Kamgar, P. Kneschaurek, and G. Dorda. *Phys. Rev. Lett.*, **32**, 1251 (1974).

[45] P. Kneschaurek, A. Kamgar, and J. F. Koch. *Phys. Rev. B*, **14**, 1610 (1976).

[46] R. G. Wheeler and H. S. Goldberg. *IEEE Trans. Electron. Devices*, **22**, 1001 (1975).

[47] F. Stern. *Phys. Rev. Lett.*, **33**, 960 (1974).

[48] T. Ando. *J. Phys. Soci. Japan*, **39**, 411 (1975).

[49] F. Stern. *Phys. Rev. Lett.*, **30**, 278 (1973).

[50] K. M. S. V. Bandara and D. D. Coon. *Appl. Phys. Lett.*, **53**, 1865 (1988).

[51] M. O. Manasreh, F. Szmulowicz, D. W. Fischer, K. R. Evans, and C. E. Stutz. *Appl. Phys. Lett.*, **57**, 1790 (1990).

[52] L. Hedin and B. I. Lundqvist. *J. Phys. C*, **4**, 2064 (1971).

[53] B. F. Levine, C. G. Bethea, G. Hasnain, V. O. Shen, E. Pelve and P. R. Abbott. *Appl. Phys. Lett.*, **56**, 851 (1990).

[54] S. D. Gunapala, B. F. Levine, D. Ritter, R. Hamm, and M. B. Panish. *Appl. Phys. Lett.*, **58**, 2024 (1991).

[55] C. G. Bethea, B. F. Levine, V. O. Shen, R. R. Abbott, and S. J. Hsieh. *IEEE Trans. Electron Devices*, **38**, 1118 (1991).

[56] L. J. Kozlowski, G. M. Williams, G. J. Sullivan, C. W. Farley, R. J. Andersson, J. K. Chen, D. T. Cheung, W. E. Tennant, and R. E. DeWames. *IEEE Trans. Electron Devices*, **38**, 1124 (1991).

[57] A. G. Steele, H. C. Liu, M. Buchanan, and Z. R.Wasilewski. *Appl. Phys. Lett.*, **59**, 3625 (1991).

[58] B. F. Levine, C. G. Bethea, G. Hasnain, J. Walker, and R. J. Malik. *Appl. Phys. Lett.*, **53**, 296 (1988).

[59] D. M. Caugley and R. E. Thomas. *Proc. IEEE*, **55**, 2192 (1967).

HIGH-TEMPERATURE DISCRETE DEVICES IN 6H-SiC: SUBLIMATION EPITAXIAL GROWTH, DEVICE TECHNOLOGY AND ELECTRICAL PERFORMANCE

M. M. Anikin, P. A. Ivanov, A. A. Lebedev,
S. N. Pytko, A. M. Strel'chuk, and A. L. Syrkin

A.F.Ioffe Physico-Technical Institute, Petersburg, RUSSIA

ABSTRACT
Open sublimation method of growing SiC epitaxial structures, dry etching and forming of ohmic contacts are described. Data on electrical characteristics of Schottky barriers, rectifier diodes, UV photodiodes and JFET's fabricated are given. The devices are of commercial quality can be operated at temperature up to $500^{\circ}C$.

In 1978 at A.F.Ioffe Institute a research and development work had been started on silicon carbide semiconductor devices with a goal of developing a technology of basic devices of semiconductor electronics encompassing treatment of the single-crystal substrate, formation of p- and n-regions and preparation of high-quality pn and np junctions. The material used was commercial 6H-SiC n-type crystals of the (0001) basal plane orientation. This paper presents results of the work done since that time and describes:
- open sublimation method of preparing pn structures;
- device processing;
- characterization of 6H-SiC devices.

1. Epitaxial Growth of pn Structures by Open Sublimation Method

The open sublimation method [1] we have developed is a modification of the sublimation sandwich method. The growth is performed in a vacuum at the temperature 1800-1900°C. The growth cell is schematized in Fig.1. The vapour source contains polycrystalline silicon carbide and silicon. Its composition is such that a silicon vapour pressure over the substrate can be varied in a wide range and maintained close to the equilibrium one. This is possible because the silicon vapour pressure over silicon at the growth temperatures cited above is about an order of magnitude higher than the equilibrium vapour pressure over silicon carbide. The growth rate is 0.1-1 µm/min. To grow epitaxial structures, highly planar and clean surfaces are required. The most perfect layers grow on the (0001)Si plane but it is not fit for growing device quality layers if etched following the standard technique which uses molten alkali as this etchant is not polishing.

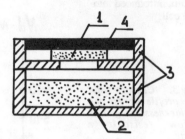

Fig.1. Growth cell.
1 - substrate
2 - vapour source
3 - graphite cruicible
4 - graphite lid

We have developed a technique of sublimation etching, which produces a polished surface regardless of the substrate orientation. The etching run is performed in the same cell where the growth occurs, by reversing the sign of the temperature gradient. Sublimation etching is used i) for removing from the surface a 15-20 μm layer, where high densities of defects are present, and for obtaining a sufficiently planar surface, and ii) for cleaning the surface just prior to the epitaxial growth (in situ etching).

 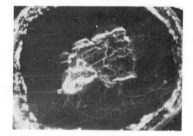

(a) (b)
Fig.2. X-ray topograms of n-n^+ structures.

Investigations of the effect of silicon vapour pressure on the structural perfection of silicon carbide epitaxial layers (see Fig.2) have shown that at below-equilibrium silicon vapour pressures over silicon carbide layers contained graphite inclusions (Fig.2,a). At pressures not much above the equilibrium one fairly perfect layers are obtained. In layers grown at high pressures inclusions of the cubic silicon carbide appear (Fig.2,b). The growth rate is decreasing (by a factor of up to 2) as the silicon vapour pressure is increased.

Layer of n-type conductivity can be obtained with the uncompensated donor concentration such that Nd-Na ~ 5×10^{15} - 10^{18} cm^{-3}. Characteristic profiles of Nd-Na across the layer thickness for growth on a low resistivity n substrate are seen in Fig.3. The data were derived from capacitance measure-ments, in several runs of etching-off the layer. It is seen from this figure that Nd-Na increases about 2 times in going from the layer surface to the layer/substrate interface, probably due to additional doping by residual nitrogen from the gas phase and to impurities diffusing from the substrate. The p-type epitaxial layers are grown in a similar way to that of the n-type layers. The doping impurity

is aluminium, introduced into the growth cell.

Fig.3. Characteristic Nd-Na profile across the layer thickness for growth on low resistivity n substrate.

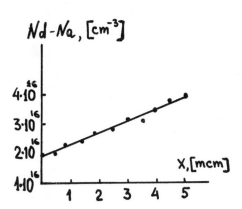

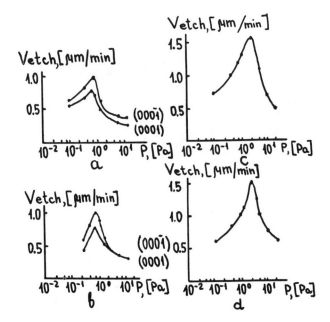

Fig.4. Etching rates in CF_4 (a,b) and SF_6 (c,d) plasmas for the $(0001)Si$ and $(000\bar{1})C$ faces of single crystal 6H-SiC vs pressure for two values of the gas flow rate $Q = 10^{-7}$ $m^3 s^{-1}$ (a,c) and $Q = 2 \times 10^{-7}$ $m^3 s^{-1}$ (b,d), excitation power $W = 10$ W/cm^3.

2. Fabrication of Devices Based on Epitaxial Structures: Profiling and Application of Ohmic Contacts

In preparation of devices based on epitaxial structures (ES) one is faced with a number of interconnected technological problems, the two most important being i) formation of a relief on the surfaces of layers grown in accordance with the device function, and ii) application of ohmic contacts to active regions of the device which should meet the requirement of small voltage drop at operating currents; methods used should be compatible with a technology of relief formation.

In the last five years profiling of silicon carbide surfaces was done using predominantly "dry" techniques, namely, plasma etch and reactive-ion etch (RIE) [2-7]. A dry method of forming devices structures in 6H-SiC was first demonstrated by the present authors (rectifier diodes) in 1984, using an inductively coupled RF plasma discharge in an atmosphere of CF_4 gas. The etching rate of 6H-SiC with CF_4 and SF_6 plasmas as a function of technological parameters of the system was investigated. In etching experiments on (0001)Si and (000$\bar{1}$)C polar planes of single crystalline SiC of 6H, 4H, 15R and 33R polytypes it was found that with pure CF_4 and SF_6 the etching rate was essentially the same for all the polytypes and for the (0001)Si plane it was lower than for the (000$\bar{1}$)C by 15-20% if CF_4 plasma was employed. With the SF_6 plasma the etching rates for the two planes coincided. Fig.4a,b presents a variation of the etching rate of the polar faces of 6H-SiC in CF_4 plasma with residual pressure inside the reactor for two values of the gas flow rate. Fig.4c,d is the same for the SF_6 plasma.

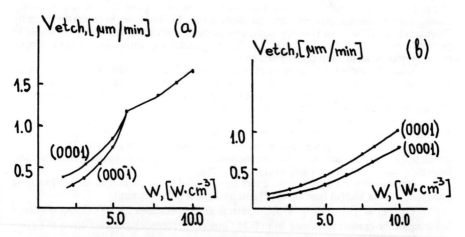

Fig.5. Etching rates in SF_6 (a) and CF_4 (b) plasmas for the (0001)Si and (0001)C faces of single crystal 6H-SiC vs discharge excitation power W.

In both SF_6 (Fig.5,a) and CF_4 (Fig.5,b) plasma the etching rate continuously increased with increasing discharge excitation power. The highest etching rate with the CF_4 plasma was 0.8 μm/min for the (0001)Si plane and 1 μm/min for the (0001)C plane. The etching rate in the SF_6 plasma reached a value of 1.7 μm/min and good selectivity was obtained with respect to an Al mask (varying from 3:1 to 40:1, depending on discharge

power and on pressure in the vacuum chamber).The profile produced with the use of the Al mask was abrupt, i.e. its elements did not show any perceptible deviations from specified dimentions due to lateral etching (Fig.6).

Fig.6. Diode mesa structure produced by RIE in SF_6 plasma with the use of an Al mask (scale: 5 μm in 1 cm).

Highly anisotropic RIE profiles obtainable with silicon carbide are in good agreement with a qualitative model proposed by Pan and Steckl [5,8] which assumes that accumulation of carbon on the surface not subjected to the ion bombardment is blocking the etching process (carbon blocking model).

It should be noted that with an capacitively-coupled (E-mode) discharge excitation used by some authors [3-8], the etching rate is somewhat lower than with an inductively-coupled (H-mode) discharge excitation [2]. In the latter case a relatively minor enrichment of the surface with carbon is observed even when using pure SF_6; it is commonly considered that just this process limits the etching rate. Furthermore, in contrast to data by Palmour et al. [7], with an H-mode excitation properties of etched surfaces were found to be unaffected by the kind of material used for the grounded specimen holder (an analogue of the cathode in an inductively-coupled system).

Using dry etching of silicon carbide it was found possible to form both diode mesas and channel regions in FETs with high depth resolution. Further research in this field is necessary in order to solve the problem of dry polishing etching of deep-etched areas and of reducing the density of radiation defects on SiC surfaces.

Main requirements to ohmic contacts are small voltage drop at operating current densities, high operating temperatures and compatibility with profiling techniques. In recent years, methods of forming ohmic contacts to SiC based on refractory metals (W, Mo) were developed [9,10]. Contacts to n- and p-materials were made both from the pure metals and from their alloys with Al, Au and others.

In many studies considerable attention was paid to interaction of the high melting point metals with SiC (of both cubic [9] and hexagonal [10] structure).

Geib et al. [9] studied W-SiC contacts at temperatures 23-900°C and observed lowering of the contact resistance from 0.24 Ohm·cm at room temperature down to 8×10^{-2} Ohm·cm at 900°C.

We studied [10] conditions in which ohmic contacts form in W-n-6H-SiC and (W/Au):Al-p-6H-SiC systems and characteristics of these contacts at elevated temperatures. Some specific details of the deposition process of W and W-based alloys and of annealing regimes resulting in formation of the contact have been described.

It was shown that the "in situ" etching affects contact formation conditions and its quality. W and its alloys were found applicable for making contacts to n- and p-6H-SiC doped in a wide range of concentrations. The contact resistance was falling-off at higher

doping levels for the W-n-6H-SiC contact and weakly varying for (W/Au):Al-p-6H-SiC. Resistance values for W and W-based contacts to 6H-SiC are listed in Table 1.

Table 1.   W-n-6H-SiC

Orientation	(0001)Si	(000$\bar{1}$)C
N_d-N_a (cm^{-3})	r^c (Ohm·cm^2)	r^c (Ohm·cm^2)
2×10^{17}	7×10^{-3}	-
8×10^{19}	-	1×10^{-2}
3×10^{18}	8×10^{-4}	5×10^{-3}
1×10^{19}	2×10^{-4}	9×10^{-4}

In studies of the contact resistance of the W-n-6H-SiC and (W/Au):Al-p-6H-SiC in the temperature range 300-950 K practically no changes of resistivity could be detected. Fig.7 shows dependences of the resistance of W-n-6H-SiC contacts on temperature for two orientations, (0001)C and (0001)Si. Differences in the properties of the polar crystal planes are revealed in higher contact resistance for the (0001)C orientation compared with the (0001)Si orientation, other conditions being equal. We have prepared ohmic contacts to n-6H-SiC based on Mo which had the resistivity of 1×10^{-4} Ohm·cm at $N_d-N_a = 3 \times 10^{18}$ cm^{-3}.

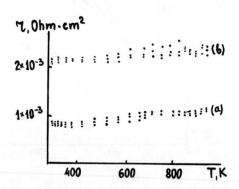

Fig.7.   Temperature dependence of the specific contact resistance of W-n-6H-SiC contacts.
(a) - (0001)Si orientation
(b) - (0001)C orientation.

We have successfully used contacts based on W, Mo and their alloys in devices such as diodes, field-effect transistors and photodetectors.

Thus, research and development work on forming devices based on silicon carbide epitaxial structures has advanced considerably and now the main problems in device fabrication can be solved in such a manner that advantages of the structures produced can be most fully exploited.

3. 6H-SiC Schottky Diodes

Because of the wide forbidden gap of 6H-SiC surface-barrier structures in silicon carbide are potentially useful for high-temperature microwave electronics. In addition, silicon carbide surface-barrier (SB) structures can be used as highly efficient UV photodetectors, practically insensitive in the visible spectral range.

A great number of works was devoted to research and development of SB structures based on 6H-SiC, we can mention here but a few [11-15]. The reverse current-voltage characteristics of SiC SB structures at room temperature can be found in [13,14]. At voltages -120 V [13] and -210 V [14] the currents before breakdown amounted to 10^{-3} A and 10^{-5} A, respectively. The metal-6H-SiC barrier, according to various authors, has a height from 1.1 [13] to 2.0 eV [15].

Mechanisms of current were studied in [11,12]. According to [11], the forward current is in agreement with the thermionic theory, but no temperature measurements were given. Observations of two current conduction mechanisms were reported [12]. One, at U = 0.7-1.2 V, is the thermal injection mechanism and another, at U < 0.7-0.8 V, is tunnelling via the impurity states within the gap. Forward current investigations in the temperature range 230-350 K reported in [12] refer to diodes made from heavily doped material (Nd-Na > 10^{18} cm^{-3}); in these diodes the exponential component of the forward current is practically unobservable because of an early onset of the series resistance.

3.1. The electrical characteristics of SiC surface-barrier structures

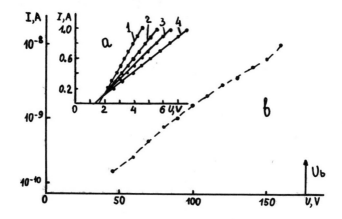

Fig.8. *Current-voltage characteristics of Schottky diodes. a) forward current at T, K: 1 - 293, 2 - 400, 3 - 483, 4 - 573; b) reverse current at room temperature. (from ref. [23])*

We prepared Schottky barriers on the surface of epitaxial n-type layers with the uncompensated donor concentration Nd-Na = $(0.5-2) \times 10^{17}$ cm^{-3} grown by open sublimation method on 6H-SiC n$^+$-substrates of (0001)Si orientation. The barrier metal was gold. The structures had an area S = 3×10^{-3} cm^2.

The breakdown voltage was 100-170 V at room temperature (Fig.8,b). In some structures up to the breakdown voltage reverse currents were less than 10^{-10} A, as indicated by an arrow in Fig.8,b. The breakdown was abrupt and irreversible, suggesting that it occured at the crystal periphery. At voltages exceeding the barrier height the forward current was linear in voltage. At room temperature the diode series resistance, a sum of the bulk and contact resistances, was equal to 2-3 Ohm (Fig.8,a).

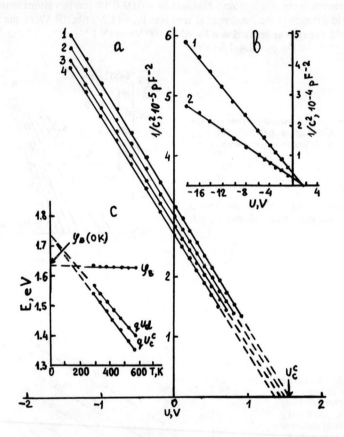

Fig.9. a) Capacitance-voltage characteristics of a 6H-SiC Schottky barrier at several temperatures T, K: 1 - 293, 2 - 373, 3 - 473, 4 - 573; b) Capacitance-voltage characteristics of two 6H-SiC Schottky barriers at room temperature; c) temperature dependence of the capacitance cutt-off voltage V_c^c, the diffusion potential V_d and φ_b - the potential barrier height of 6H-SiC Schottky barriers with N_d-N_a = $9x10^{16}$ cm^{-3} and V_c^c =1.54 V at T=293 K. (from ref. [23])

Differential capacitance-voltage characteristics in coordinates C^{-2}-U were linear throughout the investigated ranges of temperature and voltage (Fig.9). At room temperature the capacitance cut-off voltage, defined as an intercept at the voltage axis of a linear C^{-2}-U dependence extrapolated to infinite capacitance, was within 1.3-1.54 V for various structures (Fig.9) and dropped to 1.2-1.33 V at 573 K. From C-U characteristics some parameters of the SB structures at room temperature were derived: the uncompensated donor concentrate Nd-Na = $(0.7-2) \times 10^{17}$ cm^{-3} and practically independent of temperature; the space-charge layer width at zero bias was W = 0.09-0.14 μm in various structures; the electric field strength at the maximum at zero bias E_o = $(2.2-2.6) \times 10^5$ V/cm; the maximum electric field strength at break down E_b =2.3×10^6 V/cm (Nd-Na = 9×10^{16} cm^{-3}), and the potential barrier height φ_b = 1.4-1.63 eV.

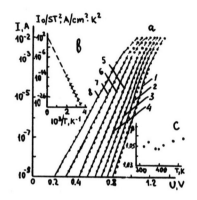

Fig.10. a) Forward current-voltage characteristics of a 6H-SiC Schottky barrier at different temperatures, T, K: 1 - 293, 2 - 317, 3 - 340, 4 - 373, 5 - 395, 6 - 423, 7 - 473, 8 - 523; b) Richardson's graph for a 6H-SiC Schottky barrier; c) temperature dependence of the ideality factor β for a 6H-SiC Schottky barrier. (from ref. [23])

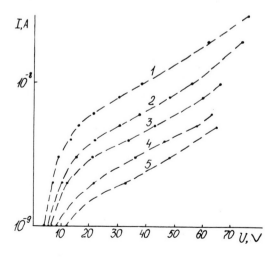

Fig.11. Reverse current-voltage characteristics of 6H-SiC Schottky barriers at different temperatures, T, K: 1 - 573, 2 - 515, 3 - 473, 4 - 431, 5 - 373. (from ref. [23])

The forward branch of the I-U characteristic for Au-SiC-6H barriers in a low current range is shown in Fig.10,a. At voltages $kT/q \ll U < \varphi_b/q$ and temperatures 293-520 K the forward current of SiC SB structures followed an exponential relationship $I=I_o\exp(qU/\beta kT)$. The exponential extended over 5 to 6 orders in current (from 10^{-9} to 10^{-3} A). The dimentionless parameter β remained constant at 1.05-1.07 for all temperatures (Fig.10,c).

In Fig.11 reverse current-voltage characteristics of Au-SiC barriers at temperatures in the range 293-550 K are given. At all the temperatures the reverse current was $\sim 10^{-8}$ A and was due to leakage at the periphery. Ideally, the operating temperature range of 6H-SiC Schottky barriers would be limited by a value of the saturation current which, using a diode theory [16], is estimated at 10^{-10} A for an SB structure with $N_d-N_a = 10^{17}$ cm^{-3} at 550 K. Reverse currents can be reduced using safety rings.

3.2. Forward current mechanism in 6H-SiC surface-barrier structures.

From capacitance measurements the variation with temperature of the capacitance cut-off voltage U_c^c, diffusion potential $U_d = U_c^c + kT/q$, and potential barrier height φ_b were determined in the temperature range 293-600 K (Fig.9,c). The temperature variation of the electrochemical potential μ [17] of electrons in the bulk semiconductor has been calculated using a formula:

$$\mu = kT \ln \{h^3(N_d-N_a)/[2(2\pi m^*kT)^{3/2}]\}, \tag{1}$$

where $m^*=0.45m_0$ is the effective mass of electrons in 6H-SiC. Lowering of the potential barrier due to image forces [18] given by:

$$\Delta\varphi_b = q[(q^3/8\pi^2\varepsilon_d^2\varepsilon_s\varepsilon_o^3)\cdot(N_d-N_a)\cdot(U_d-U-kT/q)]^{1/4} \tag{2}$$

was calculated for conditions of thermodynamic equilibrium at U=0, assuming the static dielectric permeability $\varepsilon_s=9.8$ [19] and the high-frequency dielectric permeability $\varepsilon_d=6.7$ [19]. With rising temperature in the temperature range 293-600 K U_c^c, U_d, and φ_b decrease linearly with a temperature coefficients $\alpha_c = 6.7\times10^{-4}$ V/K, $\alpha_d = 5.9\times10^{-4}$ V/K and $\alpha = 3\times10^{-5}$ eV/K. For a SiC SB structure having at room temperature $U_c^c = 1.54$ V and $N_d-N_a = 9\times10^{16}$ cm^{-3} the barrier height extrapolated to zero absolute temperature using a formula $\varphi_b(T) = \varphi_b(0 K) - \alpha T$ was found equal to $\varphi_b(0 K) = 1.63$ eV.

The forward current-voltage characteristics measured at temperatures 293-520 K were analyzed in the framework of the thermionic theory.

According to this theory, if an account taken of the barrier height lowering by applied voltage due to image forces, the forward current is

$$I = A\cdot S\cdot T^2\exp(-\varphi_b/kT)\exp\{[qU-\delta\varphi_b(U)]/kT\} = I_o\exp(qU/\beta kT), \tag{3}$$

where A is the Richardson's constant, in SB structures equal to $120m^*/m_0$ A/(cm$^2\cdot$K^2); φ_b is the potential barrier height in thermodynamic equilibrium reduced by image forces, and $\delta\varphi_b$ is a value of the barrier height lowering under applied voltage.

It is thus seen that β values exceeding unity reflect the barrier lowering due to image forces:

$$\beta^{-1} = 1 - \delta\varphi_b(U)/qU. \tag{4}$$

Calculated values of β for a Au-SiC structure having Nd-Na = 9x10^{16} cm^{-3} and U_c^c = 1.54 V at 293 K were essentially constant in the considered ranges of temperature and voltage and practically coincided with the above experimental values.

A dependence of the pre-exponential factor I_0 on inverse temperature in coordinates lg(I_0/ST^2) = f(1/T) (Richardson's graph) was linear in the temperature range 293-520 K (Fig.10,b). In the thermionic theory the pre-exponential factor is expressed as:

$$I_0 = A \cdot S \cdot T^2 \exp(\alpha/k) \exp[-\varphi_b(0,K)/kT]. \tag{5}$$

From the Richardson's graph, Aexp(α/k) and φ_b(0,K) were determined and found to coincide with theoretical values, for samples with U_c^c=1.54 V their respective values being 100 A/(cm^{2}K^2) and 1.6 eV.

The barrier height values at T=0 K derived from the current-voltage and capacitance-voltage characteristics coincided. It follows from the above comparison of the theory and experiment that in the obtained structures the forward current can be described in terms of the thermionic theory with due allowance for the effect of image forces.

It may thus be concluded that the Au-SiC surface-barrier structures can be operated in the temperature range 293-600 K. In these structures the intermediate layer between the metal and the semiconductor is practically absent and their properties are nearly "ideal".

4. Rectifier Diodes

4.1. Capacitance-voltage characteristics and deep centers.

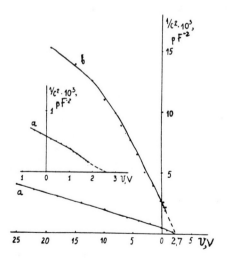

Fig.12. Capacitance-voltage characteristics of 1D pn structures at room temperatures, carrier concentration in the base Nd-Na = 1x10^{17} cm^{-3} (a) and 5x10^{16} cm^{-3} (b). (from ref. [20])

In this section p^+-n-n^+ diode structures are discussed with an n-base fabricated by the open sublimation method described above. A p-emitter was formed by the following methods: - open sublimation method with aluminium as a dopant (SE structures);
- ion implantation of Al (ID structures).

The concentration Nd-Na in the base was from 5×10^{16} to 10^{18} cm^{-3} and the concentration in the emitter was Na -Nd = 5×10^{18} - 5×10^{19} cm^{-3}.

In junctions produced with ion-implantation capacitance-voltage characteristics of the barrier plotted in $1/C^2$-U coordinates were slightly deviating from a straight line (Fig.12) showing a kink in a region of low reverse voltages. This is an indication of an existence of a compensated region near the junction [20].

In sublimation-epitaxial (SE) pn structures plots of $1/C^2$-U were straight lines in the entire voltage range (Fig.13) and yielded the same Nd-Na values as obtained for the Schottky-barrier (SB) structures described in the preceding section, i.e. the junctions in the SE structures were abrupt and asymmetric [21].

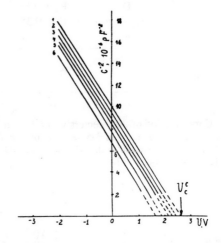

Fig.13. Capacitance-voltage characteristics of SE pn structures at different temperatures, T, K: 1 - 293, 2 - 400, 3 - 500, 4 - 600, 5 - 700, 6 - 800. (from ref. [21])

Temperature dependences of electrical parameters such as the barrier height, the electric field strength within the space-charge region of the junction, and the space-charge layer width were found to follow predictions of the Shockley theory [22].

In studies of the SB structures [23] based on 6H-SiC two types of DCs were identified denoted R and S (see Table 2). Their concentrations, Nr and Ns, were within 20% of the respective values measured on the pn structures studied, both ion-implanted and sublimation-epitaxial. This result suggests that concentrations of the R and S centers were not appreciably affected in the process of junction formation.

Besides, in the structures investigated the three more DCs designated i, L, and D, with levels located in the lower half of the forbidden gap were identified [26]. Closeness of parameter values of i-and D- centers is manifest in an overlapping of the corresponding peaks in DLTS spectra. In analyzing DLTS spectra of the SE and ID structures a technique of processing complicated spectra was used which allowed separate determinations of

ionization energies of the DCs from the low-temperature and high-temperature slopes of the DLTS peak [24]. We were also able to detect separately DLTS signals from D- and i-centers by making use of their significantly different time constants τ_r for recharging by electrons from the conduction band, the constants being in a ratio $\tau_{r,i}/\tau_{r,D} \sim 30$ [25].

Table 2. Parameters of the deep centers identified in SiC

Polytype	Level	Energy position (eV)	Capture cross-section for electrons (cm^2)	Capture cross-section for holes (cm^2)
6H	L	$E_v + 0.24$	10^{-18}	10^{-15}
	i	$E_v + 0.52$	10^{-21}	10^{-17}
	D	$E_v + 0.58$	3×10^{-20}	5×10^{-16}
	S	$E_c - 0.35$	10^{-15}	10^{-14}
	R	$E_c - 1.27$	3×10^{-13}	10^{-15}
	analogues			
4H	i	$E_v + 0.53$	-	5×10^{-17}

Concentration of the i-centers was found uniform in SE pn structures and increased in ID structures by a factor of 3-5 near the pn junction compared with its level throughout the rest of the base. Similar situation occured in ID structures prepared from 4H-SiC where an analogue to the i-center has been identified (Fig.14) [27].

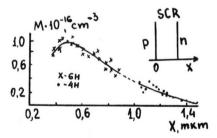

Fig.14. Distribution of i-centers in the base of an ID pn structure. (from ref. [27])

The authors of [25], by comparing capacitance and optical data for 6H-SiC pn structures obtained by various methods, have established a correspondence between the presence of the D-centers and an appearance of a yellow electroluminescent peak at $h\nu = 2.14$ eV. This yellow EL peak did not correlate with any other center. Dependences of the yellow EL intensity I on forward current density J and temperature T as well as of the temperature dependence of the time constant τ_a of the afterglow were measured. Plots of I

= f(J) (Fig.15) had two linear portions connected with a sublinear curve. With rising temperature the sublinear portion shrinked and at T=600 K the plot was practically a straight line. The plot of I = f(T) had a maximum which shifted with increasing forward current density from T = 300 K (J=1.5x10^{-3} A/cm^2) to T = 450 K (J=5 A/cm^2). With rising temperature τ_a decreased from 270 µs (300 K) to 6 µs (430 K).

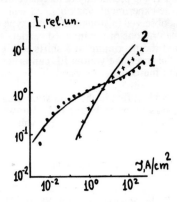
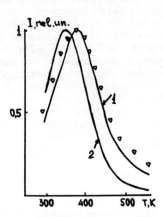

Fig.15. Dependences of the yellow electro-luminescence intensity on forward current at temperatures 300 K (1) and 630 K (2). Solid lines - calculation. (from ref. [25])

Fig.16. Dependence of the yellow electroluminescence intensity on temperature at j=0.5A/cm^2; solid lines - calculation with (1) and without (2) taking into account the temperature variation of L_p. (from ref. [26])

To calculate the temperature dependence of the EL intensity (I) and the time of afterglow (τ_a), the following expressions have been derived

$$I = \frac{M \cdot S \cdot L_p \cdot \alpha_p \cdot \alpha_n}{\alpha_n + \alpha_p}[\Delta p_0 + \frac{n_n \alpha_p - \beta_i}{\alpha_n + \alpha_p}\ln\frac{\alpha_n n_n + \beta_i + (\alpha_n + \alpha_p)\Delta p_0}{\alpha_n \cdot n_n + \beta_i}] \quad (6)$$

$$\tau_a = \frac{1}{\alpha_n \cdot n_n + \beta_i} \quad (7)$$

where M is a DC concentration; S the junction area; Δp_0 the concentration of holes injected into the base at the depletion region boundary; n_n the electron concentration in the base; $\alpha_{n(p)} = V_T \sigma_{n(p)}$; $\beta_i = N_V V_T \sigma_p \exp(-E_i/kT)$; $\sigma_{n(p)}$ the cross-section for capture of electrons (holes) by a DC; V_T the thermal velocity of electrons; E_i the DC ionization energy; k the Boltzmann's constant, and N_V the density of states in the valence band.

Characteristics of the yellow EL calculated with expressions (6) and (7) using D-center parameters provided explanation of the entire set of experimental data. Sublinearity of the part of the I = f(J) plot at 300 K is explained by a large difference in the capture

cross-sections for electrons and holes by the D-center: $\alpha_n/\alpha_p = 6 \times 10^{-5}$ (in the case $\alpha_n = \alpha_p$ the function $I = f(J)$ would have been linear for all values of the forward current density). The dependence becomes linear again when, at higher currents, a condition $\Delta p_0 \sim n_n$ is fulfilled and the D-center has a high probability of being recharged by electrons (increase of the term $\alpha_n[n_n+\Delta p_0]$). At higher temperatures the holes are more likely to be thermally ejected (β_i is large), therefore, complete filling of the DCs with holes occurs at larger values of $\Delta p_0(J)$ and an extent of the sublinear region is reduced.

The initial rise of $I = f(T)$ is caused by an additional ionization of the donor levels in the base whilst the subsequent decrease is due to the rising probability of thermal ejection of holes (β_i). The best fit between the theory and the experiment was obtained when an account was taken of the temperature variation of L_p observed in structures of this type (L_p increased by 5 to 7 times) (Fig.16). At higher concentrations of the injected carriers (i.e. with rising J) the increase in β_i is compensated and the maximum in J shifts to higher temperatures. Considering the energy position of the level, the yellow EL can originate from radiative transitions between the D-center and the conduction band, if the Frank-Condon shift of 0.35 eV is assumed for this center.

Using measured parameter values of the DC's identified in these pn structures the hole lifetime τ_p has been calculated. The S-center was found to be the only one whose concentration and parameters provided satisfactory explanation of the measured value τ_p of about $10^{-8} - 10^{-9}$ s (300 K) [28].

4.2. Forward current - voltage characteristic of the pn structures

4.2a. Low currents

Generally, the forward current - voltage (I-V) characteristic of both SE and ID pn structures up to an onset of the series resistance (currents up to $10^{-1} - 10^1$ A/cm^2) and at temperatures from 300 K to 700-800 K it consists of two exponential components of the form

$$J = J_0 \exp(qV/\varepsilon), \qquad (8)$$

where $\varepsilon = \beta kT$, $J_0 = J_0^* \exp(-E_g/kT \cdot \beta)$, $\beta \neq f(T)$.

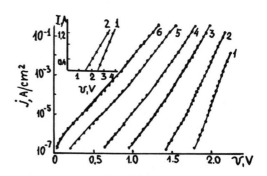

Fig.17. Forward current-voltage characteristics of a SE-type pn structure at different temperatures, T, K: 1 - 290, 2 - 370, 3 - 470, 4 - 540, 5 - 640, 6 - 720. In the insert: forward current-voltage characteristics in a range of high currents at temperatures 300 K (1) and 720 K (2) (structure area 6×10^{-3} cm^2).

In the low current region $\beta=2$; at higher currents β assumes values between 1 and 2.

In the low current region β=2; at higher currents β assumes values between 1 and 2. Current components with fractional values of β, 3/2 (Fig.17-19) and 6/5 were investigated [29,30]. These currents were interpreted as due to thermal injection of carriers into the space-charge region (SCR) of the pn junction where they recombine via a deep center (β=2), in accordance with the Sah-Noyce-Shockley model [31], or via a multivalent center (fractional values) [32], having one deep level and one or more shallow levels. Current components in SiC with fractional values of β have not been investigated earlier.

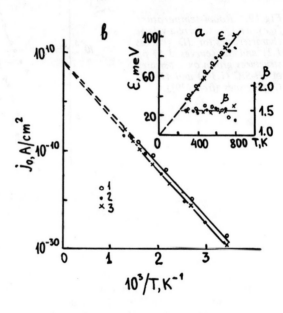

Fig.18. Temperature dependences of J, ε and β in the current-voltage relationship for three SE pn structures with β = 1.5 . (from ref. [29])

The lifetime in the SCR for deep-level recombination in a steady-state regime is 10^{-9} -10^{-10} s at room temperature [28] (likely to be somewhat less in ID structures than in SE structures). The lifetime rises with rising temperature reflecting the temperature variation of the capture cross-section. The shallow levels, which determine the lifetime at recombination via a multivalent center, have a depth from 0.1-0.2 eV. For recombination via the multivalent center, the steady-state lifetime in the SCR also rises with temperature. Fig.19 shows I-V characteristics for the current with β = 3/2 which are seen to be shifted to higher voltages for SE pn structures in 6H-SiC relative to ID structures to shorter lifetime in the former, with the curve for the 4H-SiC pn structure shifted still further along the voltage axis due to larger bandgap in 4H-SiC.

At currents exceeding 10^{-1} -10^1 A/cm^2 the voltage across the pn junction becomes comparable with the voltage drop at the series resistance of the structure, whereupon the exponential current-voltage relationship becomes distorted. Yet, a comparison of the observed recombination currents with calculated diffusion current (using measured diffusion lengths and carrier lifetimes) suggests that the recombination currents still dominate at J > 10^{-1} - 10^1 A/cm^2, i.e. in the operating current range of bipolar devices.

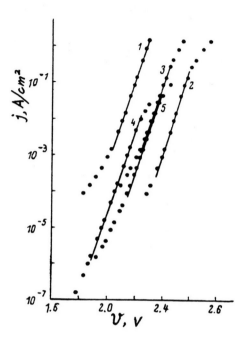

Fig.19. Room-temperature forward current-voltage characteristics for ID (curves 1,2) and SE (curves 3-5) pn structures grown on substrates of 6H-SiC (1,3-5) and 4H-SiC (2) polytypes. (from [29])

4.2b. High currents

Forward CVC's of the diodes in a current range beyond $10^0 - 10^2$ A/cm^2 are usually essentially linear (Fig.17) [33]. Limitation of the current by the structure series resistance, a sum of the substrate, the base, the emitter and contact resistances, was increasing from 1.5×10^{-3} Ohm·cm^2 to 10^{-2} Ohm·cm^2 as the structure area was increased from 7×10^{-4} cm^2 to 2×10^{-2} cm^2. In these structures currents of 1-6 A (J ~ 10^3 A/cm^2) at room temperature were reached at the forward bias values 4.5-10 V. In SE and ID diodes with an area < 2×10^{-3} cm^2 the lowest series resistance values, $(1.5-2.5) \times 10^{-3}$ Ohm·cm^2, were achieved for a given substrate thickness (450 μm), being determined primarily by the substrate resistance and current spreading in a large-area contact.

The series resistance of the structures varied with temperature in the same way as the resistivity of silicon carbide: a slight decrease from room temperature to 400-500 K followed by a 50% increase as the temperature is further raised to 800 K.

An analysis of contributions to the series resistance shows that in pn structures having the breakdown voltage up to 500V the use of low-resistivity (ρ ~ 0.1 Ohm·cm) substrates thinned to 50 μm the current would be limited mainly by resistance of the base (5×10^{-4} Ohm·cm) and of the contacts and the forward bias required to obtain currents 1-6 A at room temperature could be reduced down to 4V (for structure areas $7 \times 10^{-4} - 6 \times 10^{-3}$ cm^2).

4.3. Reverse current-voltage characteristics of the pn structures.

4.3a. The breakdown voltage

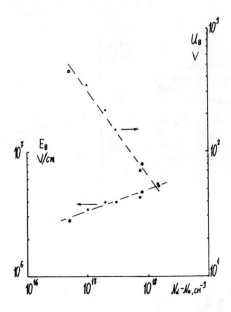

Fig.20. Voltage V_B and electric field strength E_B at breakdown as functions of the uncompensated donor concentration in the n-base.

In SE and ID structures the breakdown developed either through microplasma process and was reversible, or on the structure periphery and a recovery could be achieved with additional etching. In structures with the junction plane perpendicular to the C(0001) crystall-ographic axis (E||C) the breakdown field E_B varied from 3×10^6 to 5×10^6 V/cm (Fig.20).

The temperature coefficient of the breakdown voltage (TCBV) was investigated and found to be able to assume not only negative values, as observed earlier on 6H-SiC pn structures at E||C, but, in SE and ID structures, could also be positive or of alternating sign.

Our experimental observations suggest that the negative TCBV can be related to a presence of deep impurities [34-36]. These observations are as follows: a change of sign of TCBV from negative in dc measurements to positive in measurements with short pulses (70 ns) (Fig.21); alternating sign of TCBV at microplasma breakdown, as determined by extrapolation to zero current of the linear portion of the corresponding I-V characteristic, as well as by the voltage at which the microplasma sets in; and a change of a value or even sign of a differential TCBV with current (Fig.22). Earlier explanations of the negative TCBV relied on an assumption that the conduction band in hexagonal polytypes of SiC is split into a number of subbands, arising from a naturally occuring superlattice in the C-axis direction. Under this assumption, the negative sign of TCBV represents an intrinsic property of the breakdown in 6H-SiC pn structures at E||C which raises doubts as to the feasibility of devices operating in the avalanche breakdown regime because of their thermal instability with respect to heating by current. The problem of producing devices operating in avalanche breakdown regime can be solved, if proposed link between the breakdown

features and the presence of deep impurities is proved. Note, that E||C orientation is preferable in the growth of substrates by Lely method.

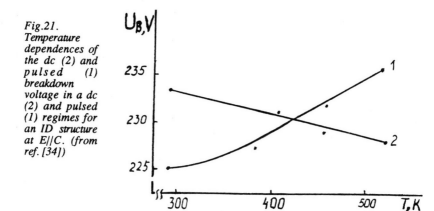

Fig.21. Temperature dependences of the dc (2) and pulsed (1) breakdown voltage in a dc (2) and pulsed (1) regimes for an ID structure at E//C. (from ref. [34])

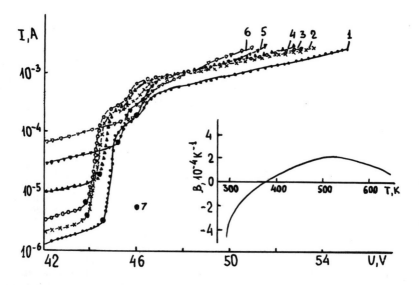

Fig.22. Reverse current-voltage characteristics of a SE pn structure at breakdown for E//C and different temperatures, T, K: 1 - 293, 2 - 335, 3 - 390, 4 - 480, 5 - 590, 6 - 670. In the insert: variation with temperature of the temperature coefficient of the breakdown voltage. (from ref. [35])

4.3b. Current in the pn structures prior to breakdown

At room temperature the reverse current before the breakdown did not exceed 10^{-9} A and was supposedly dominated by leakage through a 10^{11} -10^{12} Ohm resistance of the mesa periphery. At higher temperatures the reverse current rises, being determined, possibly, by more than one mechanism. At the highest temperatures (>700 K) the reverse I-V characteristic of both SE and ID structures assumes the features predicted in the Sah-Noyce-Shockley model, i.e. the reverse current is determined by the thermal generation of carriers in the SCR. At 800 K the reverse current is 10^{-7} - 10^{-5} A (Fig.23) [33,37], and lifetime estimates using the Sah-Noyce-Shockley model gives a value of 10^{-8} s.

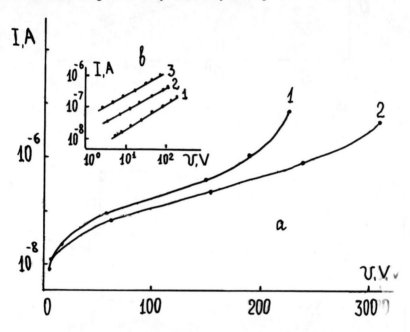

Fig.23. Reverse current-voltage characteristics (a) 1-ID diode, T=793K, S=1x10^{-3} cm^2; 2 - SE diode, T=720K, S=5x10^{-3} cm^2; (b) SE - diode, TK: 1-740, 2-780, 3-820.

4.4. Characteristics of radiationless recombination processes.

Above, from investigations of the forward and reverse I-V characteristics the lifetime estimates were given for the recombination-generation processes occuring in the space-charge region of the pn junction.

Besides, data on diffusion lengths of minority charge carriers (MCC) in quasi-neutral regions of SE pn structures in the temperature range 300-800 K [28,38] were obtained from measurements of photocurrent as a function of the SCR width under illumination with weakly absorbed intrinsic light. It has been established that the room-

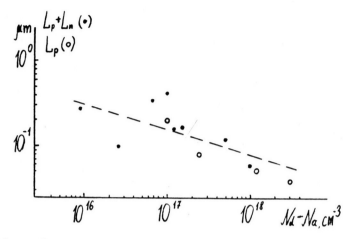

Fig.24. Minority carrier diffusion lengths: L_p+L_n, (solid circles) and L_p (open circles) versus N_d-N_a for SE structures.

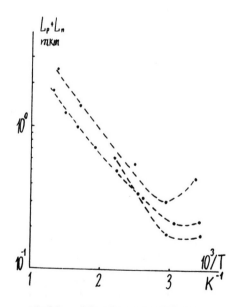

Fig.25. Minority carrier diffusion length versus reverse temperature in SE structures.

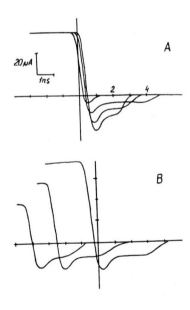

Fig.26. Transient switching characteristics of a diode: A - for constant forward current regime, B - for constant reverse current regime. (from ref. [28])

temperature diffusion length values range from 0.06 to 0.4 μm and tend to decrease with increasing Nd-Na in the n-base of the structure, from 0.3 μm at Nd-Na ~10^{16} cm^{-3} to 0.06 μm at Nd-Na ~ 3×10^{18} cm^{-3} (Fig.24). With rising temperature the diffusion length increases, except for a narrow temperature range near room temperature. From 350 to 800 K the diffusion length increases with an activation energy of 120 meV, reaching a value of 2-3 μm at 800 K (Fig.25). The temperature behavior of the diffusion length suggests that the recombination proceeds through a rather shallow level.

The room-temperature relaxation lifetime in SE structures has been derived from an analysis of transient characteristics obtained at switching the current through the structure from forward to reverse direction (Fig.26) and was found to be about 2 ns [28]. An estimate of the hole mobility in the n-base at room temperature gave a value of about 10 cm^2/V·s.

4.5. Electroluminescence spectra.

Electroluminescence (EL) spectrum of the SE pn structures is a broad band covering the entire visible spectral range (Fig.27, curve 1). The EL maximum is located in the yellow-green part of the spectrum. Above we gave characteristics of the yellow EL and described possible recombination mechanisms involved. With rising current the EL maximum shifts to the green, reflecting different dependences of the EL intensity on current of the yellow and green EL, the latter saturating at higher current densities. Above room temperature the green EL is quenched; the yellow EL intensity increases with temperature up to 300-400 K, depending on current value, and then falls off.

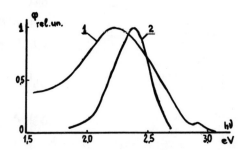

Fig.27. Room-temperature electroluminescence spectra for SE (1) and ID (2) pn structures.

In the violet an EL band is found (hv$_{max}$ = 2.92 eV at room temperature) due to a free exciton annihilation with emission of phonons. With rising temperature the intensity of the free exciton EL increases, possibly due to increasing lifetime. The maximum of the excitonic EL shifts with temperature to longer wavelengths, reflecting the temperature variation of the bandgap in agreement with data of [39] (hv$_{max}$ ~ -αT, α = 3.5×10^{-4} eV/K). As the current is raised, the excitonic EL rises in intensity faster than other bands, following a φ ~ J^n relationship, where n ~ 1. At least up to currents (10^1-10^2) A/cm^2 the φ ~ J^n dependence shows no tendency to saturation.

ID structures emit in the green (Fig.27, curve 2). Peak position of the green EL

shifts with current, at room temperature this shift is from $\lambda_{max} \sim 560$ nm at $J = 10^{-2}$ A/cm^2 to $\lambda_{max} \sim 525$ nm at $J = 10^1$ A/cm^2. Above room temperature the green EL intensity decreases with activation energy 0.2 eV, and its maximum shifts from $\lambda_{max} \sim 525$ nm at $T = 300$ K to $\lambda_{max} \sim 560$ nm at $T = 620$ K.

5. UV Photodetectors in 6H-SiC.

At the present time silicon or gallium phosphide photodiodes are predominantly used for registration of the UV radiation. However, silicon structures possess high sensitivity to the infrared and visible light, much higher than to the UV radiation and therefore can be used only with filters which considerably impairs their sensitivity, not to mention degradation of the filters under UV illumination. Currently, GaP photodiodes for voltages 4-5 V are produced only with the use of surface-barrier structures which are plagued with large leakage current.

The use of silicon carbide eliminates these drawbacks. Silicon carbide photodetectors are highly sensitive to the UV radiation (200-400 nm) and essentially insensitive to the visible light, allowing to dispense with filters. They can withstand high overloads in the input optical signal and are operable at temperatures up to 500°C.

We have developed two types of UV photodetectors in 6H-SiC: 1) UV photodiodes using SiC Schottky (SB) structures; 2) SiC photodiodes with shallow pn junctions.

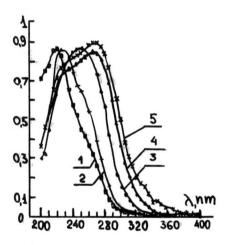

Fig.28 Quantum efficiency versus wavelength of the incident light of Schottky diode phtodetectors at different temperatures, T, K: 1-293, 2-400, 3-470, 4-520, 5-600.

In Schottky barriers the space-charge region is at the surface i.e. a passive layer of semiconductor where part of the incident radiation is lost is lacking. Therefore, photosensitivity of a Schottky-barrier photodiode can serve as a reference for sensitivity estimates of pn junction photodiodes. The present work on UV photodiodes on the base of SiC SB structures relies on a technology of Au-SiC Schottky barriers described above. The barrier was formed by vacuum evaporation of a thin semitransparent layer of gold. With an area 10^{-2} cm^2, the capacitance of the structures was about 630 pF. The leakage current was

10^{-11} A at a reverse bias of 1 V. Variation of the quantum efficiency in Au-SiC SB structures in the spectral range 200-400 nm is given in Fig.28. The monochromatic photosensitivity in the best samples reached 0.15 A/W at a wavelength of 215 nm, which corresponds to the quantum efficiency of 0.8 electron/photon.

The long-wavelength photosensitivity rises appreciably with temperature because of increasing diffusion length and decreasing bandgap. In the short waves the photosensitivity changes insignificantly.

Photodetectors with pn junctions are based on n-p^+-n^+ structures grown by sublimation sandwich method in an open system [1]. Two types of photodetectors incorporating SiC pn structures have been developed:
1. Photodiodes with a uniformly doped illuminated n-layer.
2. Photodiodes with a built-in field in the illuminated n-layer.

The built-in field was produced by doping the n-layer nonuniformly during growth. The thickness of the epitaxial n-layer was 1.0-2.0 μm. Ohmic contacts to the n-layer were deposited by vacuum evaporation of W+Au with subsequent annealing at T=1600°C. Mesas were formed using reactive ion plasma etch.

Measured spectral characteristics of the SiC pn structures are shown in Fig.29 where we compare plots of the quantum efficiency vs wavelength of the incident light for two structures: one with the uncompensated donor concentration decreasing from 3×10^{17} cm^{-3} at the illuminated surface to 9×10^{16} cm^{-3} at the space-charge boundary (curve 1), which corresponds to a built-in field strength of 0.8×10^3 V/cm; the other structure was uniformly doped to an uncompensated donor concentration of 3×10^{17} cm^{-3} (curve 2). Fig.29 shows that incorporation of the driving field enhances the quantum efficiency of the photodetectors in the maximum by a factor of 2.5-3. A ratio of intensities in the maximum is equal to the ratio of the respective sums $W+L_p+L_n$, in structures with the built-in field, L_p being the drift-diffusion length.

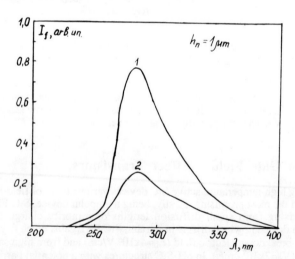

Fig.29.   Dependence of the short-circuit photocurrent of a pn junction photodiode vs wavelength of incident light for structures with (1) and (2) without the built-in field.

Spectral variation of the quantum efficiency of SiC-6H pn structures in the range 200-400 nm at different thicknesses of the illuminated n-layer are given in Fig.30. It is seen that with decreasing n-layer thickness the maximum of photosensitivity shifts markedly to shorter waves. It is also seen that the sensitivity to the UV radiation reaches a maximum when the n-layer has about the same thickness as the space-charge layer width (0.2 μm for the structure presented in Fig.30). In this case the maximum of photosensitivity occurs at a wavelength $\lambda=225$ nm. The absolute value of photosensitivity in the maximum, 0.13 A/W, corresponds to the quantum efficiency of 0.7 electron/photon.

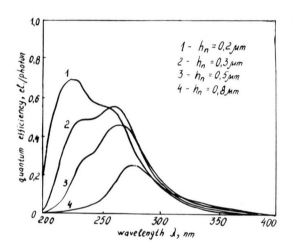

Fig.30. Spectral variation of the quantum efficiency of a pn junction photodiode for different thicknesses of the illuminated n-layer.

The pn structures investi-gated, of an area 1.7×10^{-2} cm^2, had a leakage current ~ 10^{-11} A at a reverse voltage of 1 V. Thus, the obtained results have demonstrated that SiC photodiodes are highly sensitive detectors of UV radiation having low leakage currents and can be used as UV detectors operating at high temperatures.

6. Junction Gate Field - Effect Transistors.

Among high-temperature active SiC devices various types of field-effect transistors are considered the most promising. Firstly, being a unipolar device, SiC FET is free of the drawbacks arising from small diffusion lengths of minority charge carriers in SiC. Secondly, as a device in which drift is important, SiC FET gains much from high values of the avalanche breakdown electric field ($E_b = 4 \times 10^6$ V/cm) and from high saturated electron drift velocity ($V_s = 2 \times 10^7$ cm/s). In 6H-SiC structures with a potential barrier for electrons, deep conductivity modulation of thin epitaxial n-layers can be achieved at elevated temperatures due to the fact that the depletion region having a well-defined boundary with the n-region persists in these structures up to a temperature of about 800 K [21].

Optimal doping level and dimentions of the channel in depletion-mode SiC FETs differ from those for GaAs FETs [40]. To illustrate this point, consider the maximum transconductance of a FET in terms of the Shockley model [41] where it is numerically equal to the conductance of the completely open channel :

$$g_{max} = q\mu naZ/L, \tag{9}$$

where q is the electron charge, μ is the electron drift mobility, n is the concentration of free electrons; a, Z, and L are the thickness, width, and length of the channel, respectively. Taking into account that $n \leq Nd-Na$ (Nd-Na being an uncompensated donor concentration) and $q(Nd-Na)\cdot a = \varepsilon E_p \leq \varepsilon E_b$ (where ε is a dielectric constant and E_p the maximum field in the space-charge region (SCR) when the channel is pinched-off) one obtains:

$$g_{max} \leq \varepsilon\mu E_b Z/L. \tag{10}$$

It is seen from this formula that in properly designed 6H-SiC FET the low carrier mobility could be compensated by the high electric field under the gate. If the surface donor density is chosen such that $[(Nd-Na)\cdot a]_{opt} = \varepsilon E_b/\sqrt{3}q = 1.5 \times 10^{13}$ cm^{-2}, then by lowering the donor density Nd-Na and correspondingly increasing the channel thickness a larger channel pinch-off voltage would be obtained because of $V_p = E_b a/2\sqrt{3}$. Large pinch-off voltages are undesirable, therefore, we take $V_p = 20$-40 V for a 6H-SiC FET (values characteristic of high power silicon MOS-transistors). Then the donor density will be (Nd-Na) =(15-7)$\times 10^{17}$ cm^{-3} and the channel thickness a= 0.1-0.2 μm. Note, that for GaAs MESFETs with gate length 1-2 μm the corresponding values are usually (4-10)$\times 10^{16}$ cm^{-3} and 0.3-0.7 μm, respectively (the surface donor density (Nd-Na)$\cdot$a = 3×10^{12} cm^{-2}). Further increase of the donor density in the channel of the 6H-SiC FET (above 2×10^{18} cm^{-3}) in order to lower V_p can not be recommended. The reason is that to satisfy a condition for reaching saturation of the electron velocity throughout the channel, L/a = 3 [40], the gate length of less than 0.3 μm would be required which is a resolution limit for the industrial lithographic process. Therefore, the advantages of SiC arising from its high avalanche breakdown field can be presently most fully exploited only in power FETs operating at pinch-off voltages not less than 20 V.

From the above consideration it follows that in designing 6H-SiC FETs a fairly heavily doped n-SiC (5×10^{17} cm^{-3}) must be used as a channel material [42,43].

The junction-gate field-effect transistor technology in question included the above described methods of growing pn structures, of applying ohmic contacts and forming the device topology. Epitaxial mesa structures for transistors (Fig.31) comprised p$^+$ and n layers on grown successively onto 6H-SiC (OOO1)-substrate by open sublimation epitaxy. The channel was formed with an n-layer etched off to a required depth between the drain and source ohmic contacts. The source, drain and gate ohmic contacts were prepared by metallization with Al/W. To insulate channels of separate transistors and to reduce n-layer thickness between the drain and the source the low-temperature reactive-ion etching was employed. The transistor dimentions were as follows: channel width (perpendicular to the current flow) Z=0.4 mm, channel length (drain-to-source distance) L=10 μm, epitaxial n-layer thickness between the drain and the source a=0.22 μm and thickness of the rest of the n-layer d=2 μm.

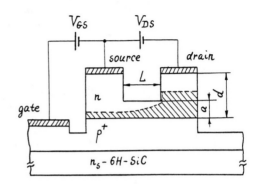

Fig.31. Schematized cross-section of the mesa-epitaxial 6H-SiC FET with a p-n gate. The hatched area is the space-charge region in a regime corresponding to the saturated drain current. (from ref. [43])

In the structures produced the electric field in the SCR of the n-p$^+$ junction was uniform and directed along the hexagonal axis of 6H-SiC, for which the avalanche breakdown field value is highest [44].

<u>Table 3. Electrical parameters of the transistors</u>

	V_T, V	I_D, mA	g, mSm	V_{DG}^m, V	P_m, W
		$V_{GS} = 0$			
300 K	-14,5	15	2,1	85	1
680 K	-16	8	1,0	85	0,5

Investigations of electrical characteristics of this FET at room temperature gave the following results. The electrical breakdown of the n-p$^+$ junction in the drain-gate region is a microplasma process caused by impact ionization in the semiconductor bulk. The maximum electric field strength in this region, 3×10^6 V/cm, is close to the limiting value for 6H-SiC along the C-axis. A corresponding value of E_p is 1.6×10^6 V/cm ((Nd-Na)=4×10^{17} cm^{-3}, a=0.22 μm, V_p=17.2 V). The threshold gate voltage V_T= -14.5 V is less than the pinch-off voltage by a value of the built-in potential of 6H-SiC pn junctions (V_{bi}=2.7 V). Output characteristics prior to the drain-gate breakdown (Fig.32,a,b) agree with the Schockley model: the drain current I_D shows complete saturation at the drain-source voltages V^s_{DS} =|V_T|-|V_{GS}| (V_{GS} is the gate-source voltage), and the transconductance g in the saturation

region of the drain current is equal numerically to the drain-source conductance G_{DS} observed at small V_{DS} values. The electrical parameters of the transistors are summarized in Table 3.

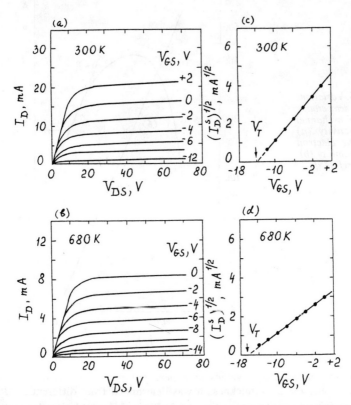

Fig.32. Output drain-source characteristics of the transistor: (a) - 300 K, (b) - 680 K; saturated drain current as a function of the gate bias: (c) - 300 K, (d) - 680 K. (from ref. [45])

As seen from Fig.32 and Table 3 the transistors fabricated operate well in in the temperature range 300-700 K. The threshold voltage and the drain-gate breakdown voltage change weakly and the transconductance drops by about a half. To elucidate the temperature variation of the transconductance, the drain-source conductance G_{DS} was studied in the temperature range 160-700K [45] using a procedure usually referred to as the G-C method. By this method variation of the derivative of G_{DS} the SCR width h of a n-p gate junction is measured, allowing one to determine the n-channel conductance $\sigma(h)$ across the channel thickness:

$$\sigma(h) = -(L/Z)\, dG_{DS}/dh. \qquad (11)$$

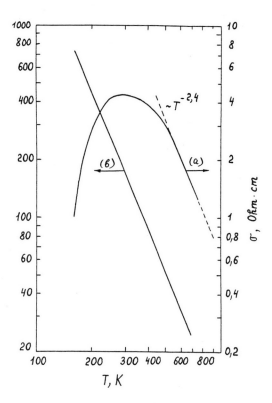

Fig.33. Variations with temperature of the n-channel conductivity (a) and the electron drift mobility (b). (from ref. [45])

For this purpose, G_{DS} values were measured in a dc regime at the chosen fixed temperatures and the SCR thickness h was determined from differential capacitance measurements on the n-p$^+$ junction $C = \varepsilon ZL/h$. Plots of G_{DS} vs h were linear throughout the temperature range investigated, giving evidence of the drift mobility being constant across the channel thickness. It was found that in the temperature range 530-700K the donors in the quasi-neutral region of the channel were completely ionized and the conductance changed in proportion to $T^{-2.4}$ (Fig.33,a), i.e. the temperature variation of the latter was determined solely by the electron drift mobility decreasing according to $\mu=180(T/300)^{-2.4}$ cm^2/Vs (T[K]). Assuming $\mu \sim T^{-2.4}$ for the entire temperature range 160-700K (Fig.33,b) the variation with temperature of the free electron concentration n(T) was calculated (Fig.34,a). An analysis of the n(T) function using carrier statistics for a partially compensated (10%) non-degenerate n-type semiconductor (Fig.34b,c) proved the supposition that in the temperature range 160-700K the electron scattering by phonons is dominant: values of the effective density of states in the conduction band, fitted by 6.8×10^{18} (T/300)$^{1.5}$ cm^{-3} (T,K), and the donor ionization energy value of 0.1 eV obtained in the analysis show a good coincidence with published galvanomagnetic data on n-type 6H-SiC samples [17].

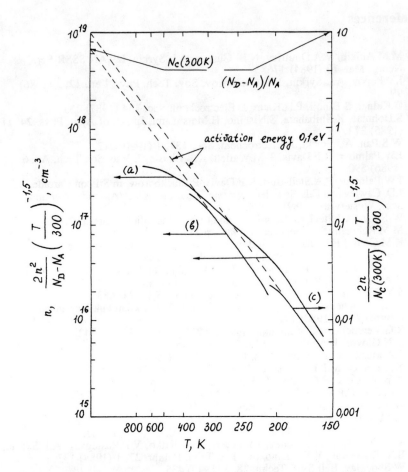

Fig.34.   The free electron concentration in the open part of the channel as a function of temperature (a). Curves (b) and (c) represent n(T) calculated using electron statistics for a partially compensated nondegenerate n-type semiconductor with the following parameters: the donor ionization energy $E_D = 100$ meV, the degree of donor compensation $Na/(Nd-Na) = 10\%$, the effective states density in the conduction band $Nc = 6.8 \times 10^{18} (T/300)^{1.5}$ cm^{-3} ($T[K]$). (from ref. [45])

Thus, the maximum in the transistor transconductance curve observed at T=300K is explained by a competition between the two processes: donor ionization in the neutral region of the channel and scattering of the electrons by phonons.

We conclude the consideration of the sublimation grown 6H-SiC FETs with a remark that Kelner et al. [46] using CVD-growth have fabricated FETs of similar design. Because of higher pinch-off voltage and shorter channel length (4 µm) their device has twice as high transconductance, equal to 15 mSm/mm.

7. References

1. M.M.Anikin, V.A.Dmitriev, N.B. Guseva, A.L.Syrkin, Izv.AN SSSR Ser. Neorg. Mat. **10** (1984) 1768.
2. I.V.Popov, A.L.Syrkin, V.E.Chelnokov, Sov. Tech. Phys. Lett. **12**, 2 (1986) 99.
3. G.Kelner, S Binoti, P.H.Klem, J. Electrochem. Soc. **134** (1987) 253
4. S.Dohmae, K.Shibahara, S.Nishino, H.Matsunami, Jpn. J. of Appl. Phys. **24**, 11 (1985) 873.
5. W.S.Pan, A.J.Steckl, J. Electrochem. Soc. **137**, 1 (1990) 212.
6. J.W.Palmour, R.F.Davis, T-M.Wallett, K.B.Bhasin, J. Vac. Sci. Tech. A**4**, 6 (1986) 590.
7. J.W.Palmour, P.A.Stell-Burt, R.F.Davis, P.Blackborow, in Silicon Carbide, ed. J.D. Cawley and C.E. Sempler, Ceramic Transactions v.2 (American Ceramik Soc., Westerville, 1989), p.478.
8. W-S Pan, A.J.Stecl, in Amorphous and Crystalline Silicon Carbide 2, ed. M.M.Rahman (Springer-Verlag,1989), p.217.
9. K.M.Geib, J.E.Mahan, C.N.Wilmsen, in Amorphous and Crystalline Silicon Carbide 2, ed. M.M.Rahman (Springer-Verlag, 1989).
10. M.M.Anikin, M.G.Rastegaeva, A.L.Syrkin, I.V.Chuiko, Extend. abstr. of 3 ICACSC (USA, Washington, 1990), poster paper N5.
11. S.Y.Wu, R.B.Campbell, Sol. State Electron. **17**, 7 (1974) 683.
12. L.A.Kosjachenko, N.M.Pan'kiv, L.V.Pivovar, V.M.Skljarchuk Ukrain. Fiz. Dzurnal **27**, 1 (1982) 101.
13. R.G.Verenchikova, V.I.Sankin, Pisma v ZhTF **14**, 19(1988) 1742.
14. G.N.Glover, J. Appl. Phys. **46**, 11 (1975) 4842.
15. S.A.Mead, W.A.Spitzer, Phys. Rev. A.**134**, 3 (1964) 713.
16. N.A.Bether, M.I.T., Radiation Lab. Rep. **4**, (1942), 42.
17. B.W.Wessels, H.C.Gatos, J.. Phys. Chem. Sol. **38**, 4 (1977) 345.
18. S.M.Sze, C.R.Crowell, D.Kahng, J. Appl. Phys. **35**, 8 (1964) 2534.
19. W.J.Choyke, L.Patric, Phys. Rev. B**2**, 6 (1970) 2255.
20. M.M.Anikin, A.A.Lebedev, I.V.Popov, V.E.Sevastjanov, A.L.Syrkin, A.V.Suvorov, V.E.Chelnokov, Pisma v Zhtf **10**, 17 (1984) 1053.
21. M.M.Anikin, A.A.Lebedev, I.V.Popov, S.N.Pjatko, V.P.Rastegaev, A.L.Syrkin, B.V.Tsarenkov, V.E.Chelnokov, Fiz. Tehn. Polupr. **22**, 1 (1988) 133.
22. W.Shockley, Bell Syst. Techn. **28**, 3 (1949) 435.
23. M.M.Anikin, A.N.Andreev, A.A.Lebedev, C.N.Pjatko, M.G.Rastegaeva, N.S.Savkina, A.M.Strel'chuk, A.L.Syrkin, V.E.Chelnokov, Fiz. Tehn. Polupr. **25**, 2 (1991) 328.
24. A.G.Kechek, N.I.Kuznetsov, A.A.Lebedev in Preprint A.F.Ioffe Inst., seperate issue N1145 (Leningrad, 1987), p.1.
25. M.M.Anikin, N.I.Kuznetsov, A.A.Lebedev, A.M.Strel'chuk, A.L.Syrkin, Fiz. Tehn. Polupr. **24**, 8 (1990) 1384.
26. M.M.Anikin, A.A.Lebedev, A.L.Syrkin, A.V.Suvorov, Fiz. Tehn. Polupr. **19**, 1 (1985) 114.
27. M.M.Anikin, A.A.Lebedev, A.L.Syrkin, A.V.Suvorov, Fiz. Tehn. Polupr. **20**, 12 (1986) 2169.

28. M.M.Anikin, A.S.Subrilov, A.A.Lebedev, A.M.Strel'chuk, Fiz. Tehn. Polupr. **25**, 3 (1991) 479.
29. M.M.Anikin, V.V.Evstrorov, I.V.Porov, V.P.Rastegaev, A.M.Strel'chuk, A.L.Syrkin, Fiz. Tehn. Polupr. **23**, 4 (1989) 647.
30. M.M.Anikin, V.V.Evstropov, I.V.Rorov, A.M.Strel'chuk, A.L.Syrkin, Fiz. Tehn. Polupr. **23**, 10 (1989) 1813.
31. C.T.Sah, R.N.Noyce, W.Shockley, Proc.IRE **45**, 9 (1957) 1228.
32. V.V.Evstropov, N.V.Kiselev, I.L.Petrovich, B.V.Tsarenkov, Fiz. Tehn. Polupr. **18**, 10 (1984) 1852.
33. M.M.Anikin, A.A.Lebedev, I.V.Popov, V.P.Rastegaev, A.M.Strel'chuk, A.L.Syrkin, J.M.Tairov, V.E.Chelnokov, Fiz. Tehn. Polupr., **22**, 2(1988) 298.
34. M.M.Anikin, S.N.Vainstein, M.E.Levinstein, A.M.Strel'chuk, A.L.Syrkin, Pisma v ZhTF **14**, 6 (1988) 545.
35. M.M.Anikin, M.E.Levinstein, I.V.Popov, V.P.Rastegaev, A.M.Strel'chuk, A.L.Syrkin, Fiz. Tehn. Polupr. **22**, 9 (1988) 1574.
36. B.S.Kondrat'ev, I.V.Popov, A.M.Strel'chuk, M.L.Tiranov, Fiz. Tehn. Polupr. **24**, 4 (1990) 647.
37. M.M.Anikin, A.A.Lebedev, I.V.Popov, A.M.Strel'chuk, A.V.Suvorov, A.L.Syrkin, V.E.Chelnokov, Fiz. Tehn. Polupr. **20**, 5 (1986) 844.
38. M.M.Anikin, A.A.Lebedev, S.N.Pjatko, V.A.Solov'ev, A.M.Strel'chuk, in Ext. Abstr.3 Int. Conf. on Amorf. and Crist. SiC (Washington, 1990), session 6, paper 6.
39. W.Choyke, L.Patric, Phys.Rev. **127**, 6 (1962) 1868.
40. P.A.Ivanov, B.V.Tsarenkov, Fiz. Tehn. Polupr. **25**, 11 (1991) 1913.
41. W.Shockley, Proc. IRE **40** (1952) 1356.
42. M.M.Anikin, P.A.Ivanov, A.L.Syrkin, B.V.Tsarenkov, V.E.Chelnokov, Pisma v ZhTF **15** (1989)36.
43. M.M.Anikin, P.A.Ivanov, A.L.Syrkin,B .V.Tsarenkov, in Extend. Abstr. of 176 Meeting Electrochem. Soc. (Hollywood Florida, 1989) p.729.
44. Ju.A.Vodakov, D.P.Litvin, V.I.Sankin, E.N.Mohov, A.D.Roenkov, Fiz. Tehn. Polupr. **19** (1985) 814.
45. M.M.Anikin, P.A.Ivanov, B.V.Tsarenkov, V.E.Chelnokov, Fiz. Tehn. Polupr. 1991 to be published.
46. G.Kelner, S.Binari, K.Sleger, J.Palmour, in Abstr. E-MRS Fall Meeting (Strasbourg, November 1990), C-3.6.

The whole paper compiled by Z. C. Feng (Emory University).